Phytochemicals in Human Health

Edited by Venketeshwer Rao, Dennis Mans and Leticia Rao

Published in London, United Kingdom

IntechOpen

Supporting open minds since 2005

Phytochemicals in Human Health
http://dx.doi.org/10.5772/intechopen.77908
Edited by Venketeshwer Rao, Dennis Mans and Leticia Rao

Contributors
António Pereira, Sérgio Martins, Ana Teresa Caldeira, Inul Ansary, Abu Taher, Soumen Saha, Parthadeb Ghosh, Anil Bhuimali, Arijit Biswas, Sinchan Adhikari, Debjoy Bhattacharjya, Pornngarm Limtrakul (Dejkriengkraikul), Warathit Semmarath, Sariya Mapoung, Setiyo Gunawan, David Febrilliant Susanto, Hakun Wirawasista Aparamarta, Arief Widjaja, Firdaus Gunawan, Sam Mashele, Hildah Mfengwana, Cristina Monteiro, Julliana Dos Santos, Stephen Rathinaraj Benjamin, Iago Almeida Da Ponte, Sudarsini Saravanabhavan, Murugan Muthuvel, Zhiyou Hao, Weisheng Feng, Meng Li, Jingke Zhang, Supatra Porasuphatana, Peeradon Tuntiteerawit, Anupon Tadee, Tichakorn Singto, Gabriela Tataringa, Ana-Maria Zbancioc, Robbert Bipat

First published in London, United Kingdom, 2020 by IntechOpen
IntechOpen is the global imprint of INTECHOPEN LIMITED, registered in England and Wales, registration number: 11086078, 7th floor, 10 Lower Thames Street, London, EC3R 6AF, United Kingdom
Printed in Croatia

British Library Cataloguing-in-Publication Data
A catalogue record for this book is available from the British Library

Additional hard and PDF copies can be obtained from orders@intechopen.com

Phytochemicals in Human Health
Edited by Venketeshwer Rao, Dennis Mans and Leticia Rao
p. cm.
Print ISBN 978-1-78985-587-6
Online ISBN 978-1-78985-588-3
eBook (PDF) ISBN 978-1-83968-413-5

For EU product safety concerns:
IN TECH d.o.o., Prolaz Marije Krucifikse Kozulić 3, 51000 Rijeka, Croatia, info@intechopen.com or visit our website at intechopen.com.

Meet the editors

Dr. Venketeshwer Rao, Professor Emeritus, Department of Nutritional Sciences, Faculty of Medicine, University of Toronto, has established a major focus in the area of diet and health. His research focuses on the role of phytochemicals in the prevention and management of human diseases including cancer, cardiovascular disease, and osteoporosis. The main area of his research is oxidative stress and antioxidant phytochemicals with particular emphasis on the role of lycopene. He is credited for bringing international awareness to the role of lycopene in human health. In addition to carotenoids, his research interests also include plant polyphenols and the role of prebiotics and probiotics in human health. He has published extensively including research papers, reviews, and books. He has a distinguished academic career spanning over 52 years. He is popularly sought by the international media to express his opinions on the subjects of nutrition and health.

Dennis R.A. Mans (October 16, 1955, Paramaribo, Suriname) graduated as a Medical Biologist in 1984 at the Utrecht University, The Netherlands. He received his PhD diploma in 1991 from the Free University of Amsterdam, The Netherlands, with a thesis on the antineoplastic agent etoposide. Between 1992 and 2000, he worked in Brazil on plants with anticancer properties. He returned to Suriname in 2000 to hold an academic position at the Anton de Kom University. In 2001, he became Head of the Department of Pharmacology, in 2006 he was appointed Associate Professor, and in 2008 Full Professor with a chair in Pharmacognosy. His scientific production amounts to more than 100 full papers and more than 200 scientific abstracts.

Dr. Leticia Rao is an Emerita Associate Professor of Medicine and former Director of the Calcium Research Laboratory, Department of Medicine, and St Michael's Hospital, Toronto, Ontario. Her degrees include BSc Chemistry, University of the Philippines; MSc Food Science, Oregon State University; and Ph.D. Biochemistry, University of Toronto. Her area of expertise is bone cell biology and osteoporosis with an emphasis on effects of drugs and nutritional supplements. She has studied the effect of gravity on bone formation in osteoblast cells in an experiment sent to space with the Space Shuttle Columbia STS 107. Her research has been presented to national and international symposia and published in peer-reviewed scientific journals. She co-authored a book entitled "Bone Building Solution" and co-edited a book on phytochemicals.

Contents

Preface

In recent years there has been a great deal of interest in naturally occurring bioactive compounds referred to as 'Phytochemicals' and their role in the prevention, management, and treatment of many human health disorders. This heightened interest in phytochemicals can be attributed to the global dietary recommendations to increase the consumption of plant-based foods, which are the main source of phytochemicals in our diet. The consumption of phytochemicals is now looked upon as being an important complementary approach to the use of traditional pharmaceutical compounds in the treatment and management of human diseases. Phytochemicals, also referred to as 'Phytonutrients', vary significantly in terms of their occurrence, chemistry, and mode of action. They must first be isolated, purified, and their physico-chemical properties established. Once identified, they can be studied for their mechanisms of action leading to the evaluation of their biological properties and their role in human health using in vitro, animal, and human clinical studies. Recognizing the importance of these compounds in humans, we undertook to edit and publish a book with chapters authored by internationally recognized scientists who are experts in their respective fields of research in this important area of the role of phytochemicals in human health. The chapters of the book include original research articles as well as up-to-date reviews. The book is intended to benefit researchers, health professionals, industrial scientists as well as government regulatory agencies. It is our hope that with a better understanding and knowledge of phytochemicals, we will be able to formulate guidelines as to their safety and their use in the management of human diseases and improvement of the quality of life.

Dr. Venketeshwer Rao
University of Toronto,
Canada

Dr. Dennis Mans
Anton de Kom University of Suriname,
Suriname

Dr. Leticia Rao
University of Toronto,
Canada

Section 1

Phytochemical Composition

Chapter 1

Anthocyanins and Proanthocyanidins in Natural Pigmented Rice and Their Bioactivities

Pornngarm Limtrakul (Dejkriengkraikul), Warathit Semmarath and Sariya Mapoung

Abstract

Natural pigmented rice is mainly black, red, and dark purple and contains a variety of flavones, tannins, phenolic, sterols, oryzanols, and essential oils. Anthocyanins and proanthocyanidins belonging to plant flavonoids are thought of as the major functional components found in black, red, and purple rice and contribute to the intense color of many fruits, vegetables, and pigmented cereals such as blueberries, grapes, red cabbages, and purple sweet potatoes. Recent data have indicated the potential for isolating and characterizing the nutrition and non-nutritive components in colored fruits, vegetables, and cereals for their potential chemopreventive and pharmaceutical agents. This chapter provides up-to-date coverage of pigmented rice in terms of the bioactive constituents, isolation, extraction and analytical methods, and related bioactivities. Special focus has been placed on the anti-inflammation, anticancer, and antiaging processes of the major components found in pigmented rice, especially with regard to germ and bran extracts.

Keywords: anthocyanins, proanthocyanidins, pigmented rice, nonpigmented rice, bioactivities of black rice, bioactivities of red rice

1. Introduction

Although white rice is consumed as a major staple food worldwide, quite a few countries in Southeast Asia (SEA) also consume pigmented cultivars such as red, black, purple, and brown rice. Rice cultivars that originated in Southeast Asia (SEA) have been classified in the species of *Oryza sativa* L., which differs from the *Oryza glaberrima* Steud. species that is cultivated in West Africa. In Thailand, the total area of cultivation has been recorded at 56.3 million Rai (22.3 million acres) with the majority being comprised of white rice cultivars (90%), while pigmented rice is only 0.1% or makes up approximately 62,000 Rai (24,506 acres) [1]. The largest cultivated area is located in the northeast of Thailand (63.10%) followed by the northern region of Thailand (21.93%), the central area (14.5%), and the south (0.47%). India and Indonesia have more cultivated area of pigmented rice than any other SEA countries, although they report a smaller proportion than that of the white rice cultivar. The total cultivated area in India has been recorded at

43.77 million acers (29.4% of the global rice area) with a production of 90 million tons [2]. The world production of rice is estimated at around 680 million tons, which is equivalent to that of wheat [3]. The color intensities of pigmented rice are obtained from the value of lightness, redness, and yellowness and seem to be correlated to the indicators of its bioactive compounds [4–6].

Recently, pigmented rice varieties have received increased amounts of attention from consumers for their high bioactive compounds that present potential nutraceutical benefits to health. It is also well known that these compounds are primarily located in the outer layer of the rice grain, which is regarded as a rice by-product. The by-products of rice processing are rice germ and bran, along with the rice hulls which protect the rice seeds during growth. These account for 20% of the rice crop. These by-products are frequently used as animal feed in developing countries. However, recently, significant amounts of data have revealed the beneficial nutritional impacts of these by-products on human health. The major bioactive compounds that are found in red, black, purple, and brown rice include gallic, protocatechuic, hydroxybenzoic, and vanillic acid, cyanidin 3-O-glucoside, peonidin-3-O-glucoside, proanthocyanidin, flavanol, catechin and epicatechin, carotenoids, and γ-oryzanol content. Several research findings have reported on the biological modulating effects of pigmented rice seeds and bran phytochemicals, including anti-inflammatory activities [7, 8], anticancer activities that have suppressed tumor growth in mice and several human cancer cell lines [9–13], the anti-metastasis properties of cancer cell invasion [14–16], antiaging effects with the reduction of oxidative stress in both in vitro and in vivo models [17, 18], the modulation of serum lipid profiles and the enhancement of mRNA expression levels of fatty acid metabolism-related genes [19], a reduction of platelet hyperactivity and hypertriglyceridemia in dyslipidemic rats [20], and skin antiaging treatments [21–24].

In this current review, we have focused on the health benefits of pigmented rice and the relevant bioactive compounds. We have tried to present the information in this chapter in a way that is easy to understand, even for readers who are not experts in this field of research. The bioactive compounds found in pigmented rice display significant immersion potential with regard to a range of beneficial health effects and also provide significant informative data that could lead to the expansion in the growing of pigmented cultivated areas in Thailand and other Southeast Asian countries. There is also the prospect of additional practical implications, not only for agriculture expansion but in the food industry as well. Several pigmented rice varieties have been used to create new nutraceuticals, and these seem to hold a promise in terms of potential cosmeceutical utilization in the new global business era.

2. Pigmented rice and bioactive compounds

The rice processing industry is well-developed and produces a number of products from rice kernels or grains (70%) along with a large quantity of rice by-products. These by-products include rice bran (8–9%), rice germ (1–2%), and rice husks (20%). Figures on rice paddy composition are presented in **Figure 1**. These by-products are frequently used as animal feed in developing countries [25], but the demand for these by-products in terms of their human nutritional impacts has increased due to their potential health benefits. Rice kernels are primarily a good source for the energy intake of carbohydrates and proteins in humans. Rice bran makes up the outer layer of the rice kernel and is mainly comprised of a pericarp,

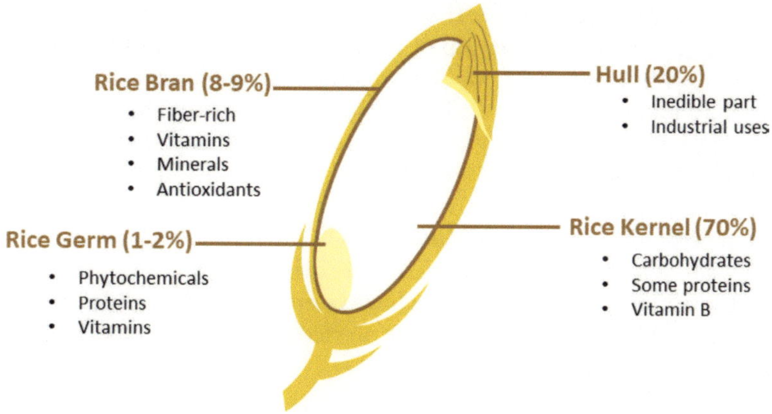

Figure 1.
Rice paddy composition.

aleurone, sub-aleurone layer, and germ. Rice bran and germ contain appreciable quantities of fiber, vitamins, minerals, unsaturated fatty acids, tocopherols, γ-oryzanol, and tocotrienols, which offer potent antioxidant content along with a range of other potential health benefits [26, 27].

2.1 Extraction methods for bioactive compounds in pigmented rice

Several common extraction techniques that are used in the process of rice extraction include the method of solvent extraction, which is a conventional technique used to extract bioactive compounds from pigmented rice, supercritical fluid extraction, and subcritical water extraction. With regard to the conventional technique, a number of organic solvents are commonly used such as acetone, methanol, ethanol, butanol, and water in certain proportions as the extraction solvent [28–30]. In our study, 50% ethanol was used as an extraction solvent at a proportion of 1:5 grain or bran liquid, and extraction was carried out at room temperature for 3–12 h. The extracts were then concentrated with a rotary evaporator until all ethanol residues were removed and then further partitioned against saturated butanol to obtain the medium polar bioactive compounds of the black rice extract [31] or red rice extract [7, 15, 16]. The bioactive compounds present in these fractions shall be described in a later section. In another study, 60% ethanol containing 0.1% HCL was used as an extraction solvent with a 1:10 feed to liquid proportion, and extraction was carried out for 3–12 h. The extracts were then concentrated and further partitioned against petroleum ether [8]. In another study, the rice bran was extracted with 70% ethanol for 30 min repeated three times and was then further partitioned with ethyl acetate at pH 2–3 [32]. The same method was used to extract soluble phenolic compounds in white rice, brown rice, and germinated brown rice [33].

Supercritical fluid extraction has been widely used for the extraction of functional active compounds from medicinal plants including rice and cereals. This was in common with the use of supercritical carbon dioxide as an extraction solvent in other successful experiments. Kim et al. [34] used the method of supercritical fluid extraction of rice bran oil from pigmented rice, which provided higher yields of polyunsaturated fatty acids than the conventional use of organic solvent extraction. In yet another study, supercritical carbon dioxide extraction was used, and yields of 17.5% oil were achieved from powdered rice bran, and a yield of 37% of γ-oryzanols was also obtained, which was characterized as 85% of the extraction efficiency [35].

Another extraction technique is the subcritical water extraction method that has been developed for the extraction of bioactive compounds from pigmented rice through the use of hot water at temperatures between 100 and 374°C under high pressure to maintain a liquid status. This technique is considered to be very friendly to the environment because no organic solvents are used, and this can potentially alleviate some of the problems associated with the conventional methods [36, 37].

There were differences in the extraction procedure and the varieties of the rice cultivars that were used to detect the amounts of bioactive compounds in different portions of rice such as in the whole grains, kernels, endosperm, husks, rice, and bran. More than 1000 published studies have been reviewed to make up the cited data based on this information. Some data on rice composition have been selectively recorded elsewhere [27].

2.2 Various bioactive compounds present in black rice

Phytochemical profiles of black rice are characterized by the presence of anthocyanins, which are a group of reddish to purple flavonoids that exist in black rice and other pigmented cereal grains. The main anthocyanins in black rice were found to be present in quantities more than 95% and were cyanidin 3-O-glucoside (2.8 mg/g) and peonidin-3-O-glucoside (0.5 mg/g) followed by flavones and flavonols (0.5 mg/g) and flavan-3-ols (0.3 mg/g) [38]. The concentrations of total anthocyanins in black rice cultivars significantly varied from one report to another, while much higher concentrations of anthocyanins were detected in Chinese black-purple rice that contained cyanidin 3-O-glucoside (6.3 mg/g) and peonidin 3-O-glucoside (3.6 mg/g) [39]. The variations of the anthocyanin content in the reports on black rice might be due to the use of different cultivars and the variety of differing growing conditions. The anthocyanidins or aglycons, the basic structure of anthocyanins, consist of an aromatic ring (A) that is bonded to a heterocyclic ring (C) that contains oxygen, which is bonded by a carbon–carbon bond to a third aromatic ring (B). When the anthocyanidins are bonded to a sugar moiety in the glycosidic linkage, they are known as anthocyanins. More than 500 different anthocyanins and 23 anthocyanidins have been reported. Anthocyanins exist as mono-, di-, or tri-O-linked glycosides and acyl glycosides of anthocyanidins in plants. The sugar moiety may be substituted by aliphatic, hydroxybenzoic, or hydroxycinnamic acids. The structural characteristics of anthocyanins make them highly reactive toward the reactive oxygen species (ROS) [27]. The basic structure of this is shown in **Figure 2**. Major flavone and flavonol glycosides present in black rice are taxifolin, quercetin, apigenin, and luteolin, which are comprised of monomeric and oligomeric constituents. The concentrations of the flavone and flavonol contents were

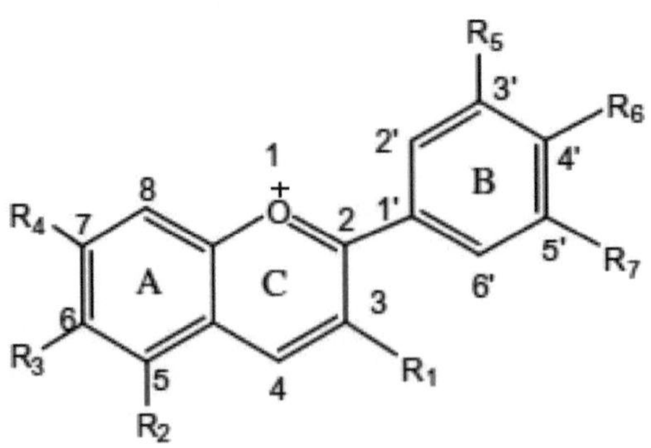

Figure 2.
General structure of anthocyanins.

significantly higher in black rice than in red, brown, or white rice. This was especially true with regard to taxifolin O-hexoside, quercetin 3-O-glucoside, and quercetin 3-O-rutinoside, which were detected only in black rice [38]. Abdel-Aal et al. [40] also reported that the mean anthocyanin content in black rice (3.276 mg/g) was about 35-fold higher than that of red rice (0.094 mg/g). Additionally, the contents of anthocyanin present in Northern Thai black rice cultivar obtained from Doi Saket, Chiang Mai, were 8.1 mg/g extract, which was considered very high when compared to the anthocyanin content found to be present in the Northern Thai red rice cultivar obtained from Dok Khamtai [31].

The total procyanidin content in black rice has been found to be present in high variations depending on the grain cultivar; however, it is noteworthy to mention that procyanidins are typically observed in red rice but not in black rice varieties [41–45]. Interestingly, some black cultivars have shown the presence of oligomeric procyanidins with a 2–10 degree of polymerization [38]. Furthermore, black and red rice were found to contain only one flavan-3-ol monomer, catechin. Additionally, a Canadian black rice variety also contained catechins at levels four times higher than epicatechin. Furthermore, the concentration of catechin was much higher in red rice (92 μg/g) than in black rice (20 μg/g) [46]. Other phytochemicals have been detected in black rice including all four derivatives of γ-oryzanol, such as 24-methylenecycloartenol, campesterol, cycloartenol, and β-sitosterol ferulates, along with lower levels of carotenoids. The main carotenoids detected in black rice were xanthophylls, lutein, and zeaxanthin, while lycopene and β-carotene could be detected but were found to be present as a minor component [38]. The value of the carotenoid content in black rice kernels is lower than the carotenoids found to be present in black rice bran. It was reported that values in a range of 33–41 μg/g of carotenoids were found in the bran extracts of four varieties of Thai black rice [47]. A range of phenolic compounds including vanillic acid, protocatechuic acid, chlorogenic acid, ferulic acid, and coumaric acid has been detected in black rice with the dominant phenolic acids being present in red and black rice bran [7, 31]. The contents of phenolic compounds, flavonoids, catechins, anthocyanins, and proanthocyanidins, are summarized in **Table 1** as examples of the phytochemicals that were detected in Doi Saket Thai black rice cultivar. The germ and bran extracts of the black and red rice varieties were found to have the greatest phytochemical content with decreasing amounts occurring in the rice hull and even less in the seeds or kernels. Additionally, the expected low levels of these phytochemicals were found in white rice as a consequence of the milling process.

2.3 Various bioactive compounds present in red rice

Red rice was characterized by a high quantity of oligomeric procyanidins (0.2 mg/g) with more than 60% of total phytochemicals found in the rice seeds. Proanthocyanidins are high molecular weight polymers or complex flavan-3-ol polymers that consist mainly of catechin, epicatechin, gallocatechin, and epigallocatechin units that can also be found in rice germ and bran, particularly in pigmented rice. The degree of polymerization varied, and the reddish colored test was associated with the presence of a class of polymeric compounds of the proanthocyanidins. These could be in the sum class of the oligomer and polymer contents of the total proanthocyanidins present in the red rice bran extract fraction. The degree of polymerization and galloylation can affect their bioactivity and proanthocyanidin profiles differently depending on the food sources [27, 48]. Proanthocyanidins can be classified into several classes depending on the degree of hydroxylation of the constitutive units and the linkages between them. Our research group has reported on the type of proanthocyanidins found in the red rice that was collected from Dok Khamtai

Compound	(mg/g extract)
Total phenolic content	117.6 ± 14.6
Vanillic acid	4.2 ± 0.4
Protocatechuic acid	2.3 ± 0.1
Gallic acid	ND
Coumaric acid	0.5 ± 0.2
Ferulic acid	1.4 ± 0.0
Chlorogenic acid	1.7 ± 0.3
Total flavonoid content	42.9 ± 2.1
Anthocyanin	8.1 ± 1.9
Catechin	ND
Proanthocyanidin	ND

Values are mean ± S.D., ND = not detectable.

Table 1.
Phytochemical content of black rice extract (polar fraction).

cultivar, Northern Thailand, as a type B proanthocyanidin. The monomeric units of proanthocyanidin in the acid hydrolysis of the red rice extract fraction were found to be catechins, epicatechins, gallocatechins, and epigallocatechins [16]. The results revealed that the proanthocyanidin types were procyanidin (catechin and/or epicatechin) and prodelphinidin (gallocatechin and/or epigallocatechin), while the degree of polymerization was recorded at approximately 4. Interestingly, the majority of proanthocyanidins in our red rice extract were of the oligomer with the same degree of polymerization that was found in grape seed extracts [49]. As has been mentioned previously, red rice has a high content of catechins and proanthocyanidins, but some of the black rice cultivars found in France and Canada have revealed the presence of catechins in their black rice cultivars (four times less than the red rice cultivars). It is worth mentioning that many other records have shown that procyanidins have been typically observed in red but not black rice varieties, including in the Northern Thai black rice cultivar obtained from Doi Saket, Chiang Mai [31]. The general structure of proanthocyanidins is shown in (**Figure 3**).

The other active compounds were γ-oryzanol and carotenoids at 27%, whereas flavones, flavonols, and anthocyanins were present in a much less quantity at less than 9% [38]. The main carotenoid detected in red rice bran was lutein, while xanthophylls and zeaxanthin were the carotenoids that were found to be present in lesser quantities. A range of phenolic acids including gallic, protocatechuic, hydroxybenzoic, vanillic, and ferulic acids in red, black, and brown rice have been detected as the dominant phenolic acids present in red and black rice bran [50, 51]. The contents of the phenolic compounds, flavonoids, catechins, anthocyanins, and proanthocyanidins, are summarized in **Table 2** as an example of the phytochemicals that were detected in Dok Khamtai Thai red jasmine rice cultivar. The contents of these bioactive compounds can be used to determine the antioxidant activities that may then provide health benefits.

2.4 Various bioactive compounds present in brown and white rice

The rice bran of whole grain brown rice (unpolished) has been acknowledged as a potential source of edible oil. Although rice bran oil is not very popular worldwide, its demand is increasing due to numerous reports on its health benefits.

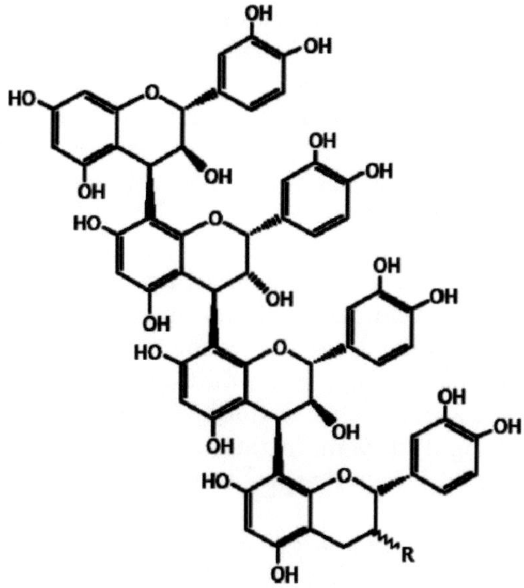

Figure 3.
General structure of proanthocyanidins [16].

Previously, rice bran obtained from brown rice has received a significant amount of attention from the nutraceutical industry as brown rice bran is recognized as the primary source of oil extraction. On this issue, agro-industrial by-products are gaining special attention from the food processing industry because rice bran oil presents a positive fatty acid profile along with the presence of other phytochemicals like ϒ-oryzanol, tocopherols, and tocotrienols. Basically, rice bran is rich in carbohydrates (34–62%), lipids (15–20%), proteins (11–15%), and dietary crude fiber (7–11%) [52]. The health benefits of rice bran include the strong antioxidant potential of rice bran oil. This is not only a consequence of the presence of significant quantities of linolenic acid (34%), oleic acid (38.4%), and other unsaturated

Compounds	(mg/g extract)
Total phenolic content	237.78 ± 17.26
Vanillic acid	1.53 ± 0.19
Protocatechuic acid	0.35 ± 0.03
Gallic acid	ND
Coumaric acid	0.2 ± 0.01
Ferulic acid	0.56 ± 0.04
Chlorogenic acid	ND
ϒ-Tocotrienol	ND
ϒ-Oryzanol	1.75 ± 0.23
Anthocyanin	ND
Catechin	6.65 ± 0.57
Proanthocyanidin	53.45 ± 3.23

Values are mean ± S.D., ND = not detectable.

Table 2.
Phytochemical content of red rice extract (polar fraction).

fatty acids but also occurs as a result of the high contents of γ-oryzanol, tocopherols, and tocotrienols that reveal strong oxidative stability along with a range of other health benefits [53, 54].

The protein content present in rice bran of brown rice is characterized as a good source of protein that is nutritionally superior and hypoallergenic in nature. Rice bran is a rich source of essential amino acids such as lysine, which seems to be present in minute quantities in other cereal grains [55, 56]. The proteins in rice bran are highly digestible and can be utilized as an effective food ingredient. Rice bran is rich in dietary fiber, and, consequently, the rice bran by-products of rice processing are now often present in food commodities and functional foods that have been marketed for the ability to add dietary fiber to the diets of consumers and to offer health benefits in terms of daily nutrition. Additionally, brown rice possesses high contents of a variety of nutrients, such as fiber, vitamins, and minerals that are lost during the process of refining and milling in the production of white rice within the rice agro-industry. Notably, brown rice possesses four times as much dietary fiber as white rice [57].

White rice is a major source of energy nourishment for the world's population, especially in Asian countries. However, the carbohydrate content in white rice accounts for 80% of its makeup, which is considered a higher amount than wheat. Wheat is a popular grain among European countries and contains a lessor proportion of carbohydrates (approximately 50–70%) [58]. For this reason, there are concerns that white rice possesses a high glycemic content and that it may not be a suitable source of carbohydrates for people who have weight problems. It is interesting to note that white rice does not contain anthocyanins and proanthocyanidins, which are the important phytochemicals that are found in black rice and red rice, respectively, particularly in portions of rice germ and bran extracts. While total flavonols and phenolic compounds are observed to be significantly high in pigmented rice, nonpigmented rice such as white rice possesses a minute quantity of flavone/flavanol content [50].

3. Pigmented rice and bioactivities that benefit health

Phytochemicals found in pigmented rice (brown, black, purple, and red rice) are not present in white rice because many valuable phytochemicals, fiber, vitamins, and nutrients are lost during the processes of refining and milling [57]. Since brown rice contains higher dietary fiber and nutrients, previous studies have revealed that when compared to a white rice diet, a brown rice diet was found to significantly reduce weight, body mass index (BMI), and the circumference of the waist and hips, as well as to lower diastole blood pressure and inflammatory biomarkers such as C-reactive proteins (CRP). Arabinoxylan and β-glucan, prebiotics that are found in brown rice, are beneficial for human gut microbiota such as *Bifidobacterium* and *Lactobacillus*. They are considered as contributing factors in producing an anti-obesity effect [57, 59]. Moreover, in terms of their antidiabetic effects, brown rice was used as an intervention for preventing type 2 diabetes. This is likely because one of their components, ϒ-oryzanol, plays an important role in controlling high-fat diet-induced ER stress in the hypothalamus, which helps in reducing the preference for fatty foods [60]. ϒ-Oryzanol in brown rice has also been found to prevent the apoptosis of pancreatic β cells and to reduce levels of blood cholesterol [61]. Dietary rice brans that give brown rice its brown color also reveal potent anticancer activities through their antioxidant activity, as well as offering antiproliferation, immune modulation, and mucosal protection [62, 63].

Natural pigmented rice, such as black and red rice, may even offer more health benefits than brown rice. Not only is natural pigmented rice higher in the beneficial antioxidant activities of black and red rice, but it also displays strong anti-inflammatory activities as well as anticancer and anti-metastasis activities. The antiaging properties of the major components found in pigmented rice may be anthocyanins and proanthocyanidins, which have been found to be especially rich in content in the germ and bran extracts of black and red rice, respectively. The details of which will be described in greater detail in the following section.

3.1 Antioxidant activities

The antioxidant activities of black and red rice and their crude extracts have been studied, and the results demonstrated that the addition of the pigmented rice could increase antioxidant capacity, both in vivo and in vitro [64–66]. In a study involving the supplementation of diets with black rice pigment fractions, the diets that attenuated atherosclerotic plaque formation in apolipoprotein E-deficient mice [66] and the anthocyanin-rich extract of the black rice might play an important role in the enhancement of atherosclerotic plaque stabilization [8]. In another study, a mixture of brown and black rice improved the lipid profiles and antioxidant status in rats [67]. Another animal study also demonstrated that black rice bran pigment effectively escalated hepatic antioxidant enzyme activities including superoxide dismutase and glutathione peroxidase in high-cholesterol-fed rats [68]. In addition to the in vivo studies, in a cell culture experiment, superoxide anions and reactive oxygen species were significantly suppressed after black rice extract exposure in HepG2 hepatocellular carcinoma [17]. When the antioxidant activities of pigmented rice were compared with those of nonpigmented rice in several studies [30, 41], the results demonstrated that the extracts from pigmented rice displayed higher antioxidant activity than did the nonpigmented rice. In another study, the radical scavenging activities of the extracts from white, black, and red rice were tested. The highest activity was observed in red rice (2.77 μmol of Trolox or vitamin E equivalents/ml), followed by black (0.92 μmol) and white (0.26 μmol) [41, 42]. Polymeric proanthocyanidins play an important role as radical-scavenging components in red rice. The relationships between the antioxidant activities and the components of pigmented rice were explored [41, 69, 70]. The antioxidant activities correlated well with the content of polyphenols and phytochemicals that contribute to the intense color of the pigmented rice. Interestingly, some studies have shown that the antioxidant activity of black rice may be reduced by up to 53% during cooking [71–73].

3.2 Anti-inflammatory properties

Inflammation is an important mechanism of immune pathogenesis, which is our body's response to tissue injury, infection, and stress. Importantly, the prolonged production of inflammatory mediators by macrophage can cause damage to the host and can contribute to the pathology of many diseases including inflamm-aging, arthritis, asthma, cancer, diabetes, and atherosclerosis. Macrophage plays a key role in response to an immediate defensive mechanism of our body against attacking foreign agents, especially with a microbial lipopolysaccharide (LPS) [74]. Macrophage is activated and produces many kinds of inflammatory mediators including nitric oxide (NO), prostaglandins, and many cytokines such as interleukin 1 (IL-1), interleukin 6 (IL-6), and tumor necrosis factor (TNF)-α [75]. Many researchers have studied in vitro and in vivo models to elucidate that natural products are able to ameliorate the inflammatory response in LPS-stimulated macrophage.

During the last decade, it has been shown that anthocyanins reduce the risks of cardiovascular diseases and cancers with inflammatory, antioxidant, and chemo-protective properties [15, 76, 77]. Some reports have demonstrated that lipophilic phytochemicals contained in pigmented rice germ and bran, such as γ-oryzanol and vitamin E derivatives, exert anti-inflammatory activities [78, 79]. On the other hand, pigmented rice contains high amounts of medium polar or hydrophilic compounds such as phenolics, bioflavonoids, anthocyanin, and proanthocyanidins that have been reported for their anti-inflammatory properties, in both in vitro and in vivo models [80–82].

Pigmented rice contains a variety of bioactive compounds with anti-inflammatory properties; however, there have been quite a few reports employing experimental designs that provide direct evidence to support using the extracts of pigmented rice. For the first time, our research group has demonstrated the molecular mechanisms underlying the anti-inflammatory effects. The anthocyanin-rich fraction of black rice extract significantly inhibited LPS that induced many pro-inflammatory mediators in RAW 264.7 macrophage white blood cells [31]. The pro-inflammatory mediators in this study were NO, TNF-α, and IL-6, and they effectively reduced the expression of two important inflammatory enzymes, the inducible NO synthase (iNOS) and the inducible cyclooxygenase-2 (COX-2). These results were regulated by an inhibition of the mitogen-activated protein kinase signaling pathway (MAPK pathway), leading to a decreased nuclear translocation of NF-κB and AP-1, two major transcription factors involved in the inflammation process. In testing the anti-inflammatory properties of anthocyanin and hydroxybenzoic acid, the major components were detected in the black rice extracts based on our extraction protocol, and similar results were obtained. A schematic diagram of the proposed mechanism of the anti-inflammatory properties of black rice anthocyanin is presented in **Figure 4**. In a study on cyanidin-3-glucoside and protocatechuic acid, no beneficial effects were found against inflammation induced by LPS [73]. Therefore, the anti-inflammatory properties of black rice might require the synergistic action of many phytochemicals, which are rich in anthocyanin and other phenolic compounds that play a role in this process. Interestingly, the same study has demonstrated that the cooking process did not alter the anti-inflammatory potential of black rice. In another study, other researchers reported that cyanidin-3-glucoside displays anti-inflammatory effects [8]. Our group also conducted a study on the anti-inflammatory effects of proanthocyanidin-rich red rice extract via the suppression of the MAPK, AP-1, and NF-κB pathways in RAW 264.7 macrophages that induced inflammation by LPS [7]. It was found that the red rice medium polar fraction that was enriched with polyphenols and proanthocyanidins exerted potent anti-inflammatory activities by inhibiting the production of TNF-α, IL-6, and NO in LPS-activated macrophage, whereas the red rice nonpolar fractions displayed no anti-inflammatory properties. All of the above results indicate that black rice that is rich in anthocyanins and red rice that is rich in proanthocyanidins exhibit therapeutic potential for the treatment of inflammatory diseases.

3.3 Anticancer properties

Cancer is one of the leading causes of morbidity and mortality worldwide. Notably, only 10% at the most of all cancers are due to genetic factors, while 90% are directly or indirectly correlated with an individual's lifestyle and dietary habits [83]. Many scientific reports have shown that a healthy lifestyle, including a diet rich in natural products, such as herbs, cereals, fruits, and vegetables, can help reduce the risk of cancer [84, 85]. Some of the phytochemicals found in these natural products

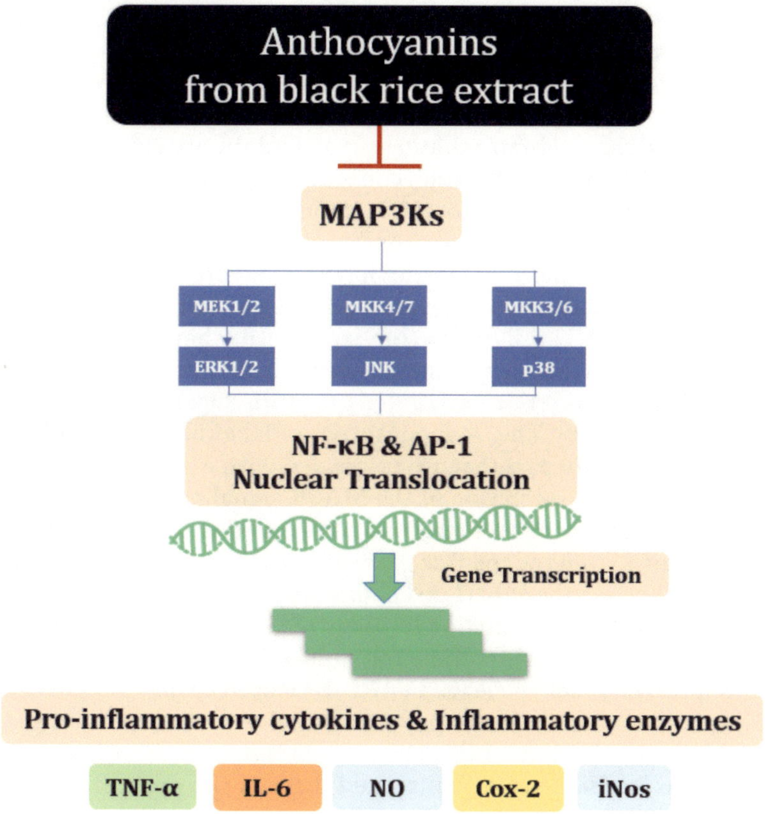

Figure 4.
Schematic diagram of anti-inflammatory properties of black rice anthocyanin.

are secondary metabolites, including phenolic compounds, bioflavonoids, terpenoids, and alkaloids. In this chapter we shall focus more on the presence of phenolic compounds and flavonoids, including anthocyanins and proanthocyanidins, as the major compounds found in pigmented rice, especially in rice germ and bran.

Active components of pigmented rice bran have demonstrated anticancer properties in in vitro cancer cell models, including those involving leukemia, colon, breast, liver, and stomach cancer cells. In a study on the anticancer potential of rice bran against the proliferation of leukemic cell lines, the antioxidant activities of the active compounds found in rice bran were noted for this beneficial effect [10]. Another investigation on the tumor suppression activities of rice bran from different pigmented and nonpigmented rice varieties reported that 70% ethanolic extract of the pigmented rice bran inhibited phorbol ester-induced tumor promotion in a better manner when compared to the nonpigmented rice bran variety [11]. In yet another study, the growth inhibitory effect of rice bran polyphenols, mainly γ-oryzanol and its derivatives, has been reported in human colorectal adenocarcinoma [86]. The anticancer activity of rice bran could be varied considerably in different rice cultivars or varieties in accordance with the different chemical profiles of the active compounds. In addition, the second study had analyzed seven varieties of rice bran for their growth inhibition potential against human colorectal cancer cells and reported on variations in the degree of growth inhibition depending upon the rice bran variety [9]. Some evidences have indicated that cyanidin-3-glucoside and peonidin-3-glucoside obtained from black rice anthocyanin can be combined with doxorubicin to inhibit cancer cell growth, while both anthocyanin compounds could inhibit cancer invasion into other tissues through the downregulation of the degradative enzymes MMP-2 and MMP-9 [14]. Interestingly, Chen et al. [87] compared the relationship of the bioactive compounds with the growth inhibitory effects of pigmented rice bran extracts. The results revealed that the light brown bran had no effect, the purple bran exhibited a minor effect on leukemia and cervical cancer cells, and the red bran exhibited strong inhibitory effects on leukemic, cervical, and stomach cancer cells. High concentrations of protocatechuic acid and anthocyanins in purple bran and proanthocyanidins in red rice bran have been singled out for their growth inhibitory effects against human cancer cells.

Many studies on anticancer properties have been reported in Thai rice cultivars. In an important study, Kum Phayao black rice cultivar was found to be highly cytotoxic to human HepG2 cells when compared with other Northern Thai purple rice cultivars [12]. In yet another study, the alcoholic extracts of black-purple rice grain cultivar Kum Doi Saket demonstrated an antimutagenic activity against aflatoxin B1 in Ames tests [88]. The therapeutic potential of black rice anthocyanin for treating inflammatory diseases that are associated with cancer has been proposed for its mechanism via the inhibition of the MAPK signaling pathway [31]. A very recent study conducted by our research group revealed that the proanthocyanidin-rich fraction isolated from the red rice germ and bran of the Kum Doi Saket cultivar grown in the northern part of Thailand significantly reduced the cell viability of HepG2 cells (IC_{50} value at 20 μg/ml) [13]. The proanthocyanidin-rich fraction could inhibit cell proliferation and induce cell apoptosis by increasing the apoptotic proteins, such as cleaved PARP-1, cleaved caspase 8, and cleaved caspase-3, and decreasing the anti-apoptotic protein survivin without p53 protein changes. A schematic diagram of this mechanism is presented in **Figure 5**. In addition, our previous studies have demonstrated that red rice grain extracts with high proanthocyanidin content displayed an anti-metastasis effect on invasive human breast carcinoma cells MDA-MB231 [16] and human fibrosarcoma HT1080 cell lines [15]. In addition, proanthocyanidins in other colored plants, such as grapes and blackberries, have demonstrated anticancer, anti-inflammatory, and antioxidant activities to a similar extent as the proanthocyanidins that are found in red rice germ and bran [7–9, 13].

3.4 Anti-inflamm-aging properties

Inflamm-aging, a state of chronic, low-level systemic inflammation, is a widespread feature of human aging and a major risk factor for disabilities and mortality in aging individuals [89, 90]. Inflamm-aging is characterized by an overall increase in plasma levels of pro-inflammatory cytokines, such as IL-6 and TNF-α, and

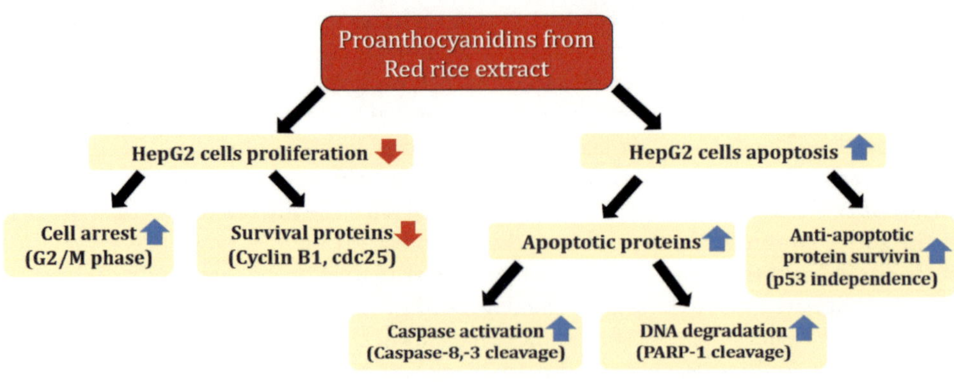

Figure 5.
Schematic diagram of anticancer properties of red rice proanthocyanidins.

subsequently can increase major inflammatory markers such as C-reactive protein (CRP) and serum amyloid A. This generalized pro-inflammatory status potentially triggers the onset of the most important age-related diseases such as cardiovascular diseases, atherosclerosis, metabolic syndrome, type 2 diabetes, obesity, neurodegeneration, sarcopenia, frailty, and cancer [91, 92].

Since the anti-inflammatory effects of phytochemical components in black rice and red rice (anthocyanins and proanthocyanidins) are able to target many inflammatory signaling pathways, such as the MAP kinase and AMP-activated protein kinase (AMPK) and mTOR pathway, the result can also decrease free radical production by their antioxidant activity, inhibiting NF-κB activation and reducing the expression of inflammatory mediators (NO, iNOS, and pro-inflammatory cytokines) [7, 31, 93, 94]. Therefore, this has made natural pigmented rice a promising candidate as an anti-inflamm-aging agent. Some relevant studies have found that a Mediterranean diet (a diet involving high consumption of vegetables, fruits, and whole grains such as pigmented rice, olive oil, and fish, but low in the intake of saturated fats and other animal fats) can modulate the multi-interconnected processes that are involved in inflammatory responses such as free radical production, NF-κB activation, and the expression of inflammatory mediators by balancing between pro- and anti-inflamm-aging activities as well as maintaining healthy gut microbiota homeostasis and epigenetic modulation of oncogenesis through specific microRNAs [95, 96].

Several studies have identified a number of actions of anthocyanins in a phytochemical diet in the context of neuroinflammation and neurodegeneration in aging individuals. It was also recently reported in an experimental model of multiple sclerosis that anthocyanins (100 mg/kg) could effectively suppress the secretion of pro-inflammatory mediators and protect cellular components against oxidative damages that were induced by demyelination [97]. Anthocyanins also protect neuronal cells from prooxidant and pro-inflammatory damage via the modulation of nuclear factor (erythroid-derived 2)-like 2 (Nrf2) and the inhibition of NF-κB pathways [98]. Moreover, anthocyanins also exhibited a similar degree of anti-inflammatory effects, and these compounds suppressed the expression and secretion of pro-inflammatory mediators in macrophages by inhibiting the nuclear translocation of NF-κB [99].

Red rice extracts that contain high levels of proanthocyanidins were also found to have neuroprotective effects and anti-inflamm-aging effects that are similar to those of anthocyanins. Previous studies have found that in primary hippocampal neuronal cells that had been treated with proanthocyanidins (14 μg/ml) and exposed to LPS, the major neuroprotective effects of proanthocyanidins were involved with a reduction of NF-κB, p38, and JNK [100]. In brief, the consumption of foods rich in polyphenols has been associated with the prevention of chronic diseases. In particular, anthocyanins, proanthocyanidins that act through various mechanisms that modulate the inflammatory signaling pathways, result in a reduction of inflammation that is often seen in aging individuals. A schematic diagram of the proposed mechanism of anti-inflamm-aging properties of black rice and red rice is presented in **Figure 6**. From the aforementioned results, it has been determined that black rice and red rice with their anti-inflamm-aging properties have a therapeutic potential that would likely need to be further investigated in geriatrics and gerontology fields.

3.5 Skin anti-aging properties

Many studies have shown that bioactive compounds found in pigmented rice, such as proanthocyanidin, catechin, vanillic acid, and oryzanol, may be useful in the cosmetic and nutraceutical industries as skin antiaging agents. As mentioned

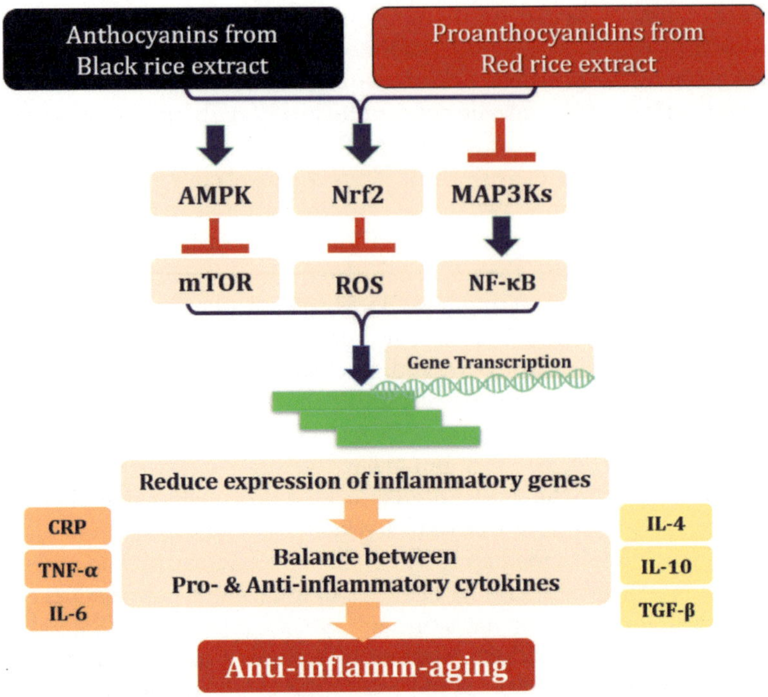

Figure 6.
Schematic diagram of anti-inflamm-aging properties of pigment rice.

in the previous section in this chapter, these bioactive compounds demonstrated antioxidant and anti-inflammatory properties. For the enhancement of the knowledge of skin antiaging properties, the bioactive compounds in the pigmented rice extract have been elucidated in a number of research laboratories. Skin aging is a process characterized by progressive physiological and structural changes in the skin. These changes could be considered as individually intrinsic and extrinsic factors, such as those associated with age, lifestyle, diet, and sunlight. Additionally, certain environmental factors can contribute to skin aging [101]. In the skin aging process, the level of degradative enzymes, such as elastase and collagenase, in skin fibroblasts are elevated, and this can lead to a loss of skin firmness and the appearance of wrinkles [102]. Mature skin in the elderly or those with sun-exposed skin can cause dark spot formations on the skin or result in the over-synthesis of melanin [103]. Hence, natural or herbal products that can exert skin benefits, including scavenging reactive oxygen species (ROS), the suppression of extracellular matrix degradation enzymes, and the inhibition of melanin synthesis, can be applied in skincare products for their beneficial skin anti-aging properties.

As pigmented rice has been reported to possess antioxidant properties, the extracts could be used for skin-anti-aging purposes. In a study by our research group, red rice extract showed anti-photoaging activity by protecting UV-induced collagen and hyaluronic acid degradation in human skin fibroblasts [21]. The red rice extract also inhibited collagenase and MMP-2 activity. In another study our group [22] has elucidated the skin antiaging properties of the main bioactive compounds in red rice extract including proanthocyanidin, catechin, hydroxybenzoic acid, vanillic acid, and oryzanol. The results showed that collagenase and MMP-2 activity were strongly inhibited by proanthocyanidin and catechin, whereas hydroxybenzoic acid, vanillic acid, and oryzanol had no effect. Both proanthocyanidin and catechin significantly induced the synthesis of collagen and hyaluronic acid, which is an important biological target for skin antiaging agents. Proanthocyanidins and γ-oryzanol could reduce the melanin content in B16-F10 melanoma cells. Some studies have proposed the use of red rice callus or stem cells as a source of materials for replenishing the aging body in a series of experiments. The results demonstrated the efficacy of red rice callus in cosmetic products on 28 volunteer subjects aged 30–55 years and proved to promote skin lightening, hydration, and elasticity. On the other hand, a study performed involving five different varieties of Thai pigmented rice demonstrated that all rice crude extracts with 50% ethanol exhibited a weak level of activity on tyrosinase inhibition [23]. This result is similar to our findings which demonstrated that proanthocyanidin and oryzanol could reduce melanin content but had no effect on mushroom tyrosinase activity [22]. However, our results have produced experimental data to support that proanthocyanidin decreased cellular tyrosinase activity leading to a decrease in melanin content. As has been mentioned previously, proanthocyanidin is highly present in red rice germ and bran and is very similar in chemical structure to the oligomers of catechin and epicatechin that are found in grape seeds and red wine. It is noteworthy to cite the findings of a study that found that the oral administration of grape seed extract was effective in lightening UV-induced pigmentation of guinea pig skin by a reduction in the number of 3,4-dihydroxyphenylalanine (DOPA)-positive melanocytes, Ki-67 positive, proliferating cell nuclear antigen (PCNA)-positive melanin-containing cells in the basal epidermal layer of the UV-irradiated skin in grape seed extract-fed guinea pigs. In addition, this study has demonstrated that grape seed extract effectively inhibited mushroom tyrosinase activity and inhibited melanogenesis without inhibiting the growth of culture B16-F10 mouse melanoma cells.

4. Conclusion

In this chapter, the by-products of rice processing, such as germ and bran, contain a wide range of biologically active compounds that can be recovered and used in a variety of approaches in nutraceuticals. This is in correlation with an increasingly deeper understanding of the predominant bioactive compounds found in pigmented rice, particularly anthocyanin and proanthocyanidin found in black and red rice, respectively. The dietary intervention and other high-value applications in functional food and cosmetic products have been attracting ever-growing attention in recent decades. The need for scientific evidence of pigmented rice bioactive compounds in different cultivars is encouraging for future perspectives within the new global business era of nutraceutical and agriculture expansion.

Most of the studies on the biological properties of black or red rice bioactive compounds have been conducted through an in vitro approach; however, only a few reports have been applied in preclinical or in animal studies. Further investigations will be needed to produce evidence on the efficacy of pigmented rice in terms of the anticancer activities and anti-inflammation properties in sub-chronic cases, especially among the aging members of the society in which sub-chronic inflammation commonly leads to noncommunicable diseases in later life. In addition, scientific studies have determined that the skin antiaging properties of pigmented rice should be useful and available in clinical studies for their efficacy and their further development in skincare products.

Acknowledgements

This work was supported by Chiang Mai University and The Excellent Center for Research and Development of Natural Products for Health, Chiang Mai University.

Conflict of interest

The authors declare no conflict of interest.

Notes/thanks/other declarations

The authors would like to thank the Faculty of Medicine of Chiang Mai University (Grant No. 096/2560), The Excellent Center for Research and Development of Natural Products for Health, the Agricultural Research Development Agency (ARDA; Public Organization), and the National Research Council of Thailand (NRCT) for their research funding for rice researches in our laboratory. Our research works that have been contributed in this chapter would not have been completed without their financial support. The authors are also grateful for the English proof editing from Mr. Russel Kirk Hollis of the English Department, Faculty of Humanities, Chiang Mai University.

Author details

Pornngarm Limtrakul (Dejkriengkraikul)[1,2*], Warathit Semmarath[1,2] and Sariya Mapoung[1,2]

1 Department of Biochemistry, Faculty of Medicine, Chiang Mai University, Chiang Mai, Thailand

2 The Excellent Center for Research and Development of Natural Products for Health, Chiang Mai University, Chiang Mai, Thailand

*Address all correspondence to: pornngarm.d@cmu.ac.th

References

[1] The Rice Department MoAaC. Thailand Rice Cultivation Areas. 2018, October 18. Available from: www.ricethailand.go.th/rkb3/Eb_024.pdf

[2] Mo A. Government of India. 2008. Available from: www.indiastat.com

[3] FAOSTAT. Food and Agriculture Organization of the United Nations. Preliminary Agricultural Production 2009 data. Available from: http://faostat.fao.org

[4] Chinprahast N, Tungsomboon T, Nagao P. Antioxidant activities of Thai pigmented rice cultivars and application in sunflower oil. International Journal of Food Science and Technology. 2016;**51**(1):46-53

[5] Chen MH, McClung AM, Bergman CJ. Bran data of total flavonoid and total phenolic contents, oxygen radical absorbance capacity, and profiles of proanthocyanidins and whole grain physical traits of 32 red and purple rice varieties. Data in Brief. 2016;**8**:6-13. DOI: 10.1016/j.dib.2016.05.001

[6] Surarit W, Jansom C, Lerdvuthisopon N, Kongkham S, Hansakul P. Evaluation of antioxidant activities and phenolic subtype contents of ethanolic bran extracts of Thai pigmented rice varieties through chemical and cellular assays. International Journal of Food Science and Technology. 2015;**50**(4):990-998

[7] Limtrakul P, Yodkeeree S, Pitchakarn P, Punfa W. Anti-inflammatory effects of proanthocyanidin-rich red rice extract via suppression of MAPK, AP-1 and NF-κB pathways in raw 264.7 macrophages. Nutrition Research and Practice. 2016;**10**(3):251-258

[8] Xia X, Ling W, Ma J, Xia M, Hou M, Wang Q, et al. An anthocyanin-rich extract from black rice enhances atherosclerotic plaque stabilization in apolipoprotein E–deficient mice. The Journal of Nutrition. 2006;**136**(8):2220-2225

[9] Forster GM, Raina K, Kumar A, Kumar S, Agarwal R, Chen M-H, et al. Rice varietal differences in bioactive bran components for inhibition of colorectal cancer cell growth. Food Chemistry. 2013;**141**(2):1545-1552

[10] Revilla E, Santa-María C, Miramontes E, Candiracci M, Rodríguez-Morgado B, Carballo M, et al. Antiproliferative and immunoactivatory ability of an enzymatic extract from rice bran. Food Chemistry. 2013;**136**(2):526-531

[11] Nam S, Choi S, Kang M, Koh H, Kozukue N, Friedman M. Bran extracts from pigmented rice seeds inhibit tumor promotion in lymphoblastoid B cells by phorbol ester. Food and Chemical Toxicology. 2005;**43**(5):741-745

[12] Banjerdpongchai R, Wudtiwai B, Sringarm K. Cytotoxic and apoptotic-inducing effects of purple rice extracts and chemotherapeutic drugs on human cancer cell lines. Asian Pacific Journal of Cancer Prevention. 2013;**14**(11):6541-6548

[13] Upanan S, Yodkeeree S, Thippraphan P, Punfa W, Wongpoomchai R, Limtrakul Dejkriengkraikul P. The proanthocyanidin-rich fraction obtained from red rice germ and bran extract induces HepG2 hepatocellular carcinoma cell apoptosis. Molecules. 2019;**24**(4):813-823. DOI: 10.3390/molecules24040813

[14] Chen P-N, Kuo W-H, Chiang C-L, Chiou H-L, Hsieh Y-S, Chu S-C. Black rice anthocyanins inhibit cancer cells invasion via repressions of MMPs and u-PA expression. Chemico-Biological Interactions. 2006;**163**(3):218-229

[15] Pintha K, Yodkeeree S, Pitchakarn P, Limtrakul P. Anti-invasive activity against cancer cells of phytochemicals in red jasmine rice (*Oryza sativa* L.). Asian Pacific Journal of Cancer Prevention. 2014;**15**(11):4601-4607

[16] Pintha K, Yodkeeree S, Limtrakul P. Proanthocyanidin in red rice inhibits MDA-MB-231 breast cancer cell invasion via the expression control of invasive proteins. Biological & Pharmaceutical Bulletin. 2015;**38**(4):571-581. DOI: 10.1248/bpb.b14-00719

[17] Chiang AN, Wu HL, Yeh HI, Chu CS, Lin HC, Lee WC. Antioxidant effects of black rice extract through the induction of superoxide dismutase and catalase activities. Lipids. 2006;**41**(8):797-803

[18] Posuwan J, Prangthip P, Leardkamolkarn V, Yamborisut U, Surasiang R, Charoensiri R, et al. Long-term supplementation of high pigmented rice bran oil (*Oryza sativa* L.) on amelioration of oxidative stress and histological changes in streptozotocin-induced diabetic rats fed a high fat diet; riceberry bran oil. Food Chemistry. 2013;**138**(1):501-508

[19] Jang H-H, Park M-Y, Kim H-W, Lee Y-M, Hwang K-A, Park J-H, et al. Black rice (*Oryza sativa* L.) extract attenuates hepatic steatosis in C57BL/6 J mice fed a high-fat diet via fatty acid oxidation. Nutrition and Metabolism. 2012;**9**(1):27

[20] Yang Y, Andrews MC, Hu Y, Wang D, Qin Y, Zhu Y, et al. Anthocyanin extract from black rice significantly ameliorates platelet hyperactivity and hypertriglyceridemia in dyslipidemic rats induced by high fat diets. Journal of Agricultural and Food Chemistry. 2011;**59**(12):6759-6764

[21] Limtrakul P, Yodkeeree S, Punfa W, Srisomboon J. Inhibition of the MAPK signaling pathway by red rice extract in UVB-irradiated human skin fibroblasts. Natural Product Communications. 2016;**11**(12):1877-1882. DOI: 10.1177/1934578X1601101226

[22] Yodkeeree S, Thippraphan P, Punfa W, Srisomboon J, Limtrakul P. Skin anti-aging assays of proanthocyanidin rich red rice extract, oryzanol and other phenolic compounds. Natural Product Communications. 2018;**13**(8):967-972. DOI: 10.1177/1934578X1801300812

[23] Teeranachaideekul V, Wongrakpanich A, Leanpolchareanchai J, Thirapanmethee K, Sirichaovanichkarn C. Characterization, biological activities and safety evaluation of different varieties of Thai pigmented rice extracts for cosmetic applications. Pharmaceutical Sciences Asia. 2018;**45**(3):140-153

[24] Saewan N, Vichit W, Prinyarux T. Anti-aging efficacy of Thai red rice callus cosmetic product. The Journal of Applied Science. 2018;**17**(Special):63-72

[25] Sharif MK, Butt MS, Anjum FM, Khan SH. Rice bran: A novel functional ingredient. Critical Reviews in Food Science and Nutrition. 2014;**54**(6):807-816

[26] Sivamaruthi BS, Kesika P, Chaiyasut C. A comprehensive review on anti-diabetic property of rice bran. Asian Pacific Journal of Tropical Biomedicine. 2018;**8**(1):79

[27] Goufo P, Trindade H. Rice antioxidants: Phenolic acids, flavonoids, anthocyanins, proanthocyanidins, tocopherols, tocotrienols, γ-oryzanol, and phytic acid. Food Science & Nutrition. 2014;**2**(2):75-104

[28] Imsanguan P, Roaysubtawee A, Borirak R, Pongamphai S, Douglas S, Douglas PL. Extraction of α-tocopherol and γ-oryzanol from rice bran. LWT-Food Science and Technology. 2008;**41**(8):1417-1424

[29] Heinemann RJ, Xu Z, Godber JS, Lanfer-Marquez UM. Tocopherols, tocotrienols, and γ-oryzanol contents in japonica and indica subspecies of rice (*Oryza sativa* L.) cultivated in Brazil. Cereal Chemistry. 2008;**85**(2):243-247

[30] Nam SH, Choi SP, Kang MY, Koh HJ, Kozukue N, Friedman M. Antioxidative activities of bran extracts from twenty one pigmented rice cultivars. Food Chemistry. 2006;**94**(4):613-620

[31] Limtrakul P, Yodkeeree S, Pitchakarn P, Punfa W. Suppression of inflammatory responses by black rice extract in RAW 264.7 macrophage cells via downregulation of NF-kB and AP-1 signaling pathways. Asian Pacific Journal of Cancer Prevention. 2015;**16**(10):4277-4283

[32] Harukaze A, Murata M, Homma S. Analyses of free and bound phenolics in rice. Food Science and Technology Research. 1999;**5**(1):74-79

[33] Tian S, Nakamura K, Kayahara H. Analysis of phenolic compounds in white rice, brown rice, and germinated brown rice. Journal of Agricultural and Food Chemistry. 2004;**52**(15):4808-4813

[34] Kim H-J, Lee S-B, Park K-A, Hong I-K. Characterization of extraction and separation of rice bran oil rich in EFA using SFE process. Separation and Purification Technology. 1999;**15**(1):1-8

[35] Chen C-R, Wang L-Y, Wang C-H, Ho W-J, Chang C-MJ. Supercritical carbon dioxide extraction of rice bran oil and column partition fractionation of γ-oryzanols. Separation and Purification Technology. 2008;**61**(3):358-365

[36] Rogalinski T, Herrmann S, Brunner G. Production of amino acids from bovine serum albumin by continuous sub-critical water hydrolysis. The Journal of Supercritical Fluids. 2005;**36**(1):49-58

[37] Haghighat Khajavi S, Ota S, Kimura Y, Adachi S. Kinetics of maltooligosaccharide hydrolysis in subcritical water. Journal of Agricultural and Food Chemistry. 2006;**54**(10):3663-3667

[38] Pereira-Caro G, Cros G, Yokota T, Crozier A. Phytochemical profiles of black, red, brown, and white rice from the Camargue region of France. Journal of Agricultural and Food Chemistry. 2013;**61**(33):7976-7986

[39] Yao Y, Sang W, Zhou M, Ren G. Antioxidant and α-glucosidase inhibitory activity of colored grains in China. Journal of Agricultural and Food Chemistry. 2009;**58**(2):770-774

[40] Abdel-Aal E-SM, Young JC, Rabalski I. Anthocyanin composition in black, blue, pink, purple, and red cereal grains. Journal of Agricultural and Food Chemistry. 2006;**54**(13):4696-4704

[41] Finocchiaro F, Ferrari B, Gianinetti A, Dall'Asta C, Galaverna G, Scazzina F, et al. Characterization of antioxidant compounds of red and white rice and changes in total antioxidant capacity during processing. Molecular Nutrition & Food Research. 2007;**51**(8):1006-1019

[42] Oki T, Masuda M, Kobayashi M, Nishiba Y, Furuta S, Suda I, et al. Polymeric procyanidins as radical-scavenging components in red-hulled rice. Journal of Agricultural and Food Chemistry. 2002;**50**(26):7524-7529

[43] Gunaratne A, Wu K, Li D, Bentota A, Corke H, Cai Y-Z. Antioxidant activity and nutritional quality of traditional red-grained rice varieties containing proanthocyanidins. Food Chemistry. 2013;**138**(2-3):1153-1161

[44] Min B, McClung AM, Chen MH. Phytochemicals and antioxidant capacities in rice brans of different color. Journal of Food Science. 2011;**76**(1):C117-C126

[45] Jang S, Xu Z. Lipophilic and hydrophilic antioxidants and their antioxidant activities in purple rice bran. Journal of Agricultural and Food Chemistry. 2009;**57**(3):858-862

[46] Qiu Y, Liu Q, Beta T. Antioxidant activity of commercial wild rice and identification of flavonoid compounds in active fractions. Journal of Agricultural and Food Chemistry. 2009;**57**(16):7543-7551

[47] Nakornriab M, Sriseadka T, Wongpornchai S. Quantification of carotenoid and flavonoid components in brans of some Thai black rice cultivars using supercritical fluid extraction and high-performance liquid chromatography-mass spectrometry. Journal of Food Lipids. 2008;**15**(4):488-503

[48] Pérez-Jiménez J, Arranz S, Saura-Calixto F. Proanthocyanidin content in foods is largely underestimated in the literature data: An approach to quantification of the missing proanthocyanidins. Foodservice Research International. 2009;**42**(10):1381-1388

[49] Labarbe B, Cheynier V, Brossaud F, Souquet J-M, Moutounet M. Quantitative fractionation of grape proanthocyanidins according to their degree of polymerization. Journal of Agricultural and Food Chemistry. 1999;**47**(7):2719-2723

[50] Irakli MN, Samanidou VF, Biliaderis CG, Papadoyannis IN. Simultaneous determination of phenolic acids and flavonoids in rice using solid-phase extraction and RP-HPLC with photodiode array detection. Journal of Separation Science. 2012;**35**(13):1603-1611

[51] Laokuldilok T, Shoemaker CF, Jongkaewwattana S, Tulyathan V. Antioxidants and antioxidant activity of several pigmented rice brans. Journal of Agricultural and Food Chemistry. 2010;**59**(1):193-199

[52] Cicero AF, Gaddi A. Rice bran oil and gamma-oryzanol in the treatment of hyperlipoproteinaemias and other conditions. Phytotherapy Research. 2001;**15**(4):277-289

[53] Wilson TA, Ausman LM, Lawton CW, Hegsted DM, Nicolosi RJ. Comparative cholesterol lowering properties of vegetable oils: Beyond fatty acids. Journal of the American College of Nutrition. 2000;**19**(5):601-607

[54] Kim JS, Godber J. Oxidative stability and vitamin elevels increased in restructured beef roasts with added rice bran oil. Journal of Food Quality. 2001;**24**(1, 1):17-26

[55] Tsuji H, Kimoto M, Natori Y. Allergens in major crops. Nutrition Research. 2001;**21**(6):925-934

[56] Kennedy G, Burlingame B. Analysis of food composition data on rice from a plant genetic resources perspective. Food Chemistry. 2003;**80**(4):589-596

[57] Sun Q, Spiegelman D, van Dam RM, Holmes MD, Malik VS, Willett WC, et al. White rice, brown rice, and risk of type 2 diabetes in US men and women. Archives of Internal Medicine. 2010;**170**(11):961-969

[58] Service USDoAAR. USDA Food Composition Databases. Available from: https://ndb.nal.usda.gov/ndb/

[59] Kazemzadeh M, Safavi SM, Nematollahi S, Nourieh Z. Effect of brown rice consumption on inflammatory marker and cardiovascular risk factors among overweight and obese non-menopausal female adults. International Journal of Preventive Medicine. 2014;**5**(4):478-488

[60] Kozuka C, Yabiku K, Takayama C, Matsushita M, Shimabukuro M,

Masuzaki H. Natural food science based novel approach toward prevention and treatment of obesity and type 2 diabetes: Recent studies on brown rice and γ-oryzanol. Obesity Research & Clinical Practice. 2013;7(3):e165-e172

[61] Mäkynen K, Chitchumroonchokchai C, Adisakwattana S, Failla M, Ariyapitipun T. Effect of gamma-oryzanol on the bioaccessibility and synthesis of cholesterol. European Review for Medical and Pharmacological Sciences. 2012;**16**(1):49-56

[62] Henderson AJ, Ollila CA, Kumar A, Borresen EC, Raina K, Agarwal R, et al. Chemopreventive properties of dietary rice bran: Current status and future prospects. Advances in Nutrition. 2012;**3**(5):643-653

[63] Tantamango YM, Knutsen SF, Beeson WL, Fraser G, Sabate J. Foods and food groups associated with the incidence of colorectal polyps: The adventist health study. Nutrition and Cancer. 2011;**63**(4):565-572

[64] Ling WH, Cheng QX, Ma J, Wang T. Red and black rice decrease atherosclerotic plaque formation and increase antioxidant status in rabbits. The Journal of Nutrition. 2001;**131**(5):1421-1426

[65] Toyokuni S, Itani T, Morimitsu Y, Okada K, Ozeki M, Kondo S, et al. Protective effect of colored rice over white rice on Fenton reaction-based renal lipid peroxidation in rats. Free Radical Research. 2002;**36**(5):583-592

[66] Xia M, Ling WH, Ma J, Kitts DD, Zawistowski J. Supplementation of diets with the black rice pigment fraction attenuates atherosclerotic plaque formation in apolipoprotein E deficient mice. The Journal of Nutrition. 2003;**133**(3):744-751

[67] Kim JY, Do MH, Lee SS. The effects of a mixture of brown and black rice on lipid profiles and antioxidant status in rats. Annals of Nutrition & Metabolism. 2006;**50**(4):347-353

[68] Cho M-K, Kim M-H, Kang M-Y. Effects of rice embryo and embryo jelly with black rice bran pigment on lipid metabolism and antioxidant enzyme activity in high cholesterol-fed rats. Applied Biological Chemistry. 2008;**51**(3):200-206

[69] Mingwei Z, Ruifen Z, Baojiang G. Hypolipidemic and antioxidative effects of black rice pericarp extract accompanied by its components analysis. Scientia Agricultura Sinica. 2006;**39**(11):2368-2373

[70] Kaneda I, Kubo F, Sakurai H. Relationship between trace metal concentration and antioxidative activity of ancient rice bran (red and black rice) and a present-day rice bran (Koshihikari). Journal of Trace Elements in Medicine and Biology. 2007;**21**(1):43-51

[71] Zaupa M, Calani L, Del Rio D, Brighenti F, Pellegrini N. Characterization of total antioxidant capacity and (poly) phenolic compounds of differently pigmented rice varieties and their changes during domestic cooking. Food Chemistry. 2015;**187**:338-347

[72] Tang Y, Cai W, Xu B. From rice bag to table: Fate of phenolic chemical compositions and antioxidant activities in waxy and non-waxy black rice during home cooking. Food Chemistry. 2016;**191**:81-90

[73] Bhawamai S, Lin S-H, Hou Y-Y, Chen Y-H. Thermal cooking changes the profile of phenolic compounds, but does not attenuate the anti-inflammatory activities of black rice. Food & Nutrition Research. 2016;**60**(1):32941

[74] Zhang X, Mosser D. Macrophage activation by endogenous danger

signals. The Journal of Pathology: A Journal of the Pathological Society of Great Britain and Ireland. 2008;**214**(2):161-178

[75] Han S, Lee JH, Kim C, Nam D, Chung W-S, Lee S-G, et al. Capillarisin inhibits iNOS, COX-2 expression, and proinflammatory cytokines in LPS-induced RAW 264.7 macrophages via the suppression of ERK, JNK, and NF-κB activation. Immunopharmacology and Immunotoxicology. 2013;**35**(1):34-42

[76] Ichikawa H, Ichiyanagi T, Xu B, Yoshii Y, Nakajima M, Konishi T. Antioxidant activity of anthocyanin extract from purple black rice. Journal of Medicinal Food. 2001;**4**(4):211-218

[77] Min S-W, Ryu S-N, Kim D-H. Anti-inflammatory effects of black rice, cyanidin-3-O-β-D-glycoside, and its metabolites, cyanidin and protocatechuic acid. International Immunopharmacology. 2010;**10**(8):959-966

[78] Mizushina Y, Kuriyama I, Yamazaki A, Akashi T, Yoshida H. Cycloartenyl trans-ferulate, a component of the bran byproduct of sake-brewing rice, inhibits mammalian DNA polymerase and suppresses inflammation. Food Chemistry. 2013;**141**(2):1000-1007

[79] Ng LT, Ko HJ. Comparative effects of tocotrienol-rich fraction, alpha-tocopherol and alpha-tocopheryl acetate on inflammatory mediators and nuclear factor kappa B expression in mouse peritoneal macrophages. Food Chemistry. 2012;**134**(2):920-925. DOI: 10.1016/j.foodchem.2012.02.206

[80] Lyu SY, Park WB. Production of cytokine and NO by RAW 264.7 macrophages and PBMC in vitro incubation with flavonoids. Archives of Pharmacal Research. 2005;**28**(5):573-581

[81] Ronchetti D, Borghi V, Gaitan G, Herrero JF, Impagnatiello F. NCX 2057, a novel NO-releasing derivative of ferulic acid, suppresses inflammatory and nociceptive responses in in vitro and in vivo models. British Journal of Pharmacology. 2009;**158**(2):569-579. DOI: 10.1111/j.1476-5381.2009.00324.x

[82] Ahmad SF, Zoheir KM, Abdel-Hamied HE, Ashour AE, Bakheet SA, Attia SM, et al. Grape seed proanthocyanidin extract has potent anti-arthritic effects on collagen-induced arthritis by modifying the T cell balance. International Immunopharmacology. 2013;**17**(1):79-87. DOI: 10.1016/j.intimp.2013.05.026

[83] Anand P, Kunnumakkara AB, Sundaram C, Harikumar KB, Tharakan ST, Lai OS, et al. Cancer is a preventable disease that requires major lifestyle changes. Pharmaceutical Research. 2008;**25**(9):2097-2116. DOI: 10.1007/s11095-008-9661-9

[84] Zheng J, Zhou Y, Li Y, Xu DP, Li S, Li HB. Spices for prevention and treatment of cancers. Nutrients. 2016;**8**(8):495-530. DOI: 10.3390/nu8080495

[85] Zhou Y, Li Y, Zhou T, Zheng J, Li S, Li HB. Dietary natural products for prevention and treatment of liver cancer. Nutrients. 2016;**8**(3):156. DOI: 10.3390/nu8030156

[86] Kong CK, Lam WS, Chiu LC, Ooi VE, Sun SS, Wong YS. A rice bran polyphenol, cycloartenyl ferulate, elicits apoptosis in human colorectal adenocarcinoma SW480 and sensitizes metastatic SW620 cells to TRAIL-induced apoptosis. Biochemical Pharmacology. 2009;**77**(9):1487-1496. DOI: 10.1016/j.bcp.2009.02.008

[87] Chen MH, Choi SH, Kozukue N, Kim HJ, Friedman M. Growth-inhibitory effects of pigmented rice bran extracts and three red bran fractions against human cancer cells: Relationships with composition and antioxidative activities. Journal of Agricultural and Food Chemistry.

2012;**60**(36):9151-9161. DOI: 10.1021/jf3025453

[88] Punvittayagul C, Sringarm K, Chaiyasut C, Wongpoomchai R. Mutagenicity and antimutagenicity of hydrophilic and lipophilic extracts of thai northern purple rice. Asian Pacific Journal of Cancer Prevention. 2014;**15**(21):9517-9522

[89] Franceschi C, Bonafe M, Valensin S, Olivieri F, De Luca M, Ottaviani E, et al. Inflamm-aging. An evolutionary perspective on immunosenescence. Annals of the New York Academy of Sciences. 2000;**908**(1):244-254

[90] Ferrucci L, Fabbri E. Inflam-mageing: Chronic inflammation in ageing, cardiovascular disease, and frailty. Nature Reviews. Cardiology. 2018;**15**(9):505-522. DOI: 10.1038/s41569-018-0064-2

[91] Franceschi C. Inflammaging as a major characteristic of old people: Can it be prevented or cured? Nutrition Reviews. 2007;**65**(12 Pt 2):S173-S176. DOI: 10.1111/j.1753-4887.2007.tb00358.x

[92] Franceschi C, Capri M, Monti D, Giunta S, Olivieri F, Sevini F, et al. Inflammaging and anti-inflammaging: A systemic perspective on aging and longevity emerged from studies in humans. Mechanisms of Ageing and Development. 2007;**128**(1):92-105. DOI: 10.1016/j.mad.2006.11.016

[93] Xia S, Zhang X, Zheng S, Khanabdali R, Kalionis B, Wu J, et al. An update on inflamm-aging: Mechanisms, prevention, and treatment. Journal of Immunology Research. 2016;**2016**:1-2. DOI: 10.1155/2016/8426874

[94] Chung HY, Kim HJ, Kim KW, Choi JS, Yu BP. Molecular inflammation hypothesis of aging based on the anti-aging mechanism of calorie restriction. Microscopy Research and Technique. 2002;**59**(4):264-272. DOI: 10.1002/jemt.10203

[95] Martucci M, Ostan R, Biondi F, Bellavista E, Fabbri C, Bertarelli C, et al. Mediterranean diet and inflammaging within the hormesis paradigm. Nutrition Reviews. 2017;**75**(6):442-455. DOI: 10.1093/nutrit/nux013

[96] Ostan R, Lanzarini C, Pini E, Scurti M, Vianello D, Bertarelli C, et al. Inflammaging and cancer: A challenge for the Mediterranean diet. Nutrients. 2015;**7**(4):2589-2621. DOI: 10.3390/nu7042589

[97] Carvalho FB, Gutierres JM, Bohnert C, Zago AM, Abdalla FH, Vieira JM, et al. Anthocyanins suppress the secretion of proinflammatory mediators and oxidative stress, and restore ion pump activities in demyelination. The Journal of Nutritional Biochemistry. 2015;**26**(4):378-390. DOI: 10.1016/j.jnutbio.2014.11.006

[98] de Pascual-Teresa S. Molecular mechanisms involved in the cardiovascular and neuroprotective effects of anthocyanins. Archives of Biochemistry and Biophysics. 2014;**599**(1):68-74. DOI: 10.1016/j.abb.2014.04.012

[99] Lee SG, Kim B, Yang Y, Pham TX, Park YK, Manatou J, et al. Berry anthocyanins suppress the expression and secretion of proinflammatory mediators in macrophages by inhibiting nuclear translocation of NF-kappaB independent of NRF2-mediated mechanism. The Journal of Nutritional Biochemistry. 2014;**25**(4):404-411. DOI: 10.1016/j.jnutbio.2013.12.001

[100] Joseph JA, Shukitt-Hale B, Brewer GJ, Weikel KA, Kalt W, Fisher DR. Differential protection among fractionated blueberry polyphenolic families against DA-, Abeta(42)- and LPS-induced decrements in Ca^{2+} buffering in primary hippocampal cells. Journal of Agricultural and Food

Chemistry. 2010;**58**(14):8196-8204. DOI: 10.1021/jf100144y

[101] Ganceviciene R, Liakou AI, Theodoridis A, Makrantonaki E, Zouboulis CC. Skin anti-aging strategies. Dermatoendocrinol. 2012;**4**(3):308-319. DOI: 10.4161/derm.22804

[102] Sardy M. Role of matrix metalloproteinases in skin ageing. Connective Tissue Research. 2009;**50**(2):132-138. DOI: 10.1080/03008200802585622

[103] Skoczynska A, Budzisz E, Trznadel-Grodzka E, Rotsztejn H. Melanin and lipofuscin as hallmarks of skin aging. Postępy Dermatologii I Alergologii. 2017;**34**(2):97-103. DOI: 10.5114/ada.2017.67070

Chapter 2

Calophyllum inophyllum: Beneficial Phytochemicals, Their Uses, and Identification

David Febrilliant Susanto, Hakun Wirawasista Aparamarta, Arief Widjaja, Firdaus and Setiyo Gunawan

Abstract

Calophyllum inophyllum Linn. is one type of mangrove plant. This plant is commonly called nyamplung. This plant is abundant in Indonesia and has many properties that can be exploited from the roots, stems, and leaves to the seeds. All parts of this plant can be useful for human needs. Its oil is generally only used as biodiesel feedstock. The aim of this chapter is to discuss the identification and the uses of phytochemicals contained in *C. inophyllum* leaves. There are various kinds of phytochemicals contained in *C. inophyllum* leaves, such as triterpenoids, steroids, flavonoids, coumarins, xanthones, fatty acids, esters, alkenes, ethers, and alicyclic compounds. They have benefits to health, such as anticancer, anti-HIV, antiviral, antitumor, anti-inflammatory, antimicrobial, antineoplastic, antiplatelet, antipsychotics, antioxidant, antiaging, antileukemic, antimalarial, anticoagulant, antifeedant, analgesic, photoprotective, molluscicidal, and piscicidal agents. Extraction is a famous method for isolating phytochemicals in *C. inophyllum* leaves, based on the solvent polarity index.

Keywords: *C. inophyllum*, human health, identification, isolation, phytochemicals

1. Introduction

The name of Calophyllum inophyllum is Kallos that is taken from the Greek word, which means beautiful and meaningful Phullon leaves. *C. inophyllum* has many name designations that vary by region country. In the UK, the tree is known as a beautiful leaf (translation from Greek), Indian laurel (because it comes from India), Alexandrian laurel, and beach *Calophyllum* (because the trees usually grow on the waterfront). Moreover, the tree is also called as tamanu (Tahiti), fetau (Samoa), damanu (Fiji Island), te itai (Kiribati Island), nyamplung (Indonesia), Penaga Laut (Malaysia), kamani (Hawaii), foraha (Madagascar), and puna (island of Lakshadweep) [1].

According to Ong [2], the distribution map of *C. inophyllum* in the world is quite extensive. This species is commonly found in areas with a tropical climate. In the world, this species is found in countries such as Australia, Cambodia, the Cook Islands, Fiji, French Polynesia, India, Indonesia, Japan, Kiribati, Laos, Madagascar, Malaysia, the Marshall Islands, Myanmar, New Caledonia, Norfolk Island, Papua New Guinea, the Philippines, Reunion, Samoa, Solomon Islands, Sri Lanka, Taiwan, Province of China, Thailand, Tonga, Vanuatu, and Vietnam.

As for exotic species (endemic to a region), it can be found in the state of Djibouti, Eritrea, Ethiopia, Kenya, Nigeria, Somalia, Tanzania, Uganda, and the USA.

C. inophyllum plant spreads almost evenly throughout Indonesia, such as in the island of Sumatra (West Sumatra, Riau, Kepulauan Riau, Lampung, and Bangka Belitung), Java (Banten, West Java, Central Java, Yogyakarta, East Java), Bali Island, East Nusa Tenggara and West Nusa Tenggara, Kalimantan (West Kalimantan, Central Kalimantan, and South Kalimantan), Sulawesi (North Sulawesi, Gorontalo, Central Sulawesi, South Sulawesi, and Southeast Sulawesi), Maluku and North Maluku Islands, and Papua [3]. *C. inophyllum* plant has a taxonomy as follows [4]:

Kingdom: Plantae
Subkingdom: Tracheobionta
Super division: Spermatophyta
Division: Magnoliophyta
Class: Magnoliopsida
Subclass: *Dilleniidae*
Order: *Theales*
Family: Clusiaceae

C. inophyllum is a plant that is grown in the earthy sand and coastal areas with a hot weather [5]. It can also grow well at an altitude of 0–800 meters above sea level such as in forests, mountains, and swamps [6]. *C. inophyllum* is a versatile crop; all parts of this plant, such as leaves, root, and fruit (**Figure 1**), can be useful for humans. The benefit of its tree, bark, and seed is as plant conservation, source of timber and non-timber forest products (NTFPs), and vegetable oil, respectively [7]. In pharmaceuticals, it is known to function as an antibacterial, anticancer, antineoplastic, anti-inflammatory, antiplatelet, antipsychotics, antiviral, photoprotective, molluscicidal, and piscicidal agent [1]. **Table 1** shows the benefits of *C. inophyllum* crops obtained from previous works.

Because all parts of this plant can be useful in treating various diseases, some researchers have conducted further research on the phytochemical content of this plant. According to Ling et al. [1], the compounds which are contained in these plants include inophynone; canophyllol; canophyllic acid; calophyllolide;

Figure 1.
Parts of C. inophyllum crop.

Part of crops	Medicinal function		
	Iskandari and Anna [8]	**Su et al. [9]**	**Ling et al. [1]**
Leaves	Inhibit the growth of larvae of *Culex quinquefasciatus* and *Aedes aegypti*, an inhibitor of the HIV virus	Treat skin rashes, swelling of the legs, caring for burns, eye irritation, dysentery, migraine, and vertigo	Treat skin diseases, arthritis, sciatica, eye irritation
Root	Antibacterial	Treat dysentery, gonorrhea, indigestion, wounds, ulcers, and others	Treating internal hemorrhage
Fruit/seed	Inhibit the growth of larvae of *Culex quinquefasciatus*, antimicrobial compounds, and toxic agents	Treating stomach pain, itching, arthritis, burns, gonorrhea, arthritis, ulcers, and ringworm	Treat wounds, leprosy, neurological diseases, burns

Table 1.
Benefits and uses of C. inophyllum crops.

inophyllolide; inophyllum B, C, P, and E; jacareubin; (+)-calanolide A; inocalophyllins A and B; calophynone; calophyllumin C; inophyllin A; and others. Su et al. [9] mentioned that according to Filho et al. [10], in various parts of the tree, C. inophyllum contains phytochemicals, including xanthones, coumarins, chromanones (flavonoids, biflavonoids), triterpenes, tripenoids, and steroids. Coumarins in *C. inophyllum* contain two components, namely, calanolides A and B. From these studies it was found that coumarin compounds in *C. inophyllum* may be effective in treating cancer and inhibiting the HIV virus.

According to Lim [11], at least nine components have been isolated from the leaves of *C. inophyllum*, including 2-hydroxyxanthone; 4-hydroxyxanthone; 1,5-dihydroxyxanthone; 1,7-dihydroxyxanthone; 1,3,5-trihydroxy-2-methoxyxanthone; 6-6-deoxyjacaerubin; flavonoids, amentoflavone; kaempferol-3-O-α-L-rhamnoside; and quercetin-3-O-α-L-rhamnoside.

Of the three studies on the leaves above, there are some differences as well as questions obtained from the leaves of *C. inophyllum* content analysis. Some of the same compounds that have been isolated from *C. inophyllum* plants are quite diverse, including derivatives of xanthones [12, 13], coumarins [9], flavonoids [13], benzodipyranones [14], triterpenoids [12, 15], and steroids [9].

2. Identification of phytochemicals in *C. inophyllum* leaves

2.1 Xanthones

Xanthones are polyphenol components in nature with molecular formula $C_{13}H_8O_2$. They consist of bonding of two benzene rings connected by a carbonyl group and one oxygen. These conjugated ring systems inhibit the free rotation carbon bond. Xanthones have a basic framework consisting of 13 carbon atoms that make up the composition of C6-C1-C6 (**Figure 2**).

Xanthones are compounds with the basic framework of two phenyls connected by bridges carbonyl and oxygen (ether). Their biosynthesis is not known clearly but allegedly still in close contact with the biosynthesis of flavonoids and stilbenoid. It can be seen from the type of oxygenation and two types of aromatic rings which are derived from the shikimate (shikimic acid) and the acetate-malonate pathways.

Xanthones compound that was isolated from *C. inophyllum* plants, there are prenylated and some are not prenylated. Most xanthone compounds isolated from these plants showed a characteristic, one of which is a hydroxy group at C1. The possible oxygenation position is shown in **Figure 2**.

Xanthones are known to have a variety of bioactive properties, notably the ability of antioxidants as can be seen in **Figure 3**. Mangosteen xanthones were isolated from *Garcinia mangostana* found against free radicals and prevent oxidative damage of low-density lipoprotein [16]. Moreover, isolated xanthones from mangosteen also can inhibit HL60 leukemia cells [17]. Also, α-mangosteen extracted from *G. mangostana* L. has antibacterial activity against vancomycin-resistant enterococci (VRE) and methicillin-resistant *Staphylococcus aureus* (MRSA) [18].

Various xanthone compounds can be isolated from *C. inophyllum* leaves, such as caloxanthone A, caloxanthone B, caloxanthone C, maclura xanthones, inoxanthone, calophynic acid, 3,4-dihydroxy xanthones [4, 12, 19], brasilixanthone B, buchanaxanthone [20], inophyxanthone A, pancixanthone A, gerontoxanthone B, jacareubin, pyranojacaereubin, 2-hydroxy xanthone, 4-hydroxyxanthone, 1,3,5-trihydroxy-2-methoxyxanthone, and xanthones [21, 22].

2.2 Coumarins

Coumarin (benzopyrones) compound is one of the members of benzopyrone components. In the coumarin structure, there is a benzene ring which is tied with pyrone ring [23] as can be seen in **Figure 4**. They can be divided into four main types: simple coumarins, pyranocoumarins, furanocoumarins, and

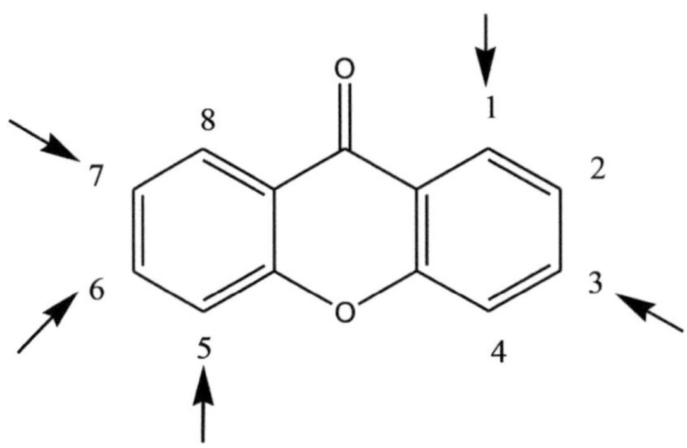

Figure 2.
Possible position oxygenation xanthone compound.

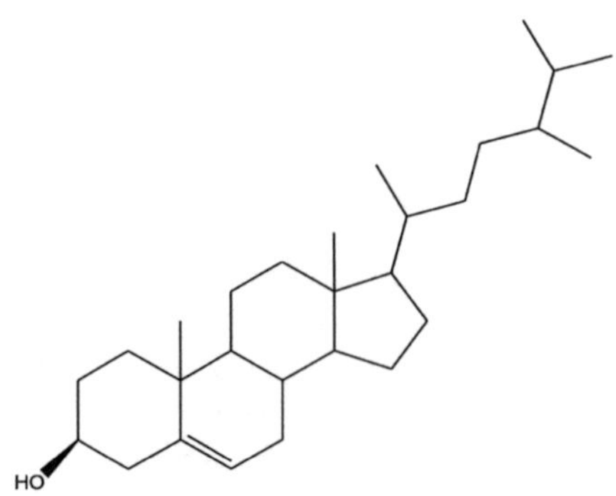

Figure 3.
Molecular structure of xanthones.

pyrone-substituted coumarins. All the reactions of coumarins focus on activation of C3,4—the double bond of the α,β-unsaturated lactone—and form a heterocyclic system [24].

Coumarins are commonly used in the agrochemical, perfume, and medical industries. They have high antitumor and antibacterial activities. Antitumor activity of 7-hydroxycoumarins against several tumor cell lines has been identified. Coumarins and their derivatives have activity as barrier against cellular proliferation in various carcinoma cell lines [25]. Besides that, they also have anticoagulant, antioxidant, antimicrobial, antiviral, anti-inflammatory, antimalarial, and analgesic activities [26].

The biosynthesis of coumarin compounds is derived from the shikimic acid pathway or still in line with the phenyl group propanoid. The skeleton benzopyran-2-on of coumarin is originating from the acid-cinnamic acid via ortho-hydrolysis. Ortho-coumaric acid produced after undergoing *cis-trans* isomerization undergoes condensation [27]. Characteristic of these compounds is their lactone group formed from the acid on the tip of propane with a hydroxy group on the phenyl group. Oxygenation coumarin compounds in the aromatic ring are also typical and are intermittent. The structure of the coumarin derivatives can be divided into four categories based on the group bound to the C_4 : 4-metilcoumarin, 4-fenilcoumarin, and 4-(n-propyl)coumarin.

2.3 Benzodipyranones

Benzodipyranones are derivative of chromone. These compounds have a skeleton similar to stilbene with two additional prenyl groups. Some benzodipyranone compounds have been isolated from the *C. inophyllum* leaves, such as (2S, 3R) and (2R, 3R)-2,3-dihydro-5-hydroxy-2,3,8,8-tetramethyl-6-(1-phenylethenyl)-4H, 8H-benzo [1,2-b: 3,4-b '] dipyran-4-one [14], inophynone, and isoinophynone [20, 28].

2.4 Terpenes and terpenoids

Terpenes are naturally derived component in the biosynthesis of isoprene C5 with molecular formula C_5H_8 ($CH_2{=}C(CH_3){-}CH{=}CH_2$) (**Figure 5**). They commonly expressed in the formula $(C_5H_8)_n$ with n states the amount of isoprene which are there, so the amount of carbon is a multiple of 5. They are classified in hemiterpenes, monoterpenes (consisting of 2 units of C5 or 10 carbon atoms), sesquiterpenes (consisting of 3 units of C5 or 15 carbon atoms), diterpenes (consisting of

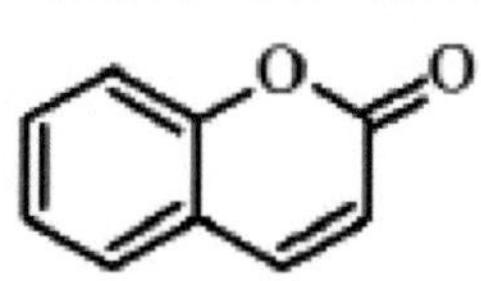

Figure 4.
Molecular structure of coumarins.

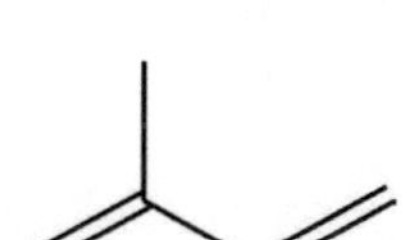

Figure 5.
Molecular structure of isoprene.

4 units of C5 or 20 carbon atoms), sesterterpenes, triterpenes (consisting of 6 units of C5 or 30 carbon atoms), tetraterpenes (consisting of 8 units of C5 or 40 carbon atoms), and polyterpenes.

Moreover, terpenoids are isoprenoid structural components which contain oxygen in its structure and can react with ketone, aldehyde, or alcohol. Chemically, they are generally soluble in fat and contained within the plant cell cytoplasm. Usually, they can be extracted with petroleum ether, ether, or chloroform and can be separated by chromatography on silica gel [29].

Terpenes are widely used as a medicine and flavor enhancers. They are commonly used in the rubber industry. They have a low molecular weight, such as essential oils that are used as natural food additives and fragrances in the perfume industry. They are also used in anticancer drug Taxol which is a diterpene. Taxol is used in the treatment of breast, ovarian, and lung cancer. One example is imberbic acid, a triterpenoid that has activity against *Mycobacterium fortuitum* and *S. aureus* [30].

Triterpenoids are a class of terpenoid compounds which consist of 30 carbon atoms or 6 units of isoprene. In plant tissue, they can be found in their native form but are also often found in the form glycoside. They are divided into cyclic and acyclic structures. The important acyclic triterpenoid is only the squalene that is considered only as an intermediate in the biosynthesis of steroids. The most widespread of triterpenoids are the pentacyclic triterpenoids. The frameworks most often found on a class of compound triterpenoids are ursam, lupan, oleanan, and friedelin [31].

Friedelin has the molecular formula $C_{30}H_{50}O$ and a molecular weight of 426,7174 g/mol (**Figure 6**). Friedelin has a melting point of 259–260°C. The structure mass spectrometry of friedelin is 426 (M^+), 411, 302, 273, 246, 231, 218, 205, 191, 179, 163, 149, 137, 125, 123, 109, 95, 81, 69, and 55. The IR spectra of friedelin in KBr was obtained using v_{max} at 1720 cm^{-1}. The form of friedelin is white crystalline-amorphous solid. Friedelin has an anti-fungal activity and has antinociceptive effects in rodents [32]. Friedelin was developed on a TLC plate by using a solvent system of 10% ethyl acetate and 90% hexane. Friedelin gave a dark spot on a TLC when exposed under UV light and iodine vapor chamber. Friedelin gave an Rf value of 0.75 with the use of a relatively nonpolar solvent system [33].

Several studies have been conducted on the benefits of friedelin. Friedelin has hepatoprotective activity [34]. It has an activity against Bacillus Calmette-Guerin (BCG) that causes tuberculosis [35]. It and some types of friedelin compound are widely used for the treatment of cancer of the bladder [36], convulsion, inflammation [37], topical ulcers, rheumatic inflammation, fever, and dysentery [38]. It is also found to have antifeedant activity in some insects [39].

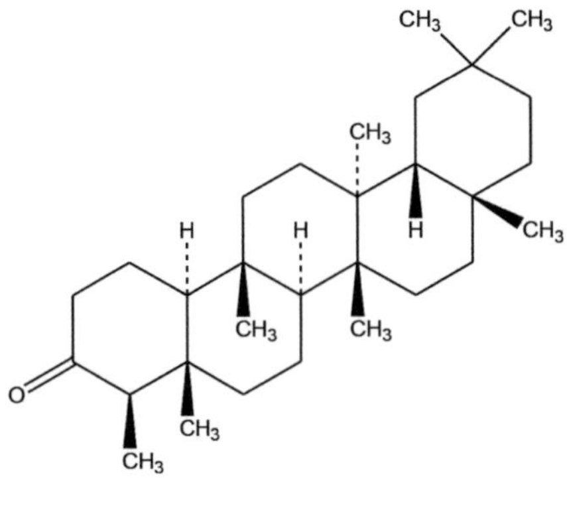

Figure 6.
Molecular structure of friedelin.

Moreover, some compound triterpenoids have been isolated from the *C. inophyllum* leaves, such as 3β, 23-epoxy-friedelane-28-OIC acid, 3-oxofriedelin-28-OIC acid, epifriedelanol, oleanolic acid [40], 3,4-secofriedelane-3,28-dioic [41], β-amyrin [20], friedelin, canophyllal, canophyllol, and canophyllic acid [4, 20, 41].

2.5 Steroid

Sterols are steroids which have a hydroxy group at C3 position as can be seen in **Figure 7**. They are found in free form or in association with glucose to form glycosides (sterolin) or as fatty acid esters (FASE). They are the natural compound that is generally composed of 27 carbon atoms [31]. They are terpenoids in which their basic framework consists of the system perhydrophenanthrene cyclopentane ring. They are a class of secondary metabolic compounds which are widely used as a drug. Steroid hormones are generally derived from natural steroid compounds, especially in plants [42]. Some steroid compounds have been isolated from the *C. inophyllum* leaves such as campesterol [20]. Campesterol also has analgesic activity.

2.6 Flavonoids

Flavonoids are the largest group of phenolic compounds found in nature, especially in tissues of higher crops. They are the product of secondary metabolites that occur from the cells and accumulate on the body crop as a toxic substance [43]. They are commonly known as flavonoids, which are water-soluble polyphenol component. They have a basic framework consisting of 15 carbon atoms where a chain of benzene (C6) is bound to a chain of propane (C3), thus forming a bond arrangement C6-C3-C6 which is particularly called phenylbenzopyran (**Figure 8**). This arrangement can produce three structures, namely, 1,3-diarilpropana (flavonoids), 1,2-diarilpropana (isoflavonoids), and 2,2-diarilpropana (neoflavonoid) [44]. Moreover, flavonoids are classified into various categories based on differences in molecular structure, such as chalcones, flavanols, catechins, flavonoes, isoflavone, dihydroflavonol, and anthocyanidins [45, 46].

According to Markham [47], flavonoids are polar compounds because they have a hydroxyl group which does not bind to sugar, so the flavonoid is quite soluble in polar solvents such as ethanol, methanol, butanol, or water. Because of the presence of sugar bound, flavonoids become more soluble in water. Conversely, the less polar aglycone, such as isoflavones, flavanones, flavones, and flavonols, which is methoxylated tends to be more soluble in solvents, such as ether and chloroform.

The largest group of flavonoids is flavones. Flavonoids have a 2-phenyl Croman order in which the ortho-position of the A ring and the carbon atom attached to the ring B of 1.3 diarilpropana is connected by bridging oxygen to form a new heterocyclic ring [47].

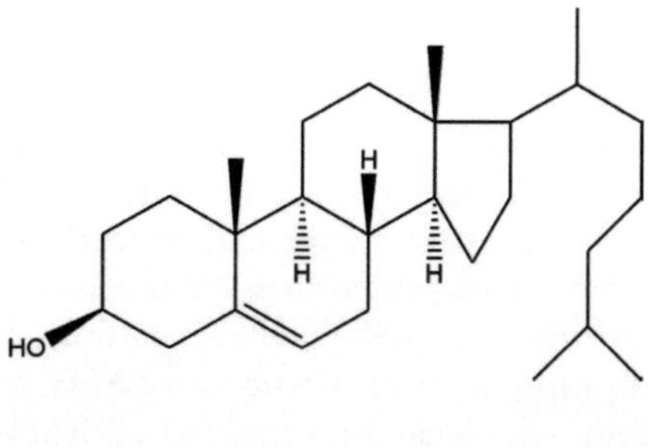

Figure 7.
Molecular structure of cholesterol.

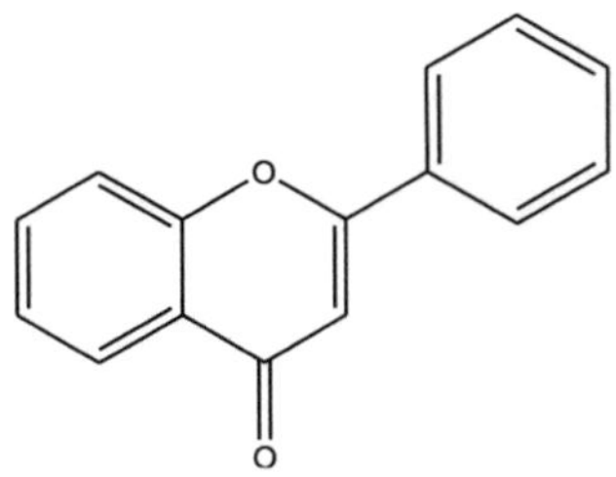

Figure 8.
Molecular structure of flavone.

Flavonoids have a variety of biological functions including pharmaceutical use and their function in plants. Examples of pigments in flowers, they provide color and attract insects for pollination. Flavonoids which are contained in the leaves have to prevent fungal infections and protect leaves from UV radiation [45]. In the aspect of pharmacology, flavonoids interact with cytochrome P450 and are used to treat heart disease. They are also known to have antioxidant activity and anti-free radicals that are useful in anticancer and antiaging. Furthermore, they also have antileukemic activity, vitamin C, 5-lipoxygenase, cyclooxygenase inhibitors, protein kinase C, tyrosine kinase, and genetic toxicity [27].

Several flavonoid compounds that have been isolated from the *C. inophyllum* leaves are bioflavonoids, neoflavonoid [48], amentoflavone [20, 40], and quercetin-3-O-α-L-rhamnoside [8, 48].

2.7 Oxygenated hydrocarbon (fatty acids)

Some of the compounds of fatty acid that has been found in the *C. inophyllum* leaves are tetradecanoic acid (myristic acid, $C_{14}H_{28}O_2$), n-hexadecanoic acid (palmitic acid, $C_{16}H_{32}O_2$), oleic acid ($C_{18}H_{34}O_2$), and octadecanoic acid (stearic acid, $C_{18}H_{36}O_2$) [49].

2.8 Esters

Some ester compounds that have been found in the *C. inophyllum* leaves are 1,2-benzenedicarboxylic acid (diisooctyl ester/phthalic acid, bis(6-methylheptyl) ester), 9,12-octadecenoic acid methyl ester, 16-octadecanoic acid methyl ester, heptadecanoic acid, and 16-methyl ester [49, 50].

2.9 Tannins

In chemistry, there are two types of tannins, namely, (1) condensed tannins or flavolan and (2) hydrolyzed tannins.

2.9.1 Condensed tannins

The condensed tannins are widespread in angiosperm plants, especially in woody plants. Another name of condensed tannins is proanthocyanidin because when they reacted with hot acid, some of the carbon-carbon connecting bond units disconnect and free monomer anthocyanidins. Most proanthocyanidin is procyanidin because when reacted with acids will produce cyanidin. Proanthocyanidin can be detected directly by dipping the plant tissue into 2 M HCl boil for half an hour that will produce a red color which can be extracted with amyl or butyl alcohol.

No.	Phytochemicals	Chemical structure	References
1.	Triterpenoids		
	3β, 23-Epoxy-friedelan-28-oic acid	$C_{30}H_{48}O_3$	[41]
	Friedelin	$C_{30}H_{50}O$	[4, 20, 28, 32, 41]
	3-Oxofriedelin-28-oic acid		[40, 41]
	Canophyllal	$C_{30}H_{48}O_2$	[20, 41]
	Canophyllol	$C_{30}H_{50}O_2$	[20, 41]
	Canophyllic acid (27-hydroxyacetate canophyllic acid)	$C_{30}H_{50}O_3$	[4, 20, 41]
	3,4-Secofriedelane-3,28-dioic acid	$C_{30}H_{50}O_4$	[19]
	Inophynone	$C_{24}H_{24}O_4$	[20, 28]
	Isoinophynone	$C_{24}H_{24}O_4$	[20, 28]
	β-Amyrin	$C_{30}H_{50}O$	[20]
	Epifriedelanol	$C_{30}H_{52}O$	[41]
	3-Oxo-27-hydroxyacetate friedelan-28-oic acid		[19]
	Oleanolic acid	$C_{30}H_{48}O_3$	[41]
	Squalene	$C_{30}H_{50}$	[50]
2.	Steroids		
	Cholesterol	$C_{27}H_{46}O$	[28]
	Campesterol	$C_{28}H_{48}O$	[20]
3.	Flavonoids		
	Biflavonoids	$C_{30}H_{20}O_{10}$	[49]
	Neoflavonoids	$C_{20}H_{18}O_8$	[49]
	Quercetin-3-O-α-L-rhamnoside (4H-1-benzopyran-4-one,2-(3,4-dihydroxyphenyl)-3,5,7-trihydroxy)	$C_{15}H_{10}O_7$	[22, 49]
	Amentoflavone	$C_{30}H_{18}O_{10}$	[20, 22, 40]
4.	Coumarins		
	Inophyllum C	$C_{25}H_{23}O_5$	[12, 40, 42]
	Inophyllum E	$C_{25}H_{22}O_5$	[12, 40]
	Inophyllum B	$C_{25}H_{24}O_5$	[4, 42]
	Inophyllum P	$C_{25}H_{24}O_5$	[4, 42]
	Calophyllic acid	$C_{25}H_{24}O_6$	[4, 20, 40]
	Isocalophyllic acid	$C_{25}H_{24}O_6$	[20, 40]
	Inophyllum G-1	$C_{25}H_{24}O_5$	[4, 42]
	Inophyllum G-2	$C_{25}H_{24}O_5$	[4, 42]
	Calocoumarin-A		[20]
	Calocoumarin-B		[20]
	Calocoumarin-C		[20]
	Apetalolide	$C_{26}H_{24}O_5$	[20]
	4-Phenylcoumarins		[20]
	Pyranocoumarins	$C_{20}H_{18}O_4$	[42]

No.	Phytochemicals	Chemical structure	References
	Calophyllolides (calophyllolide 2a, 3a, 3b, 6)	$C_{26}H_{24}O_5$	[4, 12, 42]
5.	Xanthones		
	Caloxanthone A	$C_{23}H_{22}O_6$	[4, 12]
	Caloxanthone B		[4, 12]
	Caloxanthone C		[4]
	Brasilixanthone-B	$C_{23}H_{20}O_6$	[20]
	Buchanaxanthone	$C_{14}H_{10}O_5$	[20]
	Inoxanthone	$C_{23}H_{22}O_5$	[12]
	Maclura xanthone	$C_{23}H_{22}O_6$	[12]
	Calophynic acid	$C_{35}H_{44}O_6$	[12]
	3,4-Dihydroxyxanthone	$C_{13}H_8O_4$	[12, 19]
	Inophyxanthone A		[21]
	Pancixanthone A	$C_{18}H_{16}O_5$	[21]
	Gerontoxanthone B	$C_{23}H_{22}O_6$	[21]
	Jacareubin (6-deoxyjacareubin)	$C_{18}H_{14}O_6$	[21, 22]
	Pyranojacaereubin	$C_{23}H_{20}O_6$	[21]
	2-Hydroxyxanthone	$C_{13}H_8O_3$	[22]
	4-Hydroxyxanthone	$C_{13}H_8O_3$	[22]
	1,3,5-Trihidroxy-2-methoxyxanthone		[22]
	Xanthone	$C_{13}H_8O_2$	[21]
6.	Oxygenated hydrocarbons (fatty acids)		
	Tetradecanoic acid (myristic acid)	$C_{14}H_{28}O_2$	[50]
	n-Hexadecanoic acid (palmitic acid)	$C_{16}H_{32}O_2$	[50]
	Oleic acid	$C_{18}H_{34}O_2$	[50]
	Octadecanoic acid (stearic acid)	$C_{18}H_{36}O_2$	[50]
7.	Esters		
	1,2-Benzenedicarboxylic acid (diisooctyl ester) (phthalic acid, bis(6-methylheptyl) ester) (diisooctyl phthalate)	$C_{24}H_{38}O_4$	[50]
	Methyl linoleic (9,12-octadecanoic acid methyl ester)	$C_{19}H_{34}O_2$	[50, 51]
	Methyl oleate (16-octadecanoic acid methyl ester)	$C_{19}H_{36}O_2$	[51]
	Methyl isostearate (heptadecanoic acid, 16-methyl, methyl ester)	$C_{19}H_{38}O_2$	[51]
8.	Alkenes (unsaturated compounds):		
	Azulene, 1,4-dimethyl-7-(1-methylethyl)-	$C_{15}H_{18}$	[50]
9.	Ethers		
	3-Trifluoroacetoxypentadecane (pentadecyl trifluoroacetate) (trifluoroacetic, pentadecyl ester)	$C_{17}H_{31}F_3O_2$	[50]
	1-Monolinoleoglycerol trimethylsilyl ether	$C_{27}H_{54}O_4Si_2$	[50]
10.	Alicyclic compounds		
	Cyclohexene, 3-(1,5-dimethyl-4-hexenyl)-6-methylene-,[S-(R*,S*)]	$C_{15}H_{24}$	[50]

No.	Phytochemicals	Chemical structure	References
11.	Aromatic hydrocarbon:		
	Benzene (1-methyldodecyl)	$C_{19}H_{32}$	[50]
12.	Androstan-1α-ol-17-one,23 isopropylidenedioxy-4β-methyl-	$C_{23}H_{36}O_4$	[50]
13.	Proanthocyanidin (condensed tannin)	$C_{31}H_{28}O_{12}$	[20, 49]
14.	Benzodipyranone (chromone) derivatives:		
	a. (2S,3R)-2,3-Dihydro-5-hidroxy-2,3,8,8-tetramethyl-6-(1-phenylethenyl)-4H,8H-benzo [1,2-b:3,4-b'] dipyran-4-one		[14]
	b. (2R,3R)-2,3-Dihydro-5-hidroxy-2,3,8,8-tetramethyl-6-(1-phenylethenyl)-4H,8H-benzo [1,2-b:3,4-b'] dipyran-4-one		[14]
15.	Asam inophylloidic	$C_{32}H_{46}O_6$	[12, 20]
16.	Calaustralin	$C_{25}H_{25}O_5$	[12]
17.	Shikimic acid	$C_7H_{10}O_5$	[40]
18.	Brasiliensic acid	$C_{32}H_{46}O_6$	[12]
19.	Adenanthin (7,8,12-tri-0-acetyl-3-desoxy-ingol3-one)	$C_{26}H_{34}O_9$	[51]
20.	Carbazole	$C_{12}H_9N$	[51]
21.	Diphenyl methane (1'-biphenyl, 2-methyl)	$C_{13}H_{12}$	[51]
22.	2-Phenazinamine (1,1'-biphenyl, 4-azido)	$C_{12}H_9N_3$	[51]
23.	5-Aminomethyl-dibenzosuberane (2-naphtalenecarbonitrile, 6-pentyl-)	$C_{16}H_{17}N$	[51]
24.	Phytol	$C_{20}H_{40}O$	[50, 51]
25	3,7,11,15-Tetramethyl-2-hexadecen-1-ol	$C_{20}H_{40}O$	[50]
26.	Phenol (2,4-bis(1-phenylethyl)-phenol)	$C_{22}H_{22}O$	[50]

Table 2.
Phytochemicals contained in the C. inophyllum leaves.

When dry tissues are used, the result of tannins somewhat diminished because of the occurrence of sticking tannins in place within the cell.

2.9.2 Hydrolyzed tannins

The hydrolyzed tannins are contained in dicotyledonous plants. They mainly consist of two classes; the simplest is galloylglucose. In this compound, glucose is surrounded by five or more galloyl ester groups. The second type is the core molecules of a compound gallic acid dimer, namely, hexahydroxidifenate acid that binds to glucose. Hydrolyzed tannins can be detected by determining the gallic acid or ellagic acid in ether or ethyl acetate extracts.

2.10 Other components

Some chemical compounds that have been found in the *C. inophyllum* leaves are azulene ($C_{15}H_{18}$), squalene ($C_{30}H_{50}$), 3-trifluoroacetyl pentadecane (pentadecyl trifluoroacetate), 1-monolinolein glycerol trimethylsilyl ether, cyclohexane, benzene, androstane [49], inophylloidic acid [12, 20, 52], shikimic acid [40], calaustralin, brasiliensic acid [12], adenanthin, carbazole, diphenyl methane, 2-phenazinamine, 5-aminomethyl-dibenzosuberane [50], phytol, phenol, and 3,7,11,15-tetramethyl-2-hexadecene-1-ol [49, 50]. The summary of phytochemicals in *C. inophyllum* leaves is presented in **Table 2**.

3. Isolation method of phytochemicals in *C. inophyllum* leaves

Polarity is one of the characteristics of chemical bonding, where two different atoms within the same molecule have a different electronegativity. As a result, the electrons in the bond are not shared equally by the two atoms. This causes the electric field (pole) to be asymmetric. Covalent bonding of molecules can be described as polar or nonpolar.

The polar compound is a compound formed by a single atom which has electronegativity substantially greater than the other. The more electronegative the atom, the pull of the bonding electrons is greater. The result is a bond with an uneven electron dense distribution. The nonpolar compound is a compound formed by atoms with the same or nearly the same electronegativity and forms covalent bonds, where both atoms apply traction which equals or nearly equals to the bonding electrons. Generally, the carbon-carbon and carbon-hydrogen bonds are the most common types of nonpolar bond [53].

To identify polar and nonpolar compounds from the *C. inophyllum* leaves, the first idea is separating their compounds based on the solvent used (solvent polarity index). Methanol and water are polar solvent with a polarity index of 5.1 and 9, respectively. For n-hexane or petroleum ether is nonpolar solvent with a polarity index of 0 [54]. It can be expected that polar compounds which are contained in the *C. inophyllum* leaves can be dissolved in a polar solvent and vice versa. Relative polarities of several solvents can be seen in **Table 3.**

Extraction is the separation process of material from a solid or some material from liquid with the help of the solvent. Extraction can be defined as a method of separating components of a mixture by using a suitable solvent. Solutes (dissolved substances) are separated in a manner distributed between two layers of solvents based on their solubility. Extraction is a separation of the compounds contained in the liquid material/solid using certain solvents at any given temperature.

In general, extraction techniques can be classified into two general categories:

1. Short-term extraction is extraction techniques typically used to separate a substance (liquid form), on the basis of differences in solubility of the two immiscible solvents.

2. Long-term extraction is an extraction technique normally used to separate the natural material (solid form) contained in plants or animals. It is a classic procedure to obtain the organic matter content of dry plant tissue by soaking with certain solvents (polar or nonpolar solvents) [29].

Percolation is an extraction technique that done repeatedly and performed at a room temperature. This is similar to maceration, but after soaking for a certain time, the solvent is removed and replaced with a new solvent. After filtration, the filtrate obtained is called percolate [55].

According to Mulyono [55], in terms of the extraction mechanism, known to some type of extraction, namely:

1. Single-stage extraction

Single-stage extraction is the extraction method using a single type of solvent, and extraction is only done once with a solvent.

Relative polarity	Formula	Group	Solvents
Nonpolar	R-H	Alkanes	Petroleum ethers, hexanes, ligroin
	Ar-H	Aromatics	Toluene
	R-O-R	Ethers	Diethyl ether
	R-X	Alkyl halides	Trichloromethane, chloroform
	R-COOR	Esters	Ethyl acetate
	R-CO-R	Aldehydes, ketones	Acetone, MEK
	R-NH2	Amines	Pyridine, triethylamine
	R-OH	Alcohols	MeOH, EtOH, IPA, butanol
	R-COHN2	Amides	Dimethylformamide
	R-COOH	Carboxylic acid	Ethanoic acid
Polar	H-O-H	Water	

Table 3.
Relative polarity of solvents [54].

2. Repeated extraction
Repeated extraction is the extraction method using a solvent, but the process is repeated with a number of solvents.

3. Stage extraction
Stage extraction is the extraction method using some type of solvent extraction, such as after extraction with the first solvent, followed by using other solvents, and so on.

Solvents are not or only partially soluble solids or liquids with continuous contact; the active agents move from a mixture of solids/liquid (raffinate) to the solvent (extract). After mixing the two phases, the separation process is done on the principle of gravity or centrifugal force [56].

Yunitasari [57] describes the effect of solvent on the various types of tray number from 6 to 10 for taking *C. inophyllum* oil with column extraction. From the experimental results, the authors explain that the more the number of trays, the less time is required for a solvent to extract the oil. The solvent used are between n-petroleum and n-hexane. From the experimental results, the authors explain that the maximum condition extraction was achieved by n-petroleum in the seventh tray. The amount of oil was decreasing by increasing number of tray. In the other hand, the amount of oil was increasing with number of tray while n-hexane was used.

4. Conclusions

The identification and uses of beneficial phytochemicals contained in *C. inophyllum* leaves were presented in this book chapter. It was found that all parts of *C. inophyllum* plant can be used for human needs. The information is limited to extraction and identification of mixture of phytochemical compounds that are obtained from plant extracts. The separation of individual phytochemical compounds still remains unknown. Therefore, further research on the determining of phytochemicals content in this plant is necessary.

Acknowledgements

The authors would like to convey their great appreciation for the Directorate General of Resources for Science, Technology, and Higher Education and Ministry of Research, Technology and Higher Education of the Republic Indonesia which funds the current project under the scheme No. (329/SP2H/LT/DRPM/IX/2016) called "The Education of Master Degree Leading to Doctoral Program for Excellent Graduates (PMDSU)."

Conflict of interest

We declare that we have no conflict of interest.

Author details

David Febrilliant Susanto[1], Hakun Wirawasista Aparamarta[1], Arief Widjaja[1], Firdaus[2] and Setiyo Gunawan[1*]

1 Department of Chemical Engineering, Faculty of Industrial Technology, Institut Teknologi Sepuluh Nopember (ITS), Indonesia

2 Faculty of Health, Universitas Nahdlatul Ulama Surabaya (UNUSA), Indonesia

*Address all correspondence to: gunawan@chem-eng.its.ac.id

References

[1] Ling KH, Kian CT, Hoon TC. A Guide To Medicinal Plant. Singapore: World Scientific; 2009

[2] Ong HC. Optimization of biodiesel production and engine performance from high free fatty acid *Calophyllum inophyllum* oil in Cl diesel engine. Science Direct. 2014;**81**:30-40

[3] Sudrajat. A Potential Plant for Biodiesel. Indonesia: Departemen Kehutanan; 2009

[4] Dweck AC, Meadows T. Tamanu (*Calophyllum inophyllum*) – The African, Asian, Polynesian, and Pacific Panac. USA: International Journal of Cosmetic Science; 2002

[5] Wahyuni T, Umi A, Riza Z. Pemanfaatan Hasil samping Biji Nyamplung Menjadi Biopelet sebagai Bahan bakar Pengganti minyak tanah di kawasan Pesisir. Jakarta: Pusat Pengkajian dan Perekayasaan Teknologi Kelautan dan Perikanan; 2010

[6] Baity LN, Azhar A, Eko OK. Hutan Tanaman Industri (HIT) Berbasis Nyamplung (*Calophyllum inophyllum* Linn) Sebagai Stok Energi Terbarukan dengan Sistem Zero Cutting. Bogor: Tugas akhir Institut Pertanian Bogor; 2011

[7] Wibowo S, Hendra D. Manfaat Tanaman Nyamplung dan Prospek Pengembangannya. Sumatera: Balai Penelitian Kehutanan; 2011

[8] Iskandari A. Isolasi dan Elusidasi Struktur Quercetrin dari Daun Nyamplung. Surakarta: Fakultas Matematika dan Ilmu Pengetahuan Alam Universitas Sebelas Maret; 2010

[9] Su XH, Zhang ML, Li LG, Huo CH, Gu YC, Shi QW. Chemical constituents of the plants of the genus Calophyllum. Chemistry & Biodiversity. 2008;**5**(12):2579-2608

[10] Cechinel Filho V, Meyre-Silva C, Niero R. Chemical and Pharmacological Aspects of the Genus Calophyllum. Kuching: Chem Biodiversity Centre; 2009

[11] Lim TK. Edible Medicinal and Non-Medicinal Plants Vol 2, Fruits. London New York: Springer Dordrecht Heidelberg; 2012

[12] Yimdjo MC, Azebaze AG, Nkengfack AE, Meyer AM, Bodo B, Fomum ZT. Antimicrobial and cytotoxic agents from *Calophyllum inophyllum*. Phytochemistry. 2004;**65**:2789-2795

[13] Linuma M, Tosa H, Tanaka T, Yonemori S. Two xanthones from root bark of *Calophyllum inophyllum*. Phytochemistry. 1994;**35**:527-532

[14] Khan NU, Parveen N, Singh MP, Singh R, Achari B. Two isomeric benzodipyranone derivatives from *Calophyllum inophyllum*. Phytochemistry. 1996;**42**:1181-1183

[15] Kumar V, Ramachandran S, Sultanbawa MU. Xanthones and triterpenoids from timber of *Calophyllum inophyllum*. Phytochemistry. 1976;**15**:2016-2017

[16] Williams P, Ongsakul M, Proudfoot J, Croft K, Beilin L. Mangosteen inhibits the oxidative modification of human low density lipoprotein. Free Radical Research. 1995;**23**(2):175-184

[17] Matsumoto K, Akao Y, Kobayashi E, Ohguchi K, Ito T, Tanaka T, et al. Introduction of apoptosis by xanthones from mangosteen in human leukemia cell lines. Journal of Natural Products. 2003;**66**(8):1124-1127

[18] Sakagami Y, Linuma M, Piyasena KG, Dharmaratne HR. Antibacterial activity of alpha-mangosteen against vancomycin resistant enterococci

(VRE) and synergism with antibiotics. Phytomedicine. 2005;**12**(3):203-208

[19] Laure F, Herbette G, Faure R, Bianchini JP, Raharivelomanana P. Fogliani. Structures of new secofriedelane and friedelane acids from *Calophyllum inophyllum* of French Polynesia. Magnetic Resonance in Chemistry. 2005;**43**(1):65-68

[20] Silpa S, Shrivastava B, Sharma P, Rai SS. A review article of pharmacological activities and importance of *Calophyllum inophyllum*. International Journal of Advanced Research. 2014;**2**(12):599-603

[21] Li YZ, Li ZL, Liu MS, Li DY, Zhang H, Hua HM. Xanthones from leaves of *Calophyllum inophyllum* Linn. Yao Xue Xue Bao. 2009;**44**(2):154-157. In Chinese

[22] Li YZ, Li ZL, Jua HM, Li ZG, Liu MS. Studies on flavonoids from stems and leaves of *Calophyllum inophyllum*. China Journal of Chinese Materia Medica. 2007;**32**:692-692

[23] Bezwada R. Chemistry of Comarins. Hillsborough, NJ: Indofine Chemical Company; 2008

[24] Sethna S, Shah N. The chemistry of coumarins. Chemical Reviews. 1945;**36**(1):1-62

[25] Lacy A, O'Kennedy R. Studies on coumarins and coumarin related compound to determine their therapeutic role in the treatment of cancer. Current Pharmaceutical Design. 2004;**10**:3797-3811

[26] Sahoo SS, Shukla S, Nandy S, Sahoo HB. Synthesis of novel coumarin derivatives and its biological evaluations. European Journal of Experiment Biology. 2012;**2**(4):899-908

[27] Sovia L. Senyawa Flavonoida, Fenilpropanoida, dan Alkaloida. Indonesia: Departemen Kimia Fakultas Matematika dan Ilmu Pengetahuan Alam Universitas Sumatra Utara Medan; 2006

[28] Ali MS, Mahmud S, Perveen S, Ahmad VU. Epimers from the leaves of *Calophyllum inophyllum*. Phytochemistry. 1999;**50**:1385-1389

[29] Harborne JB. Phytochemical Method. London: Chapman and Hall Ltd; 1984

[30] Katerere DRP, Gray AI, Nash RJ, Waigh RD. Antimicrobial activity of pentacyclic triterpenes isolated from African Combretaceae. Phytochemistry. 2003;**63**:81-89

[31] Kristanti AN. Buku Ajar Fitokimia Laboratorium Kimia Organik Jurusan Kimia-FMIPA, UNAIR. Surabaya: Airlangga University Press; 2008

[32] Jullyana Q, Costa EV, Tavares JF, Souza TT. Phytochemical study and antinociceptive effect of the hexanic extract of leaves from Combretum duarteanum and friedelin, a triterpene isolated from the hexanic extract, in orofacial nociceptive protocols. Revista Brasileira de Farmacognosia. 2014;**24**:60-66

[33] Gan Shu Y. Chemical Constituents from the Endemic Plant of Sarawak, Calophyllum Castaneum and their Antioxidant Activity. Malaysia: Bachelor of Science Chemistry: Faculty of Science Universiti Tunku Abdul Rahman; 2014

[34] Dzubak P, Hajduch M, Vydra D, Hustova A, Kvanica A, David M, et al. Pharmacological activities of natural triterpenoids and their therapeutic implications. Natural Product Reports. 2006;**23**:394-411

[35] Abdulahi M, Ibrahim K, Adebayo O, Amupitan JO, Fatope MO, Joseph IO. Antimycobacterial Friedelane-terpenoid from the root bark of Terminalia

Avicennioides. American Journal of Chemistry. 2011;**1**(2):52-55

[36] Simon HC. 1976 Ger. Offen. 2,508.338 (C1.461 K) 19 Feb 1976 (Chem. Abstr. 84. 169664q)

[37] Chaturvedi AK, Parmar SS, Bhatnagar SC, Mistra G, Nigam SK. Anti-convulsant and anti-inflammatory activity of natural plant coumarins and triterpenoids. Research Communications in Chemical Pathology and Pharmacology. 1974;**9**:11

[38] Subramanian SS, Nair AGR, Vedanthan TNC. Chemical examination of the aerial parts of *C. fragrans* and *C. squamatum*. The Indian Journal of Pharmacy. 1974;**36**:15

[39] Abbassy MA, El-Shazli A, El-Gayar F. A new antifeedant to *Spodoptera littoralis* Boisd. (Lepid., Noctuidae) from *Acokanthera spectabilis*) Hook. (*Apocynaceae*). Zeitschrift für Angewandte Entomologie. 1977;**83**:317 (Chem. Abstr. 87. 147016q)

[40] Prasad J, Shrivastava A, Khanna AK, Bhatia G, Awasthi SK, Narender T. Antidyslipidemic and antioxidant activity of the constituents isolated from the leaves of *Calophyllum inophyllum*. Phytomedicine. 2012;**19**:1245-1249

[41] Li YZ, Li ZL, Yin SL, Shi G, Liu MS, Jing YK, et al. Triterpenoids from *Calophyllum inophyllum* and their growth inhibitory effects on human leukemia HL-60 Cels. Fitoterapia. 2010;**81**:586-589

[42] Laure F. Screening of anti-HIV inophyllums by HPLC-DAD of *Calophyllum inophyllum* leaf extracts from French Polynesia Islands. Analytica Chimica Acta. 2008;**624**:147-153

[43] Djamal R. Tumbuhan Sebagai Sumber Bahan Obat. Pusat Penelitian: Universitas Negeri Andalas; 1988

[44] Robinson T. In: Department of Biochemistry, University of Massachusetts, editor. The Organic Constituen of Higher Plants. Their Chemistry and Interrelationships. 6th ed. New York: Burgess Publishing Company; 1991

[45] Biosintesis MP, Alami P. Terjemahan: Koensoenmardiyah. Semarang: IKIP Semarang Press; 1981

[46] Sandhar HK, Kumar B, Prasher S, Tiwari P, Salhan M, Sharna P. Review of phytochemistry and pharmacology of flavonoids. International Pharmaceutica Sciencia. 2011;**1**(1):25-41

[47] Markham KR. Cara Mengidentifikasi Falvanoid. Alih Bahasa: Kosasih Padmawinata. ITB. Bandung; 1982

[48] Sharma DK. Pharmacological properties of flavonoids including flavonolignans integration of petrocorps with drug development from plants. Journal of Scientific and Industrial Research. 2006;**65**:477-484

[49] Bandarayake WM. Bioactivities, bioactive compounds and chemical constituents of mangrove plants. Wetlands Ecology and Management. 2002;**10**:421-452

[50] Ramakhrisnan N, Malarvizhi PGC-MS. Analysis of biologically active compounds in leaves of *Calophyllum inophyllum* L. International Journal of Chemtech Research. 2011;**3**(2):806-809

[51] Saravanan P, Jaikumar K, Sheik NMM, Anand D. Phytochemical analysis of bioactive compounds from *Calophyllum inophyllum* L. leaf extract using GC-MS analysis. International Journal of Pharmacognosy and Phytochemical Research. 2015;**7**(5):956-959

[52] Mahmud S, Rizwani GR, Ahmad M, Ali S, Perveen S, Ahmad VU. Antimicrobial studies on fractions

and pure compounds of *Calophyllum inophyllum* Linn. Pakistan Journal of Pharmacology. 1998;**15**(2):13-25

[53] Fessenden JR. Kimia Organik Edisi Ketiga Jilid 1. Indonesia: Erlangga; 1986

[54] Sadek P. The HLPC Solvent Guide. United States of America: Wiley of Interscience; 2002

[55] Moelyono MW. Panduan Praktikum Analisis Fitokimia. Bandung: Laboratorium Farmakologi Jurusan Farmasi FMIPA. Universitas Padjadjaran; 1996

[56] Gamse T. Extraction: Liquid-Liquid, Solid-Liquid, High Pressure. Inffeledgasse: Graz University of Technology; 2004

[57] Yunitasari EP. Pengaruh Jenis Solvent dan Variasi Tray pada Pengambilan Minyak Nyamplung dengan Metode Ekstraksi Kolom. Semarang: Universitas Dipenogoro; 2008

Chapter 3

Analytical Methods of Isolation and Identification

Weisheng Feng, Meng Li, Zhiyou Hao and Jingke Zhang

Abstract

The chemical constituents of plants are complicated, and monomeric compounds must be obtained via extraction and isolation before structure identification, bioactivity screening, and so on. In recent years, the new technologies and methods of the extraction, isolation, and structural identification have come forth, which promote the speed of extraction and analysis of phytochemicals. The chemical structures of compounds from plants must be identified or elucidated, which may provide the necessary basis for further study on the bioactivities, structure-activity relationships, metabolisms *in vivo*, structural modification, and synthesis of the active compounds. The amount of chemical constituents isolated from plants is often minor, so the structural studies are often difficult to carry out with classical methods. Therefore, spectral analysis is mainly used. This chapter describes the isolation and identification methods during the study of phytochemicals.

Keywords: extraction, isolation, structure elucidation, spectroscopic methods

1. Introduction

The phytochemicals rich in plants have shown to be beneficial for prevention of diseases as well as long-term health. Plants are generally consumed as sources of essential compounds such as saccharides, coumarins, lignans, flavonoids, terpenoids, and steroids. The health benefits and the composition from plant have been described more and more in the literature. Because of the complexity of plant chemical constituents, pure phytochemicals must to be obtained via extraction and isolation before structure identification, bioactivity screening, and so on. In recent years, new technologies and methods of extraction occurred, which accelerate the extraction and analysis of phytochemicals.

Extraction is the first step of phytochemistry research, which is also the necessary work before the isolation of effective constituents. The purpose of extraction is to get the objective chemical constituents to the utmost extent and avoid or reduce the solution of unwanted constituents.

The separation of phytochemicals is a process of isolating the constituents of plant extracts or effective parts one by one and purifying them into monomer compounds by physical and chemical methods. Classical isolation methods, including solvent extraction, precipitation, crystallization, fractional distillation, salting-out, and dialysis, are still used commonly at present. On the other hand, modern separation technologies such as column chromatography, high performance liquid chromatography, ultrafiltration, and high performance liquid drop countercurrent chromatography also play an important role in the separation of phytochemicals [1–3].

IntechOpen

The chemical structures of plant compounds must be identified, which may provide the necessary basis for further study on the bioactivities, structure-activity relationships, metabolisms *in vivo*, structural modification, and synthesis of the active phytochemicals.

The structural studies are often difficult to carry out with classical chemical methods, such as chemical degradation and derivative synthesis, because of the minute amount of compound isolated from plants. Therefore, spectral analysis is mainly used. That is, consuming sample as little as possible to obtained structural information as much as possible by measuring and analyzing various spectra [4].

2. Extraction of phytochemicals

2.1 Solvent extraction methods

2.1.1 Principles

Solvent extraction is the commonest method to extract plant material. The main purpose is to select the suitable solvent to extract target plant materials efficiently. During the extraction, the solvent has to diffuse into the cell membrane in the first instance, in the following step it has to dissolve the solutes, then intracellular and extracellular concentration difference is formed, and finally it has to diffuse out of the cells enriched in the extracted solutes [5].

Selecting suitable solvents is the key of the solvent extraction method. Using a solvent of an appropriate polarity according to the principle of "like dissolves like" is the main point to select solvent. Thus, hydrophilic solvents are used to solubilize hydrophilic chemical constituents and vice versa. The hydrophilicity or lipophilicity of solvents and chemical constituents could be predicted by polarity. The plant compounds, such as terpenoids and steroids, possess low polarity, and could be dissolved into lipophilic solvents such as chloroform and ether, while chemical constituents, such as carbohydrates and amino acids, possess rather high polarity and could be dissolved into water and aqueous ethanol.

Solvents commonly used for extracting chemical constituents of plants are in the order of weak to strong polarity as follows: petroleum ether < carbon tetrachloride < benzene < dichloromethane < chloroform < ether < ethyl acetate < n-butanol < acetone < ethanol < methanol < water.

Water is a cheap, easy to get, and nontoxic solvent with strong polarity. It could be used to extract phytochemicals with strong polarity, such as inorganic salts, saccharides, amino acids, tannins, proteins, organic acid salts, alkaloid salts, and glycosides. Acid or alkaline water is applied sometimes to increase the solubility of certain specific components. Acid water could extract alkaline materials, such as alkaloids, via the formation of salts. Similarly, organic acids, anthraquinoids, flavonoids, coumarinoids, phenols, and other acidic materials could be extracted via the formation of salts. The disadvantage to extract chemical constituents with water is that the aqueous extract is easy to go moldy, so difficult to preserve. Additionally, water possesses high boiling point, and the water extract needs to be concentrated for a rather long time. Furthermore, the water extract contains many impurities such as proteins, pectins, tannins, mucilages, and inorganic salts, which make the extraction of target components difficult.

Hydrophilic organic solvents are strong-polarity and water miscible, such as methanol, ethanol, and acetone. Ethanol is the most commonly used hydrophilic organic solvent. Chemical constituents could be extracted by ethanol of different concentrations according to their properties. Furthermore, ethanol is inexpensive, safe, and concentrated easily. Additionally, ethanol extract is not readily moldy and glycosides are hard to be hydrolyzed in ethanol extract. Thus, ethanol is one of the most commonly used solvents in laboratories and industrial production. Methanol possesses similar property to ethanol and lower boiling point. However, methanol has rather strong toxicity, so we have to pay attention to safety when it has to be used. Acetone is a good solvent to extract lipid-soluble chemical constituents. However, acetone is easy to volatilize and flame, and it possesses certain toxicity.

Petroleum ether, benzene, chloroform, ether, ethyl acetate, dichloroform, and so on are lipophilic organic solvents and are not miscible with water. They could be applied to extract lipophilic components, such as volatile oils, fats, chlorophyll, lactones, phytosterols, some alkaloids and some aglycones (aglycones of flavonoids, anthraquinoids, steroids, and so on). These solvents possess low boiling points and are easy to concentrate. However, strong-volatility, large loss, flammability, toxicity, and high price are their disadvantages. Additionally, they are difficult to permeate into plant cell tissues.

Solvent extraction methods could be classified as cold extraction and hot extraction roughly by whether heating or not.

2.1.2 Immersion method

It is a method to dissolve out phytochemicals with appropriate solvents at room or low temperatures (<80°C). It is suitable to extract phytochemicals easily to be destroyed at high temperature. The plants with abundant starches, pectins, gums, or mucilages could also be extracted with this method. Firstly, plant powder or pieces should be loaded in the adequate container, and then the suitable solvents (water, ethanol, aqueous ethanol, and so on) are added into it to immerse the material for the given length of time. Discontinuous stirring or shaking during the process could accelerate dissolution rate. The immersion method is simple but inefficient, and the extraction ratio is also low. Furthermore, aqueous extract is easy to go moldy, so addition of appropriate preservatives is necessary.

2.1.3 Percolation method

The coarse particles of plants should be loaded in percolation apparatus and immersed with suitable solvent for 24–48 h, then collect the percolates at the bottom of percolation apparatus. New solvent should be added at the top of percolation apparatus constantly during the percolation process. It possesses higher efficiency than the immersion method because of the sustained concentration difference during the process. However, this procedure is complex and consumes rather much solvent and long time.

2.1.4 Decoction method

Load short segments, thin pieces, or coarse powder into an appropriate container, add water, and heat it to boiling; the components are then extracted. It is easy to operate; most of the constituents could be extracted in various degrees. Nevertheless, rather much nontargeted components could also be extracted, and it is not suitable to the extraction of volatile compounds and thermal unstable compounds. Furthermore, it is not suitable to extract plants with lots of starches [6].

2.1.5 Refluxing method

It is a method to extract plant chemical constituents by organic solvent using heating and refluxing. Refluxing apparatus is necessary so as not to waste solvents, and the toxicity to operators or ruin the environment is deduced. It is applicable to extraction of lipophilic phytochemicals, such as steroids, anthraquinoids, and terpenoids. It is an extraction method of high efficiency but complex, and consumes much more solvent. This method is not applicable to extract thermal unstable chemical constituents because of long time heating.

2.1.6 Constant refluxing method

It is a method developed based on the refluxing method. Soxhlet extractor is the most frequently used constant refluxing apparatus. This method avoids disadvantages of consuming too much solvent and complex operation. However, as a refluxing method, constant refluxing method is not applicable to extract thermally unstable compound either because of long time heating.

2.1.7 Supercritical fluid extraction method

In the supercritical state, the supercritical fluid is contacted with the plant tissues. By controlling different temperatures, pressures and different kinds and contents of entrainers, the supercritical fluid can selectively extract the components of different polarities, boiling points, and molecular weights successively. This method is called the supercritical fluid extraction (SFE) method [7].

The critical point of a pure substance is defined as the highest temperature and pressure at which the substance can exist in vapor-liquid equilibrium. At temperatures and pressures above this point, a single homogeneous fluid is formed, which is known as supercritical fluid (SF). SF is heavy like liquid and has low viscosity like gas meanwhile. SF possesses rather large diffusion coefficient and could dissolve many compounds well. A number of materials could be used as SFs, such as ammonia, ethane, difluoro-dichloromethane, heptane, and so on, while the most widely used SF is CO_2. The critical temperature of CO_2 (T_c = 31.26°C) is close to room temperature, and the critical pressure (P_c = 7.2 MPa) is not too high. CO_2 also has a series of other advantages, such as nontoxicity, odorless, nonflammable, chemical stability, and low cost, which allowed it to be the most commonly used solvent in SFE. CO_2 is a nonpolar substance and applicable to extract lipophilic compounds. However, its dissolvability is weak compared to strong polar substances. Hence, entrainers are always added to improve the solubility of SF CO_2 during the extraction of polar compounds. Entrainers, which are added into SF little, could enhance solubility of SF significantly. The commonly used entrainers are methanol, ethanol, water, acetone, ethyl acetate, acetonitrile, and so on.

The extraction of nonpolar and medium-polar components by SFE can avoid the sample loss and environmental pollution caused by solvent recovery in traditional extraction methods, especially for the extraction of volatile compounds with thermal instability.

The biggest advantage of SFE is that it can be performed at near-room temperature, and almost all the active ingredients in the product can be retained. There is no residual organic solvent in the process. The product has high purity and high yield. Additionally, the operation is simple and energy saving.

Compared with other conventional separation methods, SFE possesses the following advantages: (1) No residual organic solvents, fast extraction speed, simple process, high yield, and easy operation; (2) no flammable and explosive dangers, no environmental pollution. Low extraction temperature, suitable for the extraction of thermal unstable components; (3) the dissolution properties of SF are easy to improve, only the pressure needs to be changed at a certain temperature; (4) entrainers can be added to change the polarity of the extraction medium to extract polar substances; extraction medium can be recycled with low cost; (5) it could be applied combined with other chromatographic techniques, such as GC, IR, GC–MS, and HPLC, to extract, separate, and determine phytochemicals efficiently and quickly, so as to achieve the integration of extraction and quality analysis. However, supercritical extraction has some limitations: strong solubility of fat-soluble components, weak solubility of water-soluble components, high cost of equipment, resulting in higher product costs, and cleaning equipment is difficult.

Supercritical fluid extraction (SFE) technology has achieved gratifying results in the fields of medicine, chemical, food, light industry, and environmental protection. Especially, it has been widely used in phytochemical extraction field, such as the extraction of alkaloids, volatile oils, phenylpropanoids, flavonoids, organic acids, glycosides, terpenoids, and so on.

2.1.8 Ultrasonic extraction method

It is a method of solvent extraction assisted by ultrasound. Ultrasonic wave is a kind of elastic mechanical vibration wave. The vibration frequency is as high as 20 KHz in elastic medium. The ultrasonic wave could vibrate the liquid medium. When the vibration is sparse, many small holes are formed in the medium. The instantaneous closure of these small holes can cause a pressure of up to thousands of atmospheric pressures. At the same time, the local temperature can rise to 1000°C. It can cause instantaneous rupture of the cell wall of plants and the whole organism, and make the solvent permeate into the cells of plants. This accelerates the dissolution of active ingredients in plants into solvents. Ultrasonic wave extraction could shorten the extraction time and improve the extraction efficiency, but could not change the structures of chemical constituents meanwhile.

Ultrasonic extraction technology has been widely used in the extraction of natural products in recent years, for example, extraction of soy isoflavones; see [8].

2.1.9 Microwave-assisted extraction method

Microwave refers to the electromagnetic wave whose wavelength is in the range of 0.1–100 cm (the corresponding frequency is 300–300,000 MHz), which is between infrared and radio waves. Polar molecules can absorb microwave energy, then release energy in the form of thermal energy, which makes the temperature inside the medium rise rapidly, causes the rather high pressure inside, and then the components flow out and dissolve in the solvent. On the other hand, the electromagnetic field produced by microwave can make some components diffuse to the interface of the extraction solvent, accelerating their thermal movement, which not only improves the extraction efficiency but also reduces the extraction temperature [9].

Microwave-assisted extraction has the advantages of less decomposition of chemical constituents, shorter time, lower energy consumption and less environmental pollution. Microwave-assisted extraction has been widely used in a series of fields of perfume, condiments, natural pigments, herbal medicine, cosmetics, soil and environmental analysis, and so on. In China, microwave-assisted extraction technology has been used in hundreds of Chinese herbal medicine extraction, such as *Pueraria lobata*, Panax notoginseng, Ginkgo, and so on, for example, the extraction of tea polyphenols and tea caffeine from green tea leaves; see [10].

2.2 Steam distillation method

Steam distillation is suitable for the extraction of volatile components which can be distilled with steam without being destroyed and are insoluble in water. These compounds' boiling points of are mostly higher than 100°C, and they possess certain vapor pressures at about 100°C. The principle of steam distillation is that the vapor pressure of each component is equal to that of their pure state, while the existence of another liquid does not affect their vapor pressure. The total vapor pressure of the mixing system is equal to the sum of the vapor pressures of the two components. Because the total vapor pressure of the system is higher than that of any single component, so the boiling point of the mixture is lower than that of any component. It is mainly used to extract volatile oils, some alkaloids, and phenolic substances of small molecules from plants.

2.3 Sublimation method

The process that solid material converts into steam directly without melting after heating is called sublimation. The phenomenon that steam condenses into solid after cooling is called deposition. Some natural chemicals have sublimation properties, which can be extracted directly with the sublimation method, for example, the extraction of camphor from camphor wood and caffeine from tea. In addition, some small molecular alkaloids, coumarins, organic acids, and other components also have sublimation properties, such as aesculetin and benzoic acid. However, it is easy to carbonize natural products because of long heating time. The volatile tar-like substances often adhere to sublimates, which are difficult to remove and often accompanied with thermal decomposition. The yield of this method is often low, and it is not suitable for large-scale production.

2.4 Pressing method

When the content of active ingredients is relatively high and exists in the juice of plants, the juice can be extracted directly from fresh raw materials. Volatile oils can also be extracted from plant tissues by mechanical pressing, such as orange peel oil and lemon oil. It is performed at room temperature, so its components will not be decomposed by heat. However, the products obtained are impure and often contain impurities such as water, mucoid substances, and cell tissues, so they are often turbid, and it is not easy to press the volatile oil in plants entirely. Therefore, the crushed residue is often distilled by steam to extract volatile oils completely. For example, the black soybean oil from black soybean is often extracted with the low-temperature pressing method.

3. Isolation and purification of phytochemicals

The separation of phytochemicals is a process of isolating the constituents of plant extracts or effective parts one by one and purifying them into monomer compounds by physical and chemical methods. Classical isolation methods, including solvent extraction, precipitation, crystallization, fractional distillation, salting-out, and dialysis, are still used commonly at present. On the other hand, modern separation technologies such as column chromatography, high performance liquid chromatography, ultrafiltration, and high performance liquid drop countercurrent chromatography also play an important role in the separation of phytochemicals. This section describes the common methods and their specific applications in isolation of phytochemicals.

3.1 Solvent method

3.1.1 Acid and basic solvent method

It is carried out according to the different acidity and alkalinity of each component in the mixture. Water-insoluble alkaline organic components, such as alkaloids, could react with inorganic acids and form salts, which can be separated from nonalkaline and water-insoluble components. Acid components with carboxyl or phenolic hydroxyl groups can be salted by bases and dissolved in water. Components with lactone or lactam substructures can be saponified and dissolved in water and then isolated from other water-insoluble components. The total extract can be dissolved in lipophilic organic solvents (ethyl acetate is commonly used) and extracted respectively with acid water and alkali water, and then the total extract would be divided into acidic, alkaline, and neutral parts. Of course, the total extract can also be dissolved in water and extracted with organic solvents after adjusting the pH value. The alkalinity or acidity of the fractions are different and can be separated further by pH gradient extraction.

When using the acid and basic solvent method, attention should be paid to the strength of acidity or alkalinity, the contact time with the separated components, heating temperature, and time, so as to avoid the structural changes of some compounds under severe conditions or the chemical structures cannot be restored to the original states.

3.1.2 Polarity gradient extraction method

This method is to achieve the separation aim based on the different polarity of each component in plant extracts and the different partition coefficients in two-phase solvents. Generally, different two-phase solvent systems are selected according to the polarity of components in plant extracts. For example, the components with strong polarity can be separated by n-butanol-water system, the components with medium polarity can be separated by ethyl acetate-water system, and the components with weak polarity can be separated by chloroform (or ether)-water system. During the operation, the plant extract should be dissolved by water firstly, and then the solution or suspension is extracted in a separating funnel with different organic solvent which is not miscible with water based on the polarity difference. Usually, the extract was extracted with petroleum ether (or cyclohexane) firstly, then ethyl acetate (or chloroform), and finally with water saturated n-butanol, as shown in **Figure 1**. Petroleum ether layer contains lipid-soluble compounds with low polarity. Ethyl acetate layer contains medium polar compounds such as monoglycosides, flavonoids, and compounds with more polar functional groups. N-butanol layer contains compounds with strong polarity, such as oligoglycosides and other water-soluble components. Compounds in water layer possess strongest polarity, such as glycosides with more glycosyl groups, carbohydrates, amino acids, proteins, and other water-soluble compounds.

3.2 Precipitation method

It is a method based on the formation of precipitation of some phytochemicals by reaction with specific reagents, or the precipitation of some components from the solution by adding specific reagents, which can reduce the solubility of some components in the solution. The precipitation reaction must be reversible if the target components are required to form precipitation. While if the components are nontarget, the precipitation generated will be removed, so the precipitation reaction can be irreversible. According the addition of reagents or solvents, this method could be classified as follows [11].

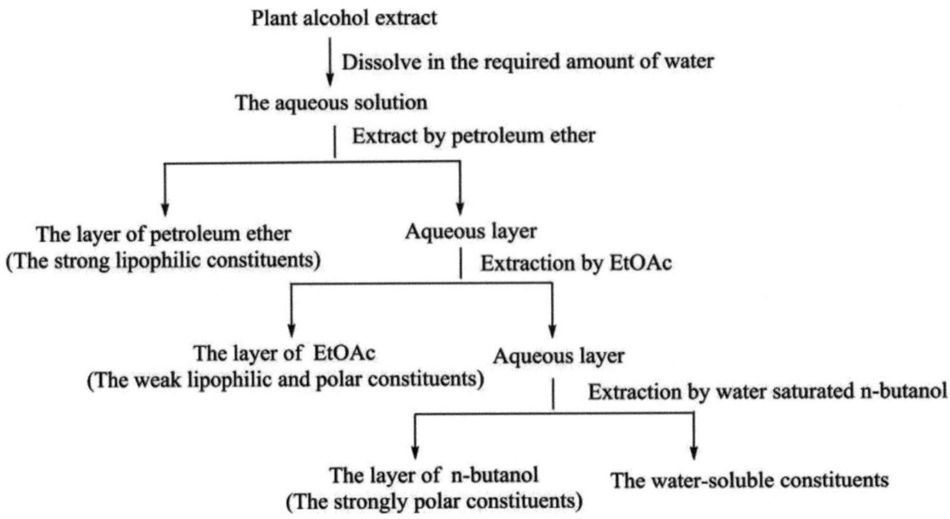

Figure 1.
Flow charts of common polarity gradient extraction method.

3.2.1 Solvent precipitation method

The solubility of some components in the mixed component solution can be changed by adding a specific solvent that can be mutually soluble with the solution, so it can be precipitated from the solution. The gradual precipitation by changing the polarity or amount of solvent added is called fractional precipitation. For example, using water as an extracting solvent to extract phytochemicals, ethanol is added to the water extracting concentrate to make its alcohol content more than 80%, and then polysaccharides, proteins, starch, gum, and so on will be precipitated and removed after filtration. The preceding procedure is called water extraction and ethanol precipitation. Crude polysaccharides from plants are often separated with this method. For example, see [12].

3.2.2 Exclusive reagent precipitation method

Some reagents could react selectively with certain chemical constituents to produce reversible precipitation, and the separation aims are achieved, which is called the exclusive reagent precipitation method. For example, alkaloid precipitation reagents such as Reynolds ammonium salt can precipitate after reacting with alkaloids, which can be used to separate alkaloids and nonalkaloids, or water-soluble alkaloids and other alkaloids. As another example, reactions of cholesterol and sterol saponins could form precipitation, which can separate them from triterpene saponins. Additionally, gelatin can precipitate tannins, which can be used to separate or remove tannins. In practical application, appropriate precipitation reagents should be selected according to the properties of target constituents and impurities in plants.

3.2.3 Salting out method

Adding inorganic salts to a certain concentration or saturated state in the water extract of plants can reduce the solubility of some components in water, thus they could be separated from water-soluble compounds. The inorganic salts commonly used for salting out are sodium chloride, sodium sulfate, magnesium sulfate, ferric sulfate, etc. For example, extractions of tetrandrine from *Daemonorops margaritae* and berberine from *Berberis poiretii* could be achieved by salting out with sodium chloride or ammonium sulfate. Some water-soluble substances, such as protoanemone, ephedrine, and matrine, are often extracted with organic solvents after adding a certain amount of salt to the water extract. For example, see [13].

3.3 Dialysis method

It is a method to let substances selectively penetrate through natural or synthetic semi-permeable membranes (or dialysis bags) under the action of concentration difference, pressure difference, or potential difference, so as to achieve the purpose of separation, classification, purification, or concentration. For example, when saponins, proteins, polypeptides, polysaccharides, and other substances in plants are separated and purified, dialysis can be used to remove inorganic salts, monosaccharides, and other impurities. On the contrary, large molecular impurities can also be left in the semi-permeable membrane, while small molecular substances can be separated and purified through the semi-permeable membrane into the solution outside the membrane [14].

3.4 Fractional distillation method

Fractional distillation is a method of separating components in liquid mixtures based on their different boiling points. It is usually categorized into atmospheric, vacuum, molecular distillation, and so on. It is mainly used for the separation of volatile oils and some liquid alkaloids in plants. For example, the boiling points of the two alkaloids in total alkaloids of *Cicuta virosa,* coniine, and conhydrine are 166–167°C for the former and 226°C for the latter, which are quite different from each other, and then they can be separated by the fractional distillation method. Generally, if the boiling point difference of compounds in liquid mixtures is above 100°C, the separation can be achieved by repeated distillation of the solution. If the boiling point difference of compounds is below 25°C, the fractionation column is needed. The smaller the boiling point difference is, the finer the fractionation device is needed [15].

3.5 Crystallization method

Crystallization is the process of solute precipitation from mother liquor with complex components, and it is an effective method to prepare pure substances. The initial crystallization is often impure and needs to crystallize again, which is called recrystallization. It is a method to separate compounds from the mixture by using the difference of solubility of each component in the solvent. Crystallization is one of the important technologies for plant chemists to prepare pure compounds.

When the content of a phytochemical is very high in one plant, crystals can be obtained by cooling or slightly concentrating the extract after extraction with appropriate solvent. For example, see [16].

Selecting suitable crystallization solvent is the key of the crystallization method. The ideal solvents for crystallization should possess the following characteristics: high solubility for the components to be purified at high temperature, low solubility at low temperature, insoluble for the impurities at high and low temperature, or soluble for the impurities at high and low temperature, moderate boiling point, no chemical reaction with the components to be crystallized, safe, low price, easy to obtain, and so on. Solvents commonly used for crystallization are methanol, ethanol, acetone, ethyl acetate, acetic acid, pyridine, etc. When crystals cannot be obtained with a single solvent, the crystallization operation can be carried out with a mixture of two or more solvents. Mixed solvents generally consist of two miscible solvents, one of which has high solubility for the component to be crystallized, and the other has low solubility. Firstly, the sample to be crystallized is heated and dissolved in as few solvents as possible with high solubility. Then the second solvent with low solubility is added to the hot solution to make it turbid. Then the first solvent is added to dissolve the sample. The solution reaches saturation at this point and crystallizes when it is cooled. The purity of crystallization can be preliminarily identified by the crystal form, color, melting point, melting range, thin layer chromatography, paper chromatography, etc.

3.6 Classical chromatographic methods

Chromatography is the most commonly used method for the separation of chemical constituents of natural products. It possesses advantages of high separation efficiency, rapidity, and simplicity. By choosing different separation principles, different operation modes, different chromatographic packings, or applying various chromatographic methods jointly, the separation and purification of various types of phytochemicals could be achieved. It can also be used for the identification of compounds.

3.6.1 Adsorption chromatography

It is a kind of chromatography based on the difference of adsorptive capacity of adsorbents to different compounds. The commonly used adsorbents include silica gel, alumina, activated carbon, polyamide, and so on. Silica gel adsorption chromatography is widely used, and it is suitable to the separation of most of the plant chemical constituents. Alumina adsorption chromatography is mainly used for the separation of alkaline or neutral lipophilic components, such as alkaloids, steroids, and terpenoids. Activated carbon is mainly used for the separation of water-soluble substances, such as amino acids, carbohydrates and some kinds of glycosides. Polyamide, which allows the separation to take place based on the formation of kinds of hydrogen bonds, is mainly used for the separation of phenols, quinones, flavonoids, anthraquinones, tannins, etc. [17].

3.6.2 Gel chromatography (exclusion chromatography, molecular sieve chromatography)

Molecular sieve is the main principle of gel chromatography, which can separate mixture compounds according to the pore size of the gel and the molecular size of the compounds. Gel is a kind of solid material with a porous network structure. The molecules of the separated substances are different in size, so their ability to enter the gel is different. When the mixture solution passes through the gel column, the molecules smaller than the gel pores can enter the gel interior freely, while the molecules with larger size than the gel pores cannot enter the gel, and only pass through the gel particle gaps. Therefore, different movement rates are emerged. The molecules with large sizes are not excluded, and the retention time is shorter. The molecules with small sizes are detained because of its diffusion into the pores, thus the retention time is longer. There are many kinds of commercial gels, dextran gel and hydroxypropyl dextran gel are used most commonly [18].

3.6.3 Ion exchange chromatography

It is to separate chemical constituents according to the difference of dissociation degrees. In this method, ion exchange resin is applied as stationary phase and water or solvent mixed with water as mobile phase. The ionic components existing in the mobile phase are absorbed by ion exchange resin after ion exchange reaction. Ion exchange chromatography is suitable for the separation of ionic compounds, such as alkaloids, amino acids, organic acids, peptides, and flavonoids. The ability of ion exchange reaction between compounds and ion exchange resins mainly depends on the compounds' dissociation degree and the amount of electric charges. If the dissociation degree of a compound is high (acidic or alkaline), it is easily exchanged on resins and difficult to elute. Therefore, when the compounds with different degree of dissociation are exchanged on the resin, the compounds with lower degree of dissociation are eluted before those with higher degree of dissociation [19].

3.6.4 Macroporous adsorption resin chromatography

It is a chromatographic method which combines the principle of adsorption and molecular sieve. Its chromatographic behavior possesses reversed-phase properties. Macroporous resin is a kind of solid macromolecule material with no dissociable group and porous structure and is insoluble in water. It is widely used in the separation and enrichment of natural compounds because of its stable physical and chemical properties (insoluble in acids, bases, and organic solvents).

In practical work, the water solution of the mixture to be separated is usually washed by water, water-containing alcohol solution with low to high concentration. The mixture can be separated into several components. The regeneration of macroporous adsorbent resin is convenient. It is often washed by 1 mol/L hydrochloric acid and 1 mol/L sodium hydroxide solution, respectively, first, then washed by distilled water to neutral, and stored in methanol or ethanol. The alcohol should be washed out with distilled water before using.

3.6.5 Partition chromatography

It is a kind of chromatography method to separate components by using different partition coefficients between stationary phase and mobile phase, which are immiscible liquids. Partition chromatography could be divided into normal phase chromatography and reverse phase chromatography. The polarity of stationary phase is stronger than that of mobile phase in normal phase partition chromatography, which is mainly used to separate polar and moderately polar molecular compounds. Carriers commonly used in normal phase distribution chromatography include silica gel, diatomite, cellulose powder, etc. Silica gel with water content of more than 17% can be used as a carrier for partition chromatography because of its loss of adsorption. It is the most widely used carrier for partition chromatography. In reverse phase partition chromatography, the polarity of mobile phase is stronger than that of stationary phase. The commonly used stationary phase is octadecylsilylated silica (ODS). The mobile phase is usually methanol-water or acetonitrile-water system, which is mainly used for the separation of nonpolar and moderately polar molecular compounds.

3.7 New technologies and methods

3.7.1 High performance liquid chromatography (HPLC)

High performance liquid chromatography (HPLC) is a rapid separation and analysis technology developed on the basis of conventional column chromatography. Its separation principle is the same as regular column chromatography, including adsorption chromatography, gel chromatography, partition chromatography, ion exchange chromatography, and other methods. HPLC columns are produced with particle fillers (particle diameter 5–20 μm) and high pressure homogenate column loading technology. The eluents are pressed into the column by a high pressure infusion pump and equipped with high sensitive detectors and automatic recording and collection devices. As a result, it is far superior to conventional column chromatography in separation speed and efficiency. It has the characteristics of high efficiency, high speed, and automation. Preparative HPLC can be used to prepare a large amount of samples of high purity. HPLC has played an increasingly important role in the separation, qualitative identification, and quantitative analysis of plant chemical constituents. During the separation of many plant chemical constituents, it is necessary to separate trace constituents from a large amount of crude extracts. Usually, in the final stage of separation, samples with high purity are prepared by high or medium pressure liquid chromatography. Constant concentration eluents are mostly used in preparative HPLC. However, gradient elution is sometimes applied for samples that are difficult to be separated. Moreover, HPLC retains the advantages of liquid chromatography, such as a wide range of application and flexibility of mobile phase change. It can be applied to chemical constituents of difficult gasification, high molecular weight, or thermal instability.

The detectors commonly used in HPLC are ultraviolet detectors and differential refractive index detectors, but both have limitations. Differential refractive index detectors are sensitive to temperature change, the detection of a small amount of substances is often not ideal, and gradient elution cannot be used. As for ultraviolet detectors, they cannot detect samples without ultraviolet absorption. In recent years, a kind of mass detector, called evaporative light scattering detector (ELSD), has been applied in HPLC. It can not only detect samples without ultraviolet absorption, but also use gradient elution. It is suitable for most nonvolatile components [20].

3.7.2 Droplet counter-current chromatography (DCCC)

DCCC is an improved liquid-liquid partition chromatography based on the counter-current partition method. The formation of droplets is required when the mobile phase passes through a liquid stationary phase column. Droplets of mobile phase contact with stationary phase effectively, and form new surfaces in thin partition extraction tubes constantly, which promote the partition of solutes in two-phase solvents, and the chemical components of mixtures are isolated in immiscible two-phase droplets due to different partition coefficients. This method is suitable for the separation of phytochemicals with strong polarity. The separation effect is usually better than counter-current partition chromatography, and there is no emulsification phenomenon. Furthermore, nitrogen is used to drive the mobile phase, so the separated substance will not be oxidized by oxygen in the atmosphere. However, the solvent system which can generate droplets must be selected in this method, the amount of sample treated is small, and special equipment is needed.

DCCC possesses good reproducibility, and can handle crude extract samples of milligram to gram grade. It can be used in either acidic or basic conditions. Because no solid separation carriers are used, the phenomenon of irreversible adsorption and band broadening of chromatographic peaks can be avoided. Compared with preparative HPLC, DCCC consumes less solvent, but the separation time is longer and the resolution is lower. For example, see [21].

3.7.3 High speed counter-current chromatography (HSCCC)

HSCCC is also a liquid-liquid partition chromatography. It is another mild form of chromatography with no solid support and hence no chance of loss of substrate by binding to the column. The only media encountered by the sample are solvent and Teflon tubing. The former is common to all forms of chromatography and the latter to most. The chemical constituents with higher partition coefficient in mobile phase are eluted first, whereas those with higher partition coefficient in stationary phase are eluted later.

HSCCC chromatography could avoid the shortcomings of irreversible adsorption and abnormal tailing of chromatographic peaks caused by solid carriers in liquid chromatography because it does not need solid carriers. The sample recovery is near 100% from a chromatography. It also has advantages of good reproducibility, high purity of separated compounds, and fast speed. It is suitable for the isolation and purification of wide kinds of phytochemicals, such as saponins, alkaloids, flavonoids, anthraquinoids, lignans, triterpenes, proteins, and carbohydrates. For example, see [22].

3.7.4 High performance capillary electrophoresis (HPCE)

It is an instrumental analysis method developed in the late 1980s combining classical electrophoresis with modern microcolumn separation technologies. In pharmaceutical analysis, the most commonly used separation modes are capillary zone electrophoresis, micellar electrokinetic capillary chromatography, and capillary gel electrophoresis. It is an efficient separation technology of large and small molecules in a hollow and thin inner diameter capillary (10–200 μm). The two ends of the capillary are immersed in a buffer solution and electrodes connected with a high voltage power supply are inserted separately. The voltage makes samples migrate along the capillary. According to the charge and volume of the separated substances, various molecules are separated under high voltage. In zone capillary electrophoresis, separation could be achieved by the movement of electrophoresis and electroosmotic flow. The strength of electroosmotic flow depends on the strength of electric field, PH value of electrolyte, composition of buffer solution, ionic strength, internal friction, and so on. Sample injection could be accomplished by pressing the sample into a capillary tube by atmospheric pressure or voltage.

HPCE has the advantages of high efficiency, microamount, economy, high automation, and wide application. However, it has the disadvantages of poor preparation ability, low sensitivity, and poor separation reproducibility. For example, see [23].

3.7.5 Affinity chromatography (AC)

Affinity chromatography is a unique chromatographic separation method based on the principle of reversible combination of high affinity and specificity between molecules. By simulating the reversible and specific interaction between biological molecules, affinity chromatography uses the adsorption medium coupled with affinity ligands as the stationary phase to adsorb target compounds. It is a development of adsorption chromatography. This method can selectively separate and analyze specific chemical constituents from complex samples. Firstly, ligands that can specifically bind to the target compounds are fixed on the filler carrier to make the chromatographic column. Then the mixture containing the target compounds is passed through the column. Only the target compounds which show affinity with the ligands can bind to the ligands and remain in the column. Finally, the adsorbed target compounds are eluted by changing the composition of the mobile phase and are separated from other chemical constituents. AC is mainly used for the separation and purification of proteins, especially enzymes, antigens, and antibodies. Its application range has been expanding along with the continuous development of technology in recent years. For example, see [24].

4. Structural identification of phytochemicals

The chemical structures of plant compounds must be identified or elucidated, which may provide the necessary basis for further study on the bioactivities, structure-activity relationships, metabolisms in vivo, structural modification, and synthesis of the active phytochemicals.

The quality of physiological active substances isolated from plants is often small, sometimes only a few milligrams, and the structural studies are often difficult to carry out with classical chemical methods, such as chemical degradation, derivative synthesis, etc. Therefore, spectral analysis is mainly used, that is, consuming sample as little as possible to obtain structural information as much as possible by measuring various spectra. Then comprehensive analysis is carried out with the assistance of literature data. If necessary, chemical means would be integrated into the former methods to determine the planar- and even the stereo-structures of the compounds.

4.1 Determination of the purity of the compounds

Before the structural investigation of an active compound, the purity must be determined, which is a prerequisite for the structural identification.

4.1.1 Measurement of physical properties

The crystals of each compound have certain shape, color, and melting point, which can be used as the basis for the preliminary determination of the purity. Generally, the crystal shape of a specific compound under the same solvent is consistent, the color is pure, and has a short melting range (generally at 1~2°C). But for compounds with double melting points or amorphous substances, the purity cannot be determined by this method.

4.1.2 Thin layer chromatography (TLC)

TLC, such as silica gel and paper chromatography, is the most commonly used method to determine the purity of compounds. Generally, a specific sample, showing an only spot (Rf value at 0.2~0.8) in three different developing agents, could be considered as a pure compound. In some cases, both normal and reverse phase chromatographic methods are needed.

4.1.3 Gas chromatography (GC) and high performance liquid chromatography (HPLC)

GC and HPLC are important methods in the purity determination of phytochemicals. GC is widely used in the analysis of volatile compounds. Both volatile and nonvolatile substances could be analyzed with HPLC, which possesses various advantages of high speed, high efficiency, sensitivity, and accuracy.

4.2 Major procedures of structural determination

The general procedures of structural determination of phytochemicals are shown roughly in **Figure 2**.

The structural identification of phytochemicals can be greatly simplified according to the researchers' habits, experiences, and skill levels of different technologies. However, the literature search almost runs through the whole process of structural

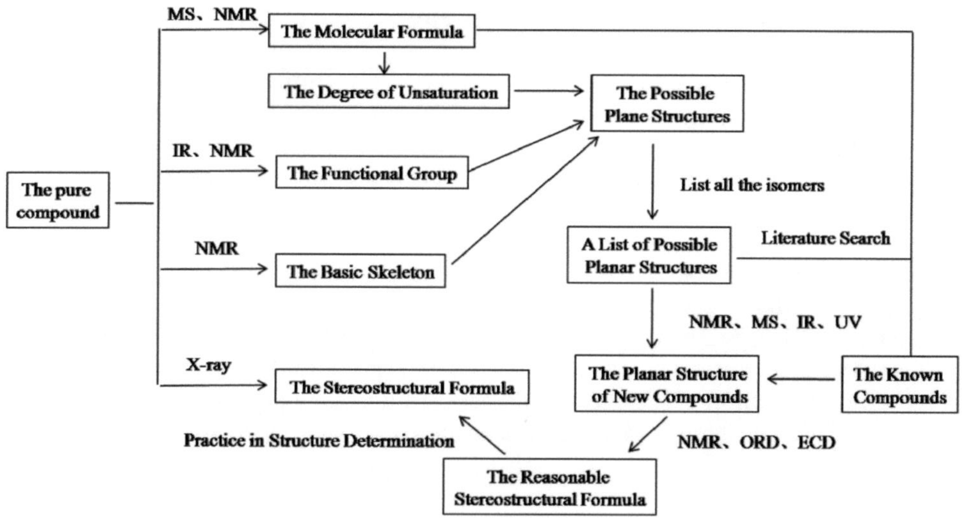

Figure 2.
The main procedures for studying the structures of phytochemicals.

research, no matter for known or new compounds. A large number of facts have been proved that taxonomically related plants, that is to say, plants of same or similar genus often contain chemical constituents of similar or even same chemical structures. Therefore, it is necessary to investigate literatures of chemical studies of the study object and the plants of its same and similar genera. It is necessary to understand not only the components from different plants of similar genera, but also their extraction methods, physicochemical properties, spectral data, and biosynthesis pathways before the extraction and separation of one specific plant. The SciFinder Scholar database is used most widely to quickly determine whether the compound was "known" or "unknown".

4.3 Spectral technologies

At present, spectrum analyses have become the main means to determine the chemical structures of plant chemicals. Particularly, with the developing of the superconducting nuclear magnetic resonance (NMR) and mass spectroscopic (MS) technologies, the speed of structural determination is greatly accelerated and the accuracy is improved. Here, the applications of infrared (IR), ultraviolet (UV), nuclear magnetic resonance (NMR), and mass (MS) spectra in the structural identification of phytochemicals are introduced briefly.

4.3.1 Ultraviolet-visible spectra (UV-Vis)

UV-vis spectrum is a kind of electron transition spectrum, which is generated after the molecules absorbing the electromagnetic waves with wavelength at the range of 200–800 nm. The valence electrons in the molecules absorb light of certain wavelengths and jump to the excited state from the ground state, and then UV spectra are recorded.

Compounds containing conjugated double bonds, α,β-unsaturated carbonyl groups (aldehydes, ketones, acids, and esters), and aromatic compounds could show strong absorption in UV spectra because of $n \rightarrow \pi^*$ or $\pi \rightarrow \pi^*$ transitions. Therefore, UV spectrum is mainly used to identify the presence of conjugated systems in the structures.

UV spectra could provide the following information: (1) the compounds show no UV absorption at 220–800 nm, indicating the compounds were aliphatic hydrocarbons, aliphatic cyclic hydrocarbons, or their simple derivatives. (2) The compounds show strong absorption at 220–250 nm, indicating that the compounds possess conjugated diene, α,β-unsaturated aldehyde, or ketone substructures. (3) The absorption at 250–290 nm is moderately strong, indicating that the compounds possess benzene rings or aromatic heterocycles. (4) Weak absorption at 250–350 nm indicates the presence of carbonyl or conjugated carbonyl groups. (5) Strong absorptions at above 300 nm indicate that the structures possess long conjugated chains.

Generally, UV spectrum can only provide part of the structural information, rather than the whole structural information of a compound, so it can only be used as an auxiliary method to identify the structures. It possesses practical value to determine the structures of phytochemicals with conjugated substructures.

4.3.2 Infrared spectra (IR)

IR is caused by the vibration-rotational energy level transition of the molecule, ranging from 4000 to 625 cm^{-1}. The region above 1250 cm^{-1} is functional group region, and the absorption of characteristic functional groups such as hydroxyl, amino, carbonyl, and aromatic rings occurs in this region. The region of 1250 to 625 cm^{-1} is fingerprint region, and the peaks appear mainly due to the stretching vibrations of C-X (X = C, O, N) single bonds, and various bending vibrations. IR is mainly used for the determination of functional groups and the types of aromatic ring substitution. In some cases, IR can also be used to determine the configuration of plant chemical constituents. For example, there is a significant difference between 960 and 900 cm^{-1} for 25R and 25S spirostanol saponins.

4.3.3 Mass spectrometry (MS)

In a mass spectrometer, mass and strength information of molecular and fragment ions is recorded after the molecules are ionized and enter into the collector under the action of electric and magnetic fields. The abscissa represents the mass-to-charge ratio (m/z) and the ordinate represents the relative intensity in a MS spectrum. Unlike IR, UV, and NMR spectra, MS is mass spectrum, which characterizes fragment ions, not an absorption spectrum. Its role is to determine weights, formulas, and fragment structures of molecules.

With the rapid development of modern techniques, new ion sources have emerged in recent years, which make MS play more important role in determining the molecular weights, elemental composition, detecting functional groups by cleavage fragments, identifying compound types, and determining carbon skeletons [25]. In the structural analysis, the information of molecular weights could be obtained on the basis of molecular ion peaks, and the molecular formula could be obtained by high-resolution mass spectrometry (HR-MS). Fragment ion peaks, combined with molecular ion peak, could be applied to conjecture chemical structures. Tandem mass spectrometry even can isolate and analyze the mixed ions again. According to the types of ion sources, common mass spectrometry could classified as electron impact mass spectrometry (EI-MS), chemical ionization mass spectrometry (CI-MS), field desorption mass spectrometry (FD-MS), fast atom bombardment mass spectrometry (FAB-MS), matrix-assisted laser desorption mass spectrometry (MALDI-MS), electrospray ionization mass spectrometry (ESI-MS), tandem mass spectrometry (MS–MS), and so on.

4.3.4 Nuclear magnetic resonance (NMR)

With the birth of Fourier transform spectrometer, the great progress of radionuclide research such as ^{1}H, ^{13}C, ^{15}N, ^{19}F, ^{31}P, and the advancement of two-dimensional and three-dimensional nuclear magnetic technology, NMR has become the most important spectroscopic method to determine chemical structures. Particularly, hydrogen spectrum and carbon spectrum are most widely used. During the operation of nuclear magnetic resonance spectrometer, compound molecules are irradiated by electromagnetic waves in a magnetic field, energy level transitions occur after the atomic nuclei with magnetic distance absorb a certain amount of energy, and then NMR spectrum is obtained by mapping the absorption strength with the frequencies of the absorption peaks. It can provide structural information about the type and number of hydrogen and carbon atoms in the molecule, the modes they are connected, the surrounding chemical environment, configuration, and conformation [26].

4.3.4.1 Commonly used deuterated reagents

Samples used to measure NMR spectra include solids, liquids, and gases. Liquid high-resolution NMR is most widely used. The solvent used in the measurement of NMR must be deuterated. The commonly used deuterated reagents to dissolve samples and their chemical shifts of their residual proton and carbon signals are shown in **Table 1**.

4.3.4.2 Proton nuclear magnetic resonance spectroscopy (1H-NMR)

Resonance absorption peaks are generated after hydrogen protons absorb electromagnetic waves of different frequencies in an external magnetic field. ^{1}H-NMR possesses high sensitivity, easy measurement, and wide application. ^{1}H-NMR spectrum can provide structural information of chemical shifts (δ), coupling constants (J) that indicate the coupling relationships between different hydrogen nucleus, and the number of protons (the peak area is proportional to the number of protons that cause the absorption).

Because of the different surrounding chemical environment, the ^{1}H nuclei possess different magnetic cloud densities and magnetic shielding effects caused by the rotation around the nucleus, and then different types of ^{1}H nuclear resonance signals appear in different regions. Tetramethylsilane (TMS) is usually used as a reference compound. Compared with the general compounds, the shielding effect of protons and carbons on the methyl groups is stronger in TMS. Therefore, regardless of the hydrogen spectrum

Solvent	δ_C	δ_H
$CDCl_3$	77.0	7.24
CD_2Cl_2	53.8	5.32
CD_3OD	49.0	3.3
Acetone-d_6	29.8, 206.0	2.04
D_2O	—	4.7
DMSO-d_6	39.5	2.49
C_6D_6	128.0	7.16
C_5D_5N	123.6135.6149.9	7.2, 7.6, 8.7

Table 1.
Chemical shifts of common deuterated solvents (TMS is an internal standard).

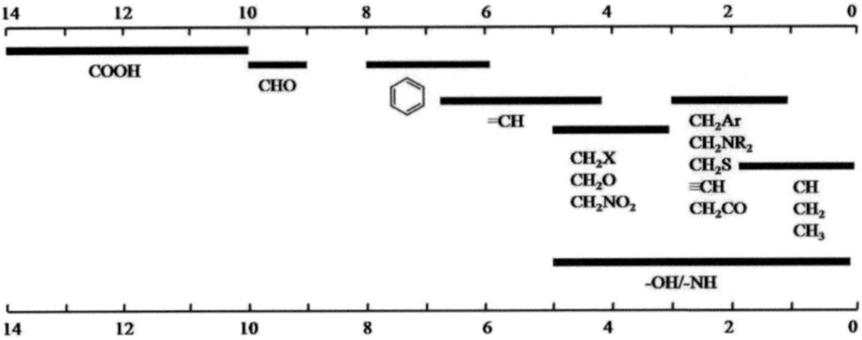

Figure 3.
^{1}H-NMR chemical shift range of common hydrogen protons.

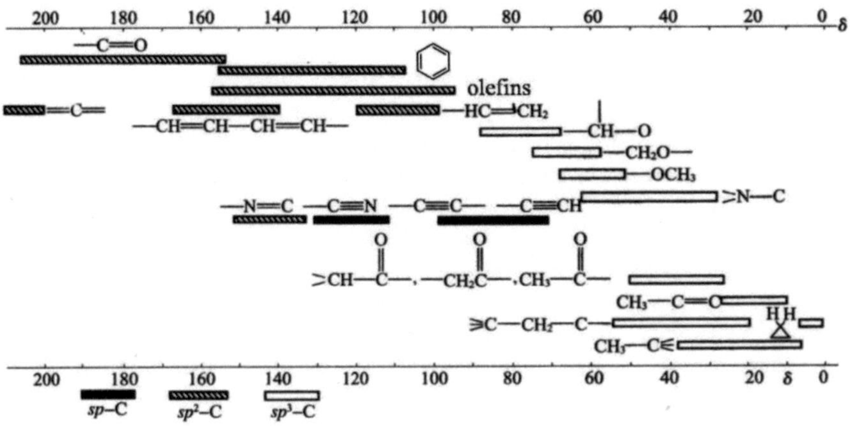

Figure 4.
^{13}C-NMR chemical shifts of common carbon signals.

or the carbon spectrum, the absorption peaks generated by the general compounds appear in the lower field than TMS, that is to say, δ values generated by common compounds is positive. The chemical shifts of the ^{1}H-NMR spectrum is mostly in the range of δ0–20. Some typical chemical shifts of ^{1}H nuclei are shown in **Figure 3** [4].

In addition to the normal ^{1}H-NMR spectrum technique, there are some auxiliary techniques that assist in structural analysis, such as selective decoupling, heavy hydrogen exchange, addition of reaction reagents, and dual irradiations.

4.3.4.3 Carbon nuclear magnetic resonance spectroscopy (^{13}C-NMR)

^{13}C-NMR spectra can provide structural information of organic compounds, including the number, types, and chemical environment of carbon atoms [27]. It is one of the important means for the structural identification of organic compounds. Especially, where there are serious signal peak overlaps in the ^{1}H-NMR spectrum, or the molecules contain several quaternary carbon atoms, ^{13}C-NMR spectra will provide crucial information for the structure identification. The chemical shifts of common carbon signals are shown in **Figure 4** [4].

Common ^{13}C-NMR techniques include proton broadband decoupling, off resonance decoupling (OFR), insensitive nuclei enhanced by polarization transfer (INEPT), and distortionless enhancement by polarization transfer (DEPT). Proton broadband decoupling and DEPT spectra are most commonly used at present.

4.3.4.3.1 Proton broadband decoupling

Proton broadband decoupling spectrum is measured after ^{1}H nuclei are saturated with broadband electromagnetic radiation. At this point, the couplings between ^{1}H and ^{13}C are completely eliminated, and all ^{13}C signals are shown as singlets, so it is very convenient to determine the chemical shift of ^{13}C signals. In addition, because of the NOE effect of ^{1}H after irradiation, the signal of ^{13}C signal connected with ^{1}H will be increased, while the quarterly carbon signal will show weak absorption peaks.

4.3.4.3.2 Distortionless enhancement by polarization transfer (DEPT)

It is an improved method of INEPT, in which a *J*-modulation is accompanied by a polarization transfer from the protons to coupled carbons, leading to significant improvement in sensitivity. In DEPT spectrum, by changing the pulse width (θ), which could be designed as 45°, 90°, and 135°, during irradiation of ^{1}H, different carbons could show different strengths and signs. The results are similar with INEPT spectrum. When θ = 45°, all CH, CH_2, and CH_3 groups display positive signals; when θ = 45°, only CH groups show positive signals; when θ = 135°, both CH and CH_3 groups show positive signals, while CH_2 groups show negative signals. Quarterly carbons show no signal peaks in DEPT spectra. An example of DEPT spectra is shown in **Figure 5**.

4.3.4.4 Two-dimensional nuclear magnetic resonance spectroscopy (2D-NMR)

Two-dimensional correlation spectroscopy (2D-COSY) is the most important and widely used in 2D-NMR spectroscopy. 2D-COSY spectra can be divided into homonuclear and heteronuclear correlation spectra. Both abscissa and ordinate represent chemical shifts in 2D-COSY. Common correlation spectrum types are show as follows.

4.3.4.4.1 ^{1}H-^{1}H COSY spectrum

It is a kind of chemical shift correlation spectrum between ^{1}H and ^{1}H. It is the coupling correlation spectrum between protons in the same coupling system. The adjacent hydrogen groups could be determined by their coupling relationships (^3J) shown in ^{1}H-^{1}H COSY spectra.

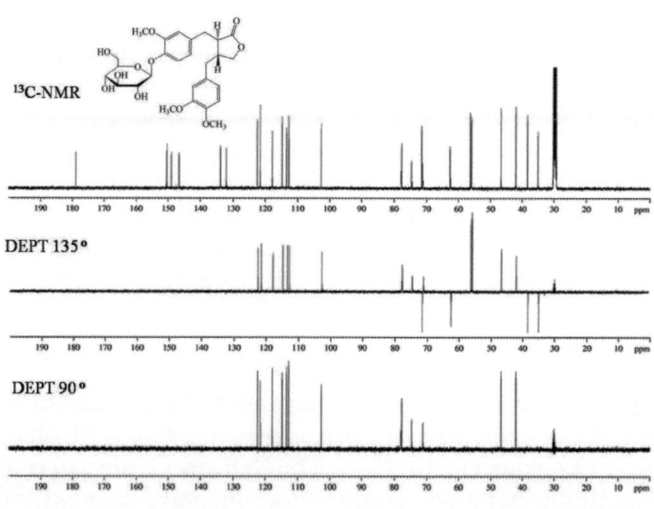

Figure 5.
The DEPT spectrum of Arctiin (CD_3OD).

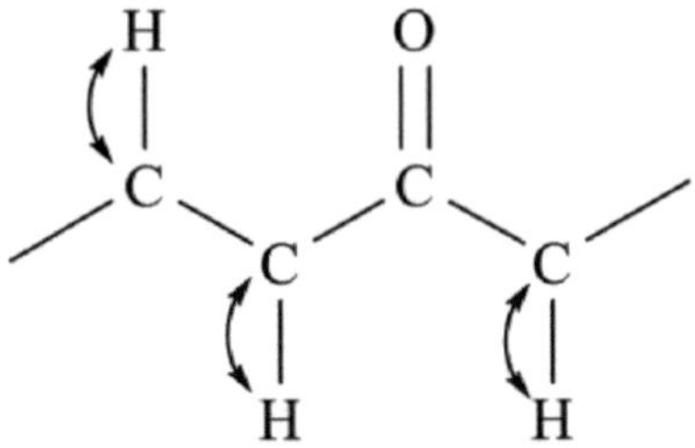

Figure 6.
Schematic diagram of correlations between ^1H and ^{13}C in the HSQC or HMQC spectrum.

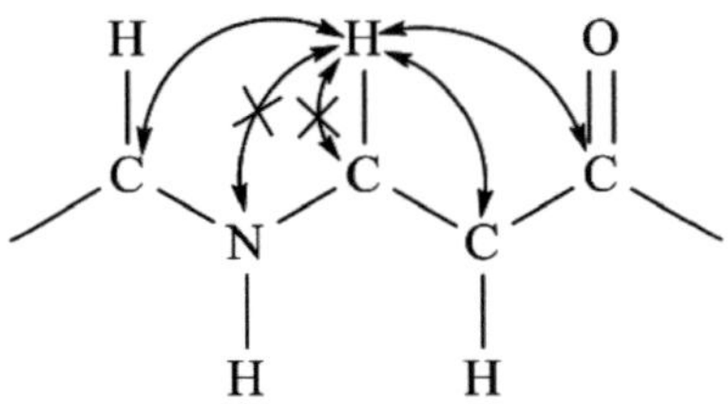

Figure 7.
Schematic diagram of correlations between ^1H and ^{13}C in the HMBC spectrum.

In addition, for compounds of aromatic systems, double bond systems, and some particular configuration systems, ^1H-^1H COSY spectra can show ^4J coupling or longer coupling relationships of hydrogen groups. It is very important for the elucidation of an unknown structure.

4.3.4.4.2 HSQC (HMQC) spectrum

^1H detected heteronuclear single quantum coherence (HSQC) and ^1H detected heteronuclear multiple quantum coherence (HMQC) can display the correlations between ^1H and ^{13}C. HSQC possesses higher sensitivity and wider application than HMQC. In the HMQC or HSQC spectrum, the signals occurred at the crosses of chemical shifts generated by corresponding carbons and protons (**Figure 6**).

4.3.4.4.3 HMBC spectrum

HMBC spectrum is short for ^1H detected heteronuclear multiple bond correlation, which associates the ^1H nucleus with ^{13}C nucleus of long-range coupling. HMBC could detect the long-range coupling of ^1H-^{13}C sensitively ($^nJ_{CH}$, n≧2). Moreover, the correlation signal peaks between protons and quaternary carbons that are two or three bonds apart could also be shown in HMBC spectra, as shown in **Figure 7**. From the HBMC spectrum, we can get the connection information of the carbon chain skeletons, the structure information of the quaternary carbons, and the structural information of the coupling systems that are cut off by heteroatoms.

4.3.4.4.4 NOESY spectrum

When two groups of protons are located at rather close spatial distances, irradiation of one group will enhance the signal strength of another, which is known as nuclear Overhauser enhancement (NOE). The NOE spectrum can determine the spatial relative position, stereoscopic configuration, and dominant conformation of some groups in the molecule, which is very important for the study of the stereostructures of organic compounds.

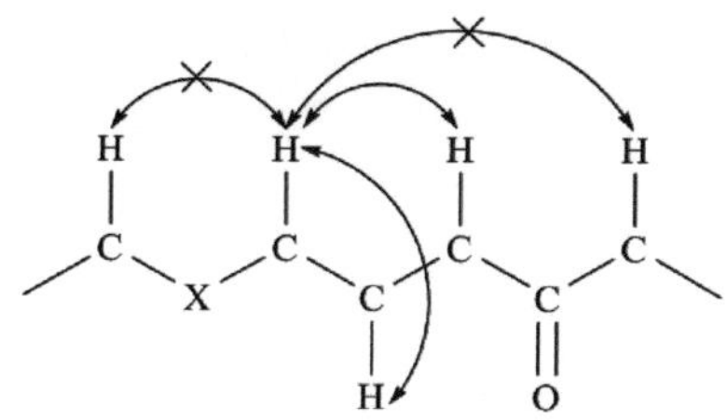

Figure 8.
Schematic diagram of correlations between ^1H and ^{13}C in the TOCSY spectrum.

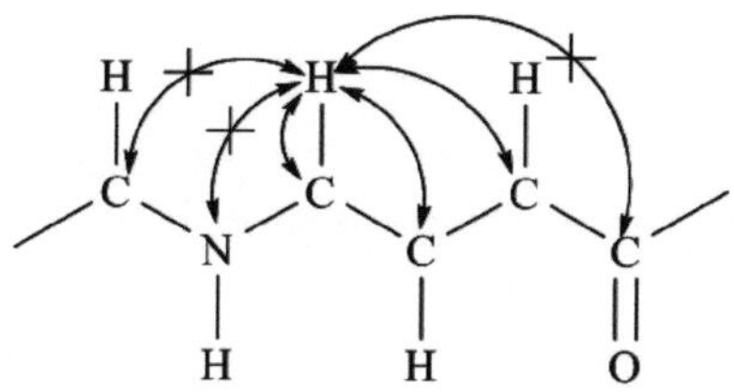

Figure 9.
Schematic diagram of correlations between ^1H and ^{13}C in the HSQC-TOCSY spectrum.

2D-NOE (NOESY) spectra could show the NOE correlations of protons. The greatest advantage of NOESY is that all the NOE information between protons of a compound could be shown in one spectrum. However, not all the cross peaks are NOE correlation signals, the residual correlation signals of COSY are often shown in NOESY spectrum as well, which should be paid attention during spectroscopic analysis.

4.3.4.4.5 Total correlation spectroscopy (TOCSY) spectrum

The TOCSY spectrum shows the correlation of the entire spin system, which is different from the ordinary ^1H-^1H COSY. The relationships between the nuclei that generated the correlation peaks are shown in **Figure 8**. Not only the correlation signals of a proton with protons connected to the adjacent carbons, but also its correlation signals with other protons in a whole spin system could be shown in the TOCSY spectrum, which provides important basis for the connection of structural fragments.

4.3.4.4.6 HSQC-TOCSY spectrum

HSQC-TOCSY is a kind of combined 2D-NMR spectrum. Comprehensive results of HSQC and HMBC are obtained by using a long pulse sequence. The correlation is shown in **Figure 9**. It is very useful for the assignment of carbon and proton signals in complex chemical structures. For example, for saponins with a series of glycosyl groups, the signals generated by glycosyl groups are often overlapped seriously in common NMR spectra, which causes difficulty to assign signals of glycosyls. HSQC-TOCSY spectrum will play an important role in this case. The spectrum includes the information of HSQC, HMBC, and ^1H-^1H COSY.

4.3.5 Optical rotary dispersion (ORD) and circular dichroism (CD)

Polarimetry is an optical method used widely in the studies of asymmetric structures, which appeared very early. The progress of the sensitive method such as ORD and CD made it possible to study stereostructures of chiral compounds more deeply. Both of them are spectra related to the optical activity of compounds, and could provide information of absolute configurations, dominant conformations, and reaction mechanisms of chiral compounds, that cannot be replaced by any other spectroscopic methods [28].

4.3.5.1 Optical rotary dispersion (ORD) spectrum

The specific rotation [α] of a chiral compound depends upon the wavelength of the monochromatic light wave. The measurement of specific rotation as a function of wavelength is called optical rotator dispersion (ORD). The common types of ORD curves are as follows.

4.3.5.1.1 Plain curves

The ORD spectrum of an optically active compound with no chromophores is plain without peaks and troughs. An ORD curve of specific rotation increases with decrease of wavelength which is called positive plain curve, while in the case of negative plain curve, negative rotation increases with decrease of wavelength (see **Figure 10**).

4.3.5.1.2 The cotton effect curve

If there is a simple chromophore in the molecule, the ORD curve is very different from plain curve. Near the absorption wavelength region of chromophore, a peak and a trough are exhibited, which is called the Cotton effect, and the spectrum drawn is called the Cotton effect curve. The spectrum with only one peak and one trough is called pure Cotton effect curve, while the spectrum with several peaks and troughs is called complex Cotton effect curve. The Cotton effect is called positive when the trough is observed at a shorter wavelength then peak. Conversely, the Cotton effect is called negative if the trough is observed at a longer wavelength than the peak. Cotton curves of $\triangle^5$-cholestenone are shown in **Figure 11,** which shows A and B possess the same structural formula, while different opposite configurations.

4.3.5.1.3 Complex Cotton effect curve

For compound with two or more different chromophores, its ORD curve may possess multiple peaks and troughs, which is called complex Cotton effect curve. Each ORD curve is the average effect of each chromophore in the molecule, and the contribution of each orientation and conformation of the molecule. Hence the Cotton effect curve is often complex.

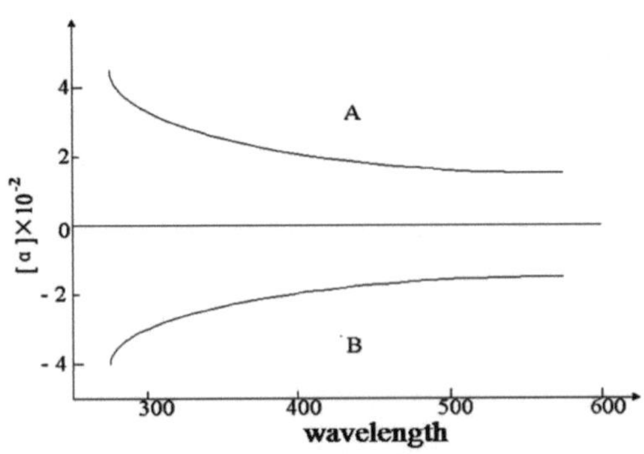

Figure 10.
ORD plain curves (A: Positive plain curve; B: Negative plain curve).

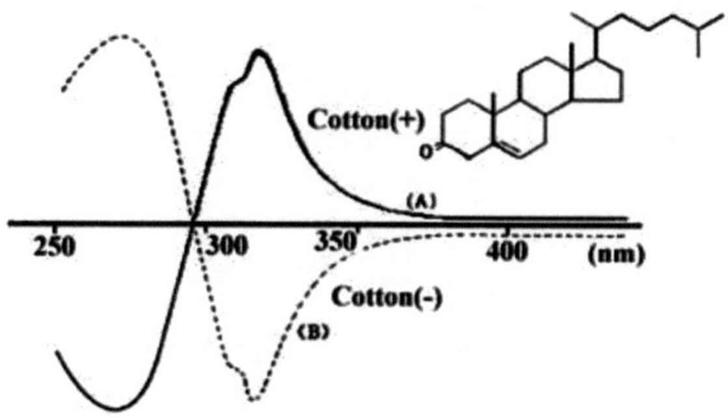

Figure 11.
The Cotton effect curves of $\triangle^5$*-cholestenone (A) natural cholesterone (+) cotton; (B) Cholesterone in the opposite absolute configuration (−) cotton.*

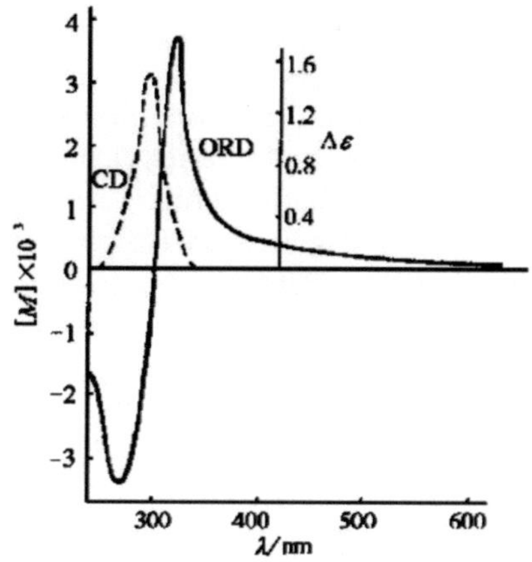

Figure 12.
The ORD and CD spectrum of (+)-camphor.

4.3.5.2 Circular dichroism (CD) spectrum

Optically active compounds have different molar absorption coefficients for left-circularly and right-circularly polarized light that make up plane polarized light, which is called circular dichroism (CD). The difference value between the two molar absorption coefficients ($\Delta\epsilon = \epsilon_L - \epsilon_R$) changes with the wavelength of the incident polarized light. With $\Delta\epsilon$ as the ordinate, the wavelength as the abscissa, the spectrum obtained is called circular dichroism spectrum. Because the absolute value of $\Delta\epsilon$ is very small, it is often replaced by molar ellipticity $[\theta]$. The relationship between $[\theta]$ and $\Delta\epsilon$ is as follows.

$$[\theta] = 3300\Delta\epsilon. \quad (1)$$

Because $\Delta\epsilon$ could be positive or negative, the circular dichroism curve also could be classified as positive and negative. In the CD spectrum showing positive Cotton effect, only a peak appears near the λ_{max} of the chromophore in the molecule. Conversely, a trough appears in the CD spectrum showing negative Cotton effect. Therefore, CD spectra are simpler and easier to analyze than ORD spectra. For example, the ORD and CD spectra of (+)-camphor are shown in **Figure 12**. CD is more widely used than ORD in the study of chiral compounds.

4.3.6 Single crystal X-ray diffraction method

Single crystal X-ray diffraction could be applied independently to analyze the structures, components, contents, configurations, conformations, solvents, and crystal forms of samples. It is widely used in the stereostructural study of natural compounds, synthetic compounds, peptides, proteins, etc. Therefore, X-ray diffraction analysis is a necessary physical method in the field of structure and function research of modern natural drugs.

Single crystal X-ray diffraction is a kind of quantitative analysis technology, which can provide three-dimensional structural information of molecules, including atomic coordinates, bond length, bond angles, dihedral angles, hydrogen bonds, salt bonds, coordinate bonds, and so on. In addition, it is also a reliable method to determine the absolute configuration of chiral drug molecules and the epimers in the stereochemical structures. For example, see [29].

5. Conclusions and future directions

In recent years, study on phytochemicals from plants becomes more and more popular due to their demonstrated health benefits. A number of plants having high contents of phytochemicals (particularly phenolic acids and flavonoids) with associated antioxidant activities have been increasingly utilized. Complementary research is also needed to enhance the potential functionalities of the phytochemicals in future, where such plants have shown to contain numerous phytochemicals that may be beneficial to human health. The compiled results indicated that many of their bioactive compounds remain to be fully isolated, identified, and characterized (alkaloids, diterpenoids, and so on).

Therefore, phytochemicals can be considered as the source of natural medicines. The compounds of plants are bioaccessible and bioavailable in humans with some demonstrated health benefits, including antioxidant, anti-inflammatory, anti-cancer, anti-microbial, hypoglycemic action, etc. Additional well-designed human intervention studies and clinical trials are needed to validate the health benefits of phytochemicals.

Author details

Weisheng Feng*, Meng Li, Zhiyou Hao and Jingke Zhang
School of Pharmacy, Henan University of Chinese Medicine, Zhengzhou, China

*Address all correspondence to: fwsh@hactcm.edu.cn

References

[1] Brusotti G, Cesari I, Dentamaro A, Caccialanza G, Massolini G. Isolation and characterization of bioactive compounds from plant resources: The role of analysis in the ethnopharmacological approach. Journal of Pharmaceutical and Biomedical Analysis. 2014;**87**:218-228. DOI: 10.1016/j.jpba.2013.03.007

[2] Alberti Á, Riethmüller E, Béni S. Characterization of diarylheptanoids: An emerging class of bioactive natural products. Journal of Pharmaceutical and Biomedical Analysis. 2018;**147**:13-34. DOI: 10.1016/j.jpba.2017.08.051

[3] Hosler DM, Mikita MA. Ethnobotany: The chemist's source for the identification of useful natural products. Journal of Chemical Education. 1987;**64**:328-332

[4] Silverstein RM, Bassler GC. Spectrometric Identification of Organic Compounds. ACS Publications; 1962

[5] Luo YM. Technology and Method of Extraction and Separation of Chemical Constituents of Traditional Chinese Medicine. Shanghai, China: Shanghai Scientific & Technical Publishers; 2016

[6] Gray AI, Igoli JO, Edrada-Ebel R. Natural products isolation in modern drug discovery programs. Natural Products Isolation. 2012;**864**:515-534. DOI: 10.1007/978-1-61779-624-1_20

[7] Tyśkiewicz K, Konkol M, Rój E. The application of supercritical fluid extraction in phenolic compounds isolation from natural plant materials. Molecules. 2018;**23**:2625. DOI: 10.3390/molecules23102625

[8] Rostagno MA, Palma M, Barroso CG. Ultrasound-assisted extraction of soy isoflavones. Journal of Chromatography A. 2003;**1012**:119-128. DOI: 10.1016/S0021-9673(03)01184-1

[9] Nayak B, Dahmoune F, Moussi K, Remini H, Dairi S, Aoun O, et al. Comparison of microwave, ultrasound and accelerated-assisted solvent extraction for recovery of polyphenols from *Citrus sinensis* peels. Food Chemistry. 2015;**187**:507-516. DOI: 10.1016/j.foodchem.2015.04.081

[10] Pan XJ, Niu GG, Liu HZ. Microwave-assisted extraction of tea polyphenols and tea caffeine from green tea leaves. Chemical Engineering and Processing. 2003;**42**:129-133. DOI: 10.1016/S0255-2701(02)00037-5

[11] Tang ZH, Guo SY, Rao LQ, Qin JP, Xu XN, Liang YZ. Optimization of the technology of extracting water soluble polysaccharides from *Morus alba* L. leaves. African Journal of Biotechnology. 2011;**10**(59):12684-12690. DOI: 10.5897/AJB10.2203

[12] Koh GY, Chou G, Liu ZP. Purification of a water extract of Chinese sweet tea plant (*Rubus suavissimus* S. lee) by alcohol precipitation. Journal of Agricultural and Food Chemistry. 2009;**57**:5000-5006. DOI: 10.1021/jf900269r

[13] Azmir J, Zaidul ISM, Rahman MM, Sharif KM, Mohamed A, Sahena F, et al. Techniques for extraction of bioactive compounds from plant materials: A review. Journal of Food Engineering. 2013;**117**:426-436. DOI: 10.1016/j.jfoodeng.2013.01.014

[14] Tahara S, Yamamoto S, Yamajima Y, Miyakawa H, Uematsu Y, Monma K. A rapid dialysis method for analysis of artificial sweeteners in foods (2nd report). Shokuhin Eiseigaku Zasshi. 2017;**58**(3):124-131. DOI: 10.3358/shokueishi.58.124

[15] Hanif MA, Nawaz H, Naz S, Mukhtar R, Rashid N, Bhatti IA, et al. Raman spectroscopy for the characterization of different fractions of hemp essential oil extracted at 130°C using steam distillation method. Spectrochimica Acta Part A: Molecular and Biomolecular Spectroscopy. 2017;**182**:168-174. DOI: 10.1016/j.saa.2017.03.072

[16] Escribà-Gelonch M, Hessel V, Maier MC, Noël T, Neira d'Angelo MF, Gruber-Woelfler H. Continuous-flow in-line solvent-swap crystallization of vitamin D_3. Organic Process Research and Development. 2018;**22**:178-189. DOI: 10.1021/acs.oprd.7b00351

[17] Poole CF. Chromatographic test methods for characterizing alkylsiloxane-bonded silica columns for reversed-phase liquid chromatography. Journal of Chromatography B. 2018;**1092**:207-219. DOI: 10.1016/j.jchromb.2018.06.011

[18] Porath J. From gel filtration to adsorptive size exclusion. Journal of Protein Chemistry. 1997;**16**:463-468

[19] Gerberding SJ, Byers CH. Preparative ion-exchange chromatography of proteins from dairy whey. Journal of Chromatography. A. 1998;**808**:141-151

[20] Ji S, Wang S, Xu H, Su Z, Tang D, Qiao X, et al. The application of on-line two-dimensional liquid chromatography (2DLC) in the chemical analysis of herbal medicines. Journal of Pharmaceutical and Biomedical Analysis. 2018;**160**:301-331. DOI: 10.1016/j.jpba.2018.08.014

[21] Cardoso CAL, Wilegas W. Droplet counter-current chromatography of indole alkaloids from *Tabernaemontana hilariana*. Phytochemical Analasis. 1999;**10**:60-63. DOI: 10.1002/(SICI)1099-1565(199903/04)10:2<60:AID-PCA446>3.0.CO;2-Y

[22] Gu M, Ouyang F, Su ZG. Comparison of high-speed counter-current chromatography and high-performance liquid chromatography on fingerprinting of Chinese traditional medicine. Journal of Chromatography. A. 2004;**1022**:139-144. DOI: 10.1016/j.chroma.2003.09.038

[23] Horie H, Kohata K. Analysis of tea components by high-performance liquid chromatography and high-performance capillary electrophoresis. Journal of Chromatography. A. 2000;**881**(1-2):425-438. DOI: 10.1016/S0021-9673(99)01345-X

[24] Cao H, Liu XJ, Ulrihc NP, Senguptad PK, Xiao JB. Plasma protein binding of dietary polyphenols to human serum albumin: A high performance affinity chromatography approach. Food Chemistry. 2019;**270**:257-263. DOI: 10.1016/j.foodchem.2018.07.111

[25] Kumar K, Siva B, Rama Rao N, Suresh Babu K. Rapid identification of limonoids from *Cipadessa baccifera* and *Xylocarpus granatum* using ESI-Q-ToF-MS/MS and their structure-fragmentation study. Journal of Pharmaceutical and Biomedical Analysis. 2018;**152**:224-233. DOI: 10.1016/j.jpba

[26] Monakhova YB, Kuballa T, Lachenmeier DW. Chemometric methods in NMR spectroscopic analysis of food products. Journal of Analytical Chemistry. 2013;**68**:755-766

[27] Bakiri A, Hubert J, Reynaud R, Lanthony S, Harakat D, Renault JH, et al. Computer-aided ^{13}C NMR chemical profiling of crude natural extracts without fractionation. Journal of Natural Products. 2017;**80**:1387-1396. DOI: 10.1021/acs.jnatprod.6b01063

[28] Wang J, Yang XD. Determination of absolute configuration of chiral compounds based on chiroptical

spectroscopic methods: From instrument characterization to computational chemistry. University Chemistry. 2016;**31**:37-44. DOI: 10.3866/PKU.DXHX201603022

[29] Zhang D, Guo J, Zhang M, Liu X, Ba M, Tao X, et al. Oxazole-containing diterpenoids from cell cultures of *Salvia miltiorrhiza* and their anti-HIV-1 activities. Journal of Natural Products. 2017;**80**:3241-3246. DOI: 10.1021/acs.jnatprod.7b00659

Chapter 4

The Phytochemical Composition of Medicinal Plants: Brazilian Semi-Arid Region (Caatinga)

Iago Almeida da Ponte, Murugan Muthuvel, Sudarsini Saravanabhavan and Stephen Rathinaraj Benjamin

Abstract

Carnauba wax, the most important vegetable wax under the economic and extracted from the leaves of the carnauba (*Copernicia prunifera* (Miller) H. E. Moore), is extensively applied in food due to its physiochemical characteristics with a majority of esters. *p*-Methoxycinnamic acid diesters obtained from the ceriferous powder of carnauba wax (PCO-C) have been associated with biological actions. However, being a versatile product, many types of research have been carried out seeking to expand the possibilities of applications of this raw material. Furthermore, different experimental studies on the pharmacological activities have also been undertaken in recent years and have tested various biological activities, such as hypolipidemic, hypocholesterolemic and hypoglycemic effects in mice. Therefore, in this book chapter, it is reviewing the development of a process of extraction of 4-hydroxycinnamic acid diesters of carnauba wax powder and investigates their biological actions and physical and chemical characteristics.

Keywords: Caatinga, *Copernicia prunifera*, *p*-methoxycinnamic acid, phytochemistry, biotechnological uses

1. Introduction

Caatinga is a Brazilian biome with a semi-arid climate, vegetation with small leaves and adapted to dry periods, as well as great biodiversity. This biome is found in areas of northeastern Brazil, in the states of Maranhão, Piaui, Ceará, Rio Grande do Norte, Paraiba, Pernambuco, Alagoas, Sergipe, Bahia and part of Minas Gerais. This whole area covers about 844,000 km^2, or 11% of the Brazilian territory [1]. This ecosystem is very important from the biological point of view because it has unique fauna and flora, formed by vast biodiversity, rich in genetic resources and vegetation consisting of species, woody, herbaceous, cactus, and bromeliads. It has 932 species of plants, 148 mammals and 510 birds, for example, and many of these species occur only in the Caatinga.

The main characteristics of the Caatinga are: Strong presence of shrubs with twisted branches and deep roots; Presence of cacti and bromeliads; Shrubs usually lose their leaves almost completely in times of drought (property used to prevent evaporative water loss). The leaves of this vegetation type are small in size; Caatinga soil has low fertility and is stony. Caatinga biodiversity supports various economic activities aimed at agroforestry and industrial purposes. Despite its importance, Caatinga vegetation is a type of vegetation adapted to the aridity of the soil and the scarcity of water in the region. They are classified depending on the natural conditions of the areas and different characteristics like strata: arboreal: with species ranging between 8 and 12 m in height; shrub: with species ranging between 2 and 5 m in height; herbaceous: with species below 2 m in height [2]. *Copernicia prunifera* (Miller) H. E. Moore (Arecaceae) family, a typical desert flora animal categories and exclusively located of areas through the Caatinga biome [3]. It is also known as "Tree of life", *carnauba*, *carandauba*, *carnaba*, *carnaubeira*, *caranaiba*, *carnaúva*, among others.

The carnauba is a palm tree very common in the northeast region whose main feature is its height, which can reach 15 m. The stem is straight and cylindrical, with a diameter that can vary from 10 to 20 cm and has thorns at the bottom. The tree provides fruits from November to March. They are greenish when young and turn purple when they mature. Its fruits are well used to feed farm animals. According to Brazilian specialized guidelines characterize the "pó de olho" and "pó de palha" wax powder as category A and B, respectively [4–6]. Meanwhile, the apical leaves have found lower chlorophyll content, type A wax has a pigmentation that shifts from white to light yellow and has a higher incentive than category B, which has a greenish-gray pigmentation.

Carnauba wax is derived besides the leaves regarding the *Copernicia prunifera* tree (**Figure 1**) and is made principally out of long-chain wax esters (80%), 20% contained fatty acids, fatty alcohols, and hydrocarbons [7–9]. Carnauba wax has the most maximum melting point conditions of all vegetable waxes and has been utilized in an assortment of items, including cosmetic and food products, nourishment items, and the paper area [9]. Additionally, this material is widely used in folk medicine, including the treatment of rheumatism and syphilis. However, carnauba

Figure 1.
Copernicia prunífera *tree (from Fortaleza, Brazil).*

wax is utilized as a stabilizer to different waxes, for example, beeswax to improve the melting point, taking into account expanded utilization of these waxes [10].

Despite that, a great deal of research has been carried out into an attempt to extend the chances of potential outcomes of utilizations of this crude material. With this goal, analysis has been finished streamlining customary applications and examining advancements, for example, the utilization of wax for the microencapsulation of flavors and as a wellspring of molecules following up on the avoidance and treatment of, diabetes, dyslipidemia, and others. The wax is an item to show the level of local consumption with extraordinary potential for use all through the Brazilian food production chain. Along these lines, it is imperative to experts in the food area to more likely comprehend this crude material so as to misuse its maximum capacity. In this way, this review talks about the utilization of carnauba wax in food ranging from the nutritional, phytochemical evaluation, ethnobotanical and biotechnological applications.

2. Nutritional and chemical composition

Carnauba wax consists of complex mixture regarding long-chain fatty acids, free alcohols, esters, aromatic acids, aliphatic acids, triterpene diols, cinnamic acids, proteins, and hydroxy acids and ω-hydroxycarboxylic free acids [10–14]. Recently, one triterpene carnaubadiol was also isolated and identified present in the leaves were reported. The inorganic compounds existing such as aluminum, copper, magnesium, zinc, manganese, calcium, iron, and sodium [15]. Recent studies continued to assess more genetic resource of carnauba wax while revealing a more extensive variety in the nutritional composition as described in the following sections.

2.1 Pectin

Paim et al. [16] extracted the pectin from the aqueous pulp extracts (APE) of *Copernicia prunifera* analyzed by chromatographic and spectroscopic methods. From this study, the pectin substance acquired from the pulp of unripe fruits of *C. prunifera* demonstrated an estimation of 2.9%. Additionally, the pectin was observed by using the absorption spectra by demonstrating several carbonyl groups in the form of esterified and carboxylate compounds. Furthermore, the thin layer chromatography (TLC) technique identified galactose, galacturonic acid patterns, and arabinose compounds, respectively. By using, ^{13}C NMR spectroscopy analysis method, various forms of polymers were recognized in the pectic polysaccharides chain compounds including D-galacturonic acid (major signs), D-galactose (lower signs) and the peak molar mass (*M*pk) was determined by gel permeation chromatography of 0.6 × 105 g mol^{-1}. All these studies highlighted that pectin presence of higher molecular weight and a igher degree of esterification displaying improved performances.

2.2 Triterpenes

Almeida et al. [11], explored phytochemical investigation of hexane and ethanolic extracts carnauba wax (types 1 and 4) was analyzed and identified 16 dammarane-type triterpenes, with 13 newly categorized as (24R*)-methyldammara-20,25-dien-3α-ol and a mixture of alkyl (24R*)-methyldammar25-en-20-ol-3β-carboxylates, and 3 triterpenes such as carnaubadiol,(24R*)-methyldammara-20,25-dien-3β-ol and (24R*)-24-methyldammara-20,25-dien-3-one. Furthermore, fatty alcohols such as docosanol, eicosanol, and hexacosanol, tetracosanol as well as four sterols (campesterol, cholesterol, sitosterol and stigmasterol) were detected and identified. These finding isolated compounds were characterized by using Infrared (IR) spectra and confirmed by classical chromatographic techniques such as gas chromatography-flame ionization detections (GC-FID), ^{1}H and ^{13}C nuclear magnetic resonance (NMR) methods.

^{1}H and ^{13}C NMR spectroscopy techniques have been applied for structural elucidation of dammarane triterpenoids [basic skeleton as carnaubadiol] in carnauba wax powder obtained from the leaves of *Copernicia cerifera* [13]. Totally four types of triterpenes were identified from hexane extract of carnauba wax. Four of these compounds were, structure1, (24RŁ)-24-methyldammara-21,25- diene-3ˇ-ol, structure of 2 and 3 was distinguished as (24RŁ)-24-methyldammara-25-ene-3-one. Furthermore, the structure of 4, illustrated as (E)-25-hydroperoxydammar-23-ene-3ˇ,20-diol. The chemical composition analyzed after successive column chromatography using silica gel hexane followed by ethanol at room temperature, respectively.

2.3 Proteins

Cruz et al. [12] isolated the wax protein from "Carnauba" wax and the samples accomplished by SDS-Tricine-gel electrophoresis technique. It showed relative molecular masses of 26,000 (β-1,3-glucanase) and 24,000 Da (class III chitinase), respectively. However, these proteins have been involved in the resistance systems of plants against insects and pathogens. In addition, the authors found that proteins segregated from the different portions of carnauba wax have antifungal enzymatic action. These chemicals, chitinase and β-1,3-glucanases, can hinder early development of organisms and modify hyphal (threadlike fibers like mycelium of parasites) morphology of growths developing within the proteins.

2.4 Ethnobotanical study of carnauba wax

2.4.1 Antioxidant activity

Phenolic bioactive and polyphenolic compounds occur normally and significant segments of the human diet due to their antioxidant capacity that decreases oxidative stress-inducing cellular damage associated with severe pathologies such as cardiovascular, neurodegenerative diseases and cancers [17]. The simplest bioactive phytochemicals containing a single substituted phenolic ring, like cinnamic acid and caffeic acid. Cinnamic acid is a naturally proceeding organic acid in plants, has low toxicity and a broad spectrum of biological activities. However, cinnamic acid derivatives comprise a series of trans-3-phenylpropenoic acids which differ in their substituents on the aromatic ring. The presence of a benzene ring and a low unsaturated hydrocarbon chain determines its low polarity and solubility in water. The most common cinnamic acid derivatives in plants are *p*-coumaric, caffeic, and chlorogenic acids and hydroxybenzoic and hydroxycinnamic acids, respectively [18].

Claisa et al. [19] studied the antioxidant activity by ABTS and FRAP methods and *in-vivo* cellular antioxidant activity assay. The antioxidant activity of ethyl acetate and hexane extracts of *p*-methoxy cinnamic diester (**Figure 2**) (PCO-C) showed the values 107.27 ± 3.92 μM Trolox/g and 73.3 ± 1.83 μM iron sulfate/g, respectively. From these results showed significant antioxidant activity values due to the presence of derivative of cinnamic acid compounds [20]. In addition, the *in-vivo* antioxidant activity showed lower ROS values in PCO-C alone (50 and 250 μg/mL). Accordingly, PCO-C did not produce any cellular oxidation significantly it produces low level of ROS, because of the oxidation of lymphocytes endured with

H_2O_2. However, PCO-C had a best antioxidant effect in high dose level (250 μg/mL) similar of Trolox (80 μM) and found an oxidation inhibition capacity in human peripheral blood lymphocytes (HPBLs). According to the authors, it is revealed that antioxidant activates arise from *p*-methoxy cinnamic diesters presence of phenolic compounds PCO-C. Therefore, the presence of these excellent antioxidant potentials of produce reflects its ability to deliver bioactive substances that neutralize reactive oxygen species (ROS) and scavenge free radicals produced by oxidative stress.

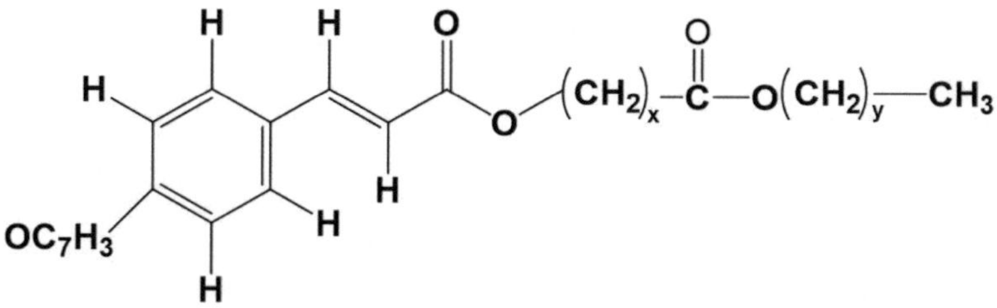

Figure 2.
Chemical structure of p-methoxy cinnamic acid diesters.

Rufino et al. [21] reported that significant antioxidant activities of polyphenol-rich extracts from tropical fruits and dry fruits especially carnauba, both DPPH, ABTS, FRAP, β-Carotene oxidation methods and total phenolic contents were performed. The antioxidant activity of the methanolic and ethanolic extracts of fresh fruits of carnauba found decreased values DPPH values 3549 ± 184 g fruit/g DPPH˙ and increased ABTS values 10.7 ± 0.2 μmol Trolox/g, FRAP values 15.5 ± 0.4 μmol Fe_2SO_4/g and high β-Carotene bleaching values found 87.7 ± 2.7(% O.I) and extractable polyphenols values 830 ± 28.3 mg GAE/100 g, respectively. Additionally, the bioactive compounds values (mg/100 g fresh matter) such as, vitamin-C (78.1 ± 2.6), total anthocyanins (4.1 ± 0.1), yellow flavonoids (66.4 ± 2.3), and total carotenoids (0.6 ± 0.2), chlorophyll (4.2 ± 0.2), respectively. According to the authors, this study provides an adaptation of ABTS, DDPH, FRAP and β-carotene bleaching methods, along with an evaluation of the compounds related to antioxidant potential. The results showed promising perspectives for the exploitation of non-traditional tropical fruit species with considerable nutritional properties and antioxidant capacity.

2.4.2 Anti-microbial and anti-fungal activities

The prevention of the decomposition and assurance of the food safety can be attained by the use of compounds that act as preservatives of foodstuffs, by presenting antimicrobial properties, preventing the degradation by enzymatic and non-enzymatic reactions. The identification of new sources of compounds and increased demand for the prospection that has antimicrobial properties. However, its realization depends on some conditions, including solubility of the food and pH. Cinnamic acid derivatives (CAD) such as trans, hydroxy and methoxy cinnamic acid, 4-chlorocinnamate, cinnamic acid derived from oxazoline ions, antimicrobial and antifungal activities. These substances showed strong activity against Gram-positive bacteria (*Staphylococcus aureus* and *Bacillus cereus*) and Gram-negative bacteria (*Escherichia coli* and *Proteus vulgaris*).

Gonçalves et al. [22] studied the effects on different concentration of carnauba wax (1, 2, 3 and 4.5%) on the brown rot, produced by *Monilinia fructicola* (G. Wint.) and Rhizopus rot, developed by *Rhizopus stolonifer* (Erhenb.:Fr.) *in vitro* evaluation on infection of nectarine and plums. The authors, distinguished that no mycelial development of *M. fructicola* at any wax concentrations in post-contamination tests, however, *R. stolonifer* was totally restrained via by carnauba wax at all concentrations except at 1%. Additionally, *in vitro* evaluation for both *M. fructicola* and *R. stolonifera* no germination occurred of spores at any carnauba wax concentrations. There was 50% inhibition observed in spore germination for *M. fructicola* by utilizing 9% carnauba wax concentration and covered with nectarines 90% for *R. stolonifera*. The carnauba wax concentrations (4.5% and 9%) were applied to the protections with essentially reduced frequencies of both diseases in nectarines and plums. Nevertheless, the utilization of wax control was ineffective after infection by both diseases.

According to Jo et al. [23] studied quality and microbial safety of Fuji apples coated with CSW/LO (Carnauba-shellac wax nanoemulsion containing lemongrass oil). In this work, carnauba wax incorporated into shellac wax (Carnauba-shellac wax) with essential oils like lemongrass oil coating formulations and their effects on the coating and shelf life of the Fuji apples were evaluated. Total soluble solid content to titratable acid ratio, hardness, weight loss and color, sensory quality and microbial growth of fresh Fuji apples were studied during 5 months of storage at room temperature. According to the authors, results showed that carnauba extracts incorporated to shellac wax-based coatings together with lemongrass oil successfully maintained the firmness and color of coated freshly harvested apples in comparison with uncoated control samples, which presented severe texture softening. During storage conditions, the hardness of the uncoated apples exhibited the lowest conditions by 3.3 N and the weight loss was found by 7.7%. Interestingly, the weight loss was found to be 5.2% and the hardness of the coated apples did not change at any conditions, respectively. The total soluble solids and titratable acidity revealed that not significantly different between coated and uncoated apples.

Hence, the application of CSW-LO coated apples had better sensory scores with the sensory acceptability threshold for any attributes evaluated. In addition, the total aerobic bacteria population on the coated apples were deteriorated (1.4 log CFU/g) compared with uncoated apples after 5 months of storage. Additionally, the population of yeast and molds of the uncoated apples were found 2.2 log CFU/g after 5 months of storage, although yeast and molds were not detected on the coated apples, respectively. The results achieved demonstrate the feasibility of the addition of carnauba wax coating formulations for increasing the nutritional value of fresh apples without compromising their fresh-like quality attributes.

2.4.3 Antifungal activity

Different kinds of antimicrobial proteins have been purified from plants such as, β-1,3-glucanases, chitinases, ribosome-inactivating proteins, thionins, and defensins. In this case, β-1,3-glucanase and chitinase separated from type B wax of *Copernicia cerifera*, has revealed antifungal activity against phytopathogenic fungi medium [12]. Based on the results, the yeast *Saccharomyces cerevisiae* showed the patterns of growth for *Fusarium oxysporum* and *S. cerevisiae* in the presence of different fractions obtained from "Carnauba" wax and in control medium. Plant chitinases and β-1,3-glucanases are known as antifungal hydrolases since they inhibit fungal growth in model experiments by using on agar plates and in liquid media. The presence of isolated proteins by using SDS-Tricine-gel electrophoresis, and showed inhibit early growth of all fungi in their fractions in agar plates. Based on these results, defense proteins like chitinase and glucanases which appear to inhibit the early growth of all fungi and cause hyphal morphological alterations for fungi growing in the presence of these proteins (relative molecular masses of 26,000 and

24,000 Da) as compared with growth on control medium. According to the authors, *Copernicia cerifera* wax contains defense proteins ability to inhibit fungal growth. Moreover, the fungal cell walls together with β-1,3-glucans, recommend a protective role for these hydrolases.

2.4.4 Hypercholesterolemic activity

Paim et al. [24] first time studied *in vivo* study of the antihypercholesterolemic effect of the aqueous pulp extracts (APE) from the *C. prunifera* (APE 150 and 300 mg/Kg b.w./day) were directed to hyperlipidemic mice for 90 days. It showed that APE was promising results with lipidemic alterations were effective in both models causing significant changes in the values of total cholesterol, low-density lipoprotein cholesterol (LDL-C), HDL-C and triglycerides in serum. Nevertheless, it showed no renal toxicity and liver toxicity parameters (enzyme AST) and renal metabolites (urea and creatinine) to animals. Additionally, APE in high doses showed no renal and liver toxicity to animals. Despite the fact that the histological results bring about liver of mice treated with APE shows that doses (150 and 300 mg/Kg b.w./day) were not ready to alter the inflammatory procedure contrasted with the standard diet (SD) fed mice, all things considered, that better reaction opposing the hypercholesterolemic diet (HD). Besides, it was recognized the reduced intensity of inflammation in higher dose receiving present in the group. According to these results revealed that aqueous fruit pulp extracts of carnauba reduced hypercholesterolemia showing a potential preventive effect against cardiovascular diseases without side effects cause.

Furthermore, in this investigation, Filho et al. [25] revealed that the extract of PCO-C (100 mg/kg) found that productive in decreasing total cholesterol (TC) and triglyceride (TG) levels in both dyslipidemia induction models in hypercholesterolemic mice. This effect ascribed to the presence of high dietary and crude fiber content and antioxidant potential of PCO-C. Histological investigations demonstrated that PCO-C has no hepatotoxic impact and diminishes hepatic steatosis in animals that expended hyperlipidemic ration. In this manner, it was inferred that PCO-C separated from *Copernicia prunifera* may be helpful in the treatment of hyperlipidemia and atherosclerosis. Additionally, the authors highlighted that the results obtained in animals treated with PCO-C were pivotal compound had therapeutic potential in the prevention and treatment of diseases related with the metabolism of carbohydrates and lipids.

2.4.5 Hypoglycemic activity

Rodrigues et al. [26] studied that oral administration of *Copernicia cerifera* in glibenclamide diabetic mice at doses of 100 and 150 mg/kg bodyweight for 21 days.

According to the authors, the findings of this study indicated that 10% isopropanol in heptane leaf extract of Carnauba powder extract had antidiabetic activity when using therapeutic doses (100 and 150 mg/kg body weight (b.w.)). However, after treatment with 150 mg/kg b.w dose was found to be effective in significantly controlling blood glucose levels ($p < 0.05$), when compared to the reference drug glibenclamide. The observed hypoglycemic activity could be associated with the phytochemicals present in carnauba wax powder. These finding results suggested that PCO-C leads to diabetes by protecting beta-cells from oxidative damage. Indeed, the presence of the antioxidant effect of PCO-C may improve the pancreatic beta-cells to inhibit glucagon secretion and release more insulin levels. Finally, this study clearly shows that the leaf extract of carnauba wax powder possesses possible hypoglycemic activity in alloxan-induced diabetic mice.

2.4.6 *Antiprotozoal activity*

Almeida et al. [22] identified antiprotozoal metabolites from the *Copernicia prunifera* (Miller) showed *in vitro* action against promastigote and amastigote types of *Leishmania infantum*, trypomastigote forms of *Trypanosoma cruzi*. Among the separated dammarane-type triterpenoids, from the hexane and ethanolic extracts 'carnauba' wax (type 1 and 4) indicated antiprotozoal activity against promastigotes of Leishmania *infantum*, which showed the values of IC_{50} of 46.2 mM in tested extract 1. Besides, considering the positive controls miltefosine and benznidazole, the obtained results recommended that the impact of tested extract 1 against *L. infantum* is less noticeable than that observed against trypomastigotes of *Trypanosoma cruzi*. The intracellular amastigotes of *L. infantum* were sensitive to three types of triterpenoids, with IC_{50} estimations of 7.8, 37.6 and 51.9 μM, individually. Regardless of triterpenoid 2 and 3 exhibited absence of activity against the extracellular promastigotes, they killed the intracellular structures with selective index (SI) esteems more than 5.3 and 3.8, respectively, proposing a conceivable commitment of macrophages at the end of parasites. Notwithstanding, the tested extract 1 and 2 were less effective than standard drug miltefosine, which showed an IC_{50} of 16.4 μM. Finally, this study provided useful information about the antiprotozoal activity of 'carnauba' (*C. prunifera*) wax as well as the identification of compounds responsible to this potential.

2.5 Pharmaceutical processing

Carnauba wax has a wide scope of utilizations and, as a result, is industrially accessible in an assortment of blends. Carnauba wax utilized in fruit and vegetable coating is constantly connected as a microemulsion made with unsaturated fatty acids and an essential counterion [27]. These produce an anionic emulsifier where the carnauba wax is dispersed. In addition, various types of unsaturated fats utilized incorporate oleic, linoleic, palmitic, myristic or lauric acids. The fundamental counterion may be hydroxides of sodium, potassium salts, ammonium, morpholine [28] and triethanolamine [29]. Since carnauba wax is just utilized as a fruit coating in the mix with different substances, the adequacy and consistence of different substances should likewise be considered.

Nart et al. [30] studied carnauba wax demonstrates a pivotal reinforcement to support the sustained release of high soluble medications in relationship with Ethocel™ (EC) and Kollicoat® SR 30D utilizing reservoir and matrix systems, respectively. However, melt granulation of the medication with carnauba wax was connected as an intermediary of the key to sustained release mini-tablets, utilizing captopril (6.25 mg/mini-tablet) and metformin hydrochloride (15.0 mg/mini-tablet) as profoundly soluble model drugs. In addition, investigating the impacts of carnauba wax as a granulating excipient in the arrangement of mini-scale tablets, unmistakably the excipient diminished the contact of the drug particles with the disintegration medium, decreasing the release rate of the drugs and submitting the disintegration of the smaller than usual tablets. In this manner, it was seen that the melt granulation technique with carnauba wax improved the rheology of the considered drugs. The carnauba wax added to diminish the diffusion rate of the drug to the medium by expanding the hydrophobicity and lessening the disintegration rate of the structure of the measurements, impeding water dispersion a while later. The blend of carnauba wax with the EC at 50% indicated promising profiles for sustained release formulations.

Neto et al. [31] developed methionine microencapsulated with lipid matrix using carnauba wax by the melt emulsification technique. Different compositions of carnauba wax: methionine (MEM 2:1 and MEM 4:1) were prepared and compared with pure methionine. In addition, scanning electron micrograph results showed no invade by ruminal microorganisms of both formulations after *in situ* testing. Taking into account that carnauba can apply an impact of protective on amino acids by covering their degradation in the rumen due to its hydrophobic distinguishing. In addition, it is a characteristic result of low degradability because of its concoction structure in unsaturated fats, and it is easy to obtain. Notwithstanding, carnauba wax sustained its thermal degradation temperatures and typical melting after the microencapsulation procedure, this diminishing in thermal stability of methionine is not because of its collaboration with the wax however most likely is because of the forces of intermolecular level (presences of hydrogen bridges) among the methionine particles. Finally, the formulation MEM 4:1 showed that promising results of the lower level of thermal degradation and higher yield and efficiency of microencapsulation.

2.5.1 Post-harvest storage

The valuable role of carnauba wax is outstanding for improving shelf life and supporting postharvest quality of a few fruits, for example, mango [32], avocado [33] and mamey sapote organic product [34]. Barmen et al. [35] studied pomegranate (*Punica granatum* L., cv. Mridula) fruits were treated with putrescine, carnauba wax and putrescine + carnauba wax combination prior at 2°C cold storage temperature. Further, carnauba wax is additionally stated to reduce the improvement of chilling injury (CI) manifestations. Respiration rate of stored fruits has been discovered expanded with the progression of the capacity period under every one of the medicines. Up to the fifteenth day of capacity, there was no critical contrast in breath rate in the organic products treated with polyamine like putrescine (PUT), carnauba wax and their mix. The low breath rate in carnauba wax treated organic product ascribed because of diminished gas exchange and thusly low oxygen accessibility to the natural product tissues for breath. The utilization of carnauba wax gave higher maintenance of fruit solidness, most likely because of the less drying out happened and furthermore to a slower degradation of cell divider segments. In this way, in control group and carnauba wax treated pomegranate fruits, the expansion in juice recuperation after 30th day of storage capacity may be ascribed to CI intervened activities of cell degrading enzymes such as pectin methylesterase and polygalacturonase. In addition, the utilization of carnauba wax covering in blend with PUT may have applied synergistic impact which aided in keeping up higher juice recuperation by diminishing loss of moisture from the fruits.

Germano et al. [36] studied, a galactomannan-carnauba wax-based coating improved the guava fruit in postharvest quality and storability over preservation of firmness in ambiental conditions (25°C). The authors prepared edible coating galactomannan (0.75%) and carnauba wax (0.9%) were treated with guava fruits (Paluma). At day 15, coated and refrigerated (FR) guava fruits were showed a climacteric rise in 59.3 mg CO_2 kg^{-1} h^{-1} and however, coated guava at ambient (FA) showed a diminished value 168.6 mg CO_2 kg^{-1} h^{-1} at 15 days of storage and firmness of 14.3 N attributed to lower lipid peroxidation and cell wall hydrolysis. In addition, no increase values of control refrigerated fruit (CR) and no further evaluation of control-uncoated 'Paluma' guava stored at ambient (CA) for 9 days. Additionally, coating improves increased antioxidant enzymes CAT and SOD activities refrigerated samples presented 35% lower H_2O_2 levels ($p < 0.05$) while compared to uncoated control samples. However, symptoms of chilling injury (CI) inhibition of softening and respiratory peaks are exhibited in refrigerated uncoated fruits. According to the authors, galactomannan carnauba wax coating was effective in guava postharvest quality and maintaining firmness and color, also preventing chilling symptoms under refrigerated conditions, respectively.

3. Conclusion and future perspectives

Despite the current recognition of *Copernicia prunifera* as a quintessential Brazilian plant with growing interest of research and a boost in its commerce and industrial application for the formulation of therapeutic products, it can be safely postulated that its therapeutic potentials have not been fully explored. At present, researches and commercial interest on carnauba wax are skewed toward its cosmeceutical, food and pharmaceutical applications which have dwarfed research interest in its potential as a remedy for other diseases. Thus, further researches on its pharmacological activity recommended with the end-goal of unraveling the pharmacodynamics, pharmacokinetics and clinical relevance. In addition, toxicity risk assessment studies of both the bioactive extracts and isolated constituents need to be given more attention. Nevertheless, further studies and long-term human trials should be carried out in order to clarify the relationship between the consumption of 4-methoxy cinnamic acid diesters and its derivatives such as isolated pectin and their benefic impact on the human body, there seems to be certainly a promising future for new investigations.

Acknowledgements

All the authors of the manuscript thank and acknowledge support of their respective Universities and Institutes. We are also grateful to Mr. Arunkumar Thanga Periyasamy, Srivilliputtur, Tamilnadu, India, for his valuable suggestions and encouragements.

Conflict of interest

The authors declare no conflicts of interest to disclose in relation to this book chapter.

Acronyms and abbreviations

PIP	precipitation index permits
TLC	thin layer chromatography
Mpk	peak molar mass
CSW	carnauba-shellac wax
LO	lemongrass oil
APE	aqueous pulp extracts
LDL-C	low-density lipoprotein cholesterol
TC	total cholesterol
TG	triglyceride
HD	hypercholesterolemic diet
GC-FID	gas chromatography-flame ionization detections
IR	infrared spectroscopy
PCO-C	4-methoxy cinnamic acid diesters
b.w	body weight

DPPH	2,2-diphenyl-1-picrylhydrazyl
EC	effective concentration
ABTS	2,2′-azino-bis(3-ethylbenzothiazoline-6-sulfonic acid)diammonium salt
FRAP	ferric reducing antioxidant power
NMR	nuclear magnetic resonance spectroscopy
SDS	sodium dodecyl sulfate
AST	aspartate transaminase
ROS	reactive oxygen species
PG	polygalacturonase
PUT	putrescine
CI	chilling injury
PPO	polyphenol oxidase
SI	selective index

Author details

Iago Almeida da Ponte[1], Murugan Muthuvel[2], Sudarsini Saravanabhavan[3] and Stephen Rathinaraj Benjamin[1*]

1 Department of Biotechnology and Molecular Biology, State University of Ceará, Fortaleza, Brazil

2 Department of Pharmaceutical Chemistry, Faculty of Pharmacy, Sriramachandra Institute of Higher Education and Research, Chennai, Tamilnadu, India

3 Department of Ayurveda and Siddha, Trichy Ayurvedic Union Pvt. Ltd., Tiruchirappalli, Tamilnadu, India

*Address all correspondence to: biotechstephen@gmail.com

References

[1] Giulietti A, Harley R, Queiroz L, Barbosa M, Bocage Neta AL, Figueiredo MA. Plantas endêmicas da Caatinga. Vegetação e Flora da Caatinga. 2002;**1**:103-115

[2] Coe HHG, De Sousa LOF. The Brazilian "Caatinga": Ecology and vegetal biodiversity of a semiarid region. In: Dry Forests: Ecology, Species Diversity and Sustainable Management. New York: Nova Science Publisher; 2014

[3] Taylor NP, Zappi D. Distribuição de espécies de Cactaceae na Caatinga. Vegetação e Flora da Caatinga. 2002;**1**:123-125

[4] Carnaúba R. Técnico de identidade e de qualidade para a classificação da cera de Brasil. Ministério da Agricultura, Pecuária e Abastecimento Gabinete do Ministro Instrução Normativa No 34, de 30 de Novembro de 2004

[5] Brasil. Regras Para Análise de Sementes—Ministério da Agricultura, Pecuária e Abastecimento [Internet]. Instrução Normativa No. 34. 2009. 339p. Available from: http://www.agricultura.gov.br/assuntos/laboratorios/arquivos-publicacoes-laboratorio/regras-para-analise-de-sementes.pdf/view

[6] de Carvalho FPA, Gomes JMA. Eco-efficiency in the production of carnauba wax in the city of Campo Maior, Piaui, 2004. Revista de Economia e Sociologia Rural [Internet]. 2008;**46**(2):421-453. Available from: http://www.scielo.br/scielo.php?script=sci_arttext&pid=S0103-20032008000200006&lng=pt&tlng=pt

[7] Steinle JV. Carnauba Wax: An expedition to its source. Industrial & Engineering Chemistry [Internet]. 1936;**28**(9):1004-1008. Available from: http://pubs.acs.org/doi/abs/10.1021/ie50321a003

[8] Valmalle G, Karleskind AR. Application de la Chromatographie en Phase Gazeuse à Haute Temperature à l'Ana-lyse des Cires et à Leur Dosage dans des Melanges Industriels. Revue Francaise des Corps Gras. 1977;**24**:203-209

[9] Lawrence JF, Iyengar JR, Page BD, Conacher HBS. Characterization of commercial waxes by high-temperature gas chromatography. Journal of Chromatography A [Internet]. 1982;**236**(2):401-403. Available from: https://linkinghub.elsevier.com/retrieve/pii/S0021967300848918

[10] Müller MG, Lindner JA, Briesen H, Sommer K, Foerst P. On the properties and application of beeswax, carnauba wax and palm fat mixtures for hot melt coating in fluidized beds. Advanced Powder Technology [Internet]. 2018;**29**(3):781-788. Available from: https://linkinghub.elsevier.com/retrieve/pii/S0921883117305150

[11] De Almeida BC, Araújo BQ, Barros EDS, Freitas SDL, Maciel DSA, Ferreira AJS, et al. Dammarane triterpenoids from Carnauba, *Copernicia prunifera* (Miller) H. E. Moore (Arecaceae), wax. Journal of the Brazilian Chemical Society. 2017;**28**(8):1371-1376

[12] Cruz MAL, Gomes VM, Fernandes KVS, Machado OLT, Xavier-Filho J. Identification and partial characterization of a chitinase and a β-1,3-glucanase from *Copernicia cerifera* wax. Plant Physiology and Biochemistry [Internet]. 2002;**40**(1):11-16. Available from: https://linkinghub.elsevier.com/retrieve/pii/S0981942801013407

[13] de Brito Cysne J, Braz-Filho R, Assunção MV, de Andrade Uchoa DE, Silveira ER, Pessoa ODL. 1H and13C NMR spectral assignments of four dammarane triterpenoids from

carnauba wax. Magnetic Resonance in Chemistry [Internet]. 2006;**44**(6):641-643. Available from: http://doi.wiley.com/10.1002/mrc.1792

[14] Harron AF, Powell MJ, Nunez A, Moreau RA. Analysis of sorghum wax and carnauba wax by reversed phase liquid chromatography mass spectrometry. Industrial Crops and Products [Internet]. 2017;**98**:116-129. Available from: https://linkinghub.elsevier.com/retrieve/pii/S0926669016305957

[15] Wang L, Ando S, Ishida Y, Ohtani H, Tsuge S, Nakayama T. Quantitative and discriminative analysis of carnauba waxes by reactive pyrolysis-GC in the presence of organic alkali using a vertical microfurnace pyrolyzer. Journal of Analytical and Applied Pyrolysis [Internet]. 2001;**58-59**:525-537. Available from: http://linkinghub.elsevier.com/retrieve/pii/S0165237000001558

[16] Paim RTT, Benjamin SR, Rondina D, Marques MMM, de Viana DA, da Gonzaga MLC, et al., Antihypercholesterolemic effects of fruit aqueous extract of *Copernicia prunifera* (Miller) H. E. Moore in mice diet-induced hypercholesterolemia. Evidence-Based Complementary and Alternative Medicine [Internet]. 2017;**2017**:1-15. Available from: https://www.hindawi.com/journals/ecam/2017/6376173/

[17] Marques MPM, Batista de Carvalho LAE, Valero R, Machado NFL, Parker SF. An inelastic neutron scattering study of dietary phenolic acids. Physical Chemistry Chemical Physics: PCCP [Internet]. 2014;**16**(16):7491-7500. Available from: http://www.ncbi.nlm.nih.gov/pubmed/24626795

[18] Sova M. Antioxidant and antimicrobial activities of cinnamic acid derivatives. Mini Reviews in Medicinal Chemistry [Internet]. 2012;**12**(8):749-767. Available from: http://www.ncbi.nlm.nih.gov/pubmed/22512578

[19] Freitas CAS, Vieira ÍGP, Sousa PHM, Muniz CR, Gonzaga MLDC, Guedes MIF. Carnauba wax p-methoxycinnamic diesters: Characterisation, antioxidant activity and simulated gastrointestinal digestion followed by in vitro bioaccessibility. Food Chemistry. 2016;**196**:1293-1300

[20] Jakovetić SM, Jugović BZ, Gvozdenović MM, Bezbradica DI, Antov MG, Mijin DŽ, et al. Synthesis of aliphatic esters of cinnamic acid as potential lipophilic antioxidants catalyzed by lipase B from *Candida antarctica*. Applied Biochemistry and Biotechnology [Internet]. 2013;**170**(7):1560-1573. Available from: http://link.springer.com/10.1007/s12010-013-0294-z

[21] Rufino M do SM, Alves RE, de Brito ES, Pérez-Jiménez J, Saura-Calixto F, Mancini-Filho J. Bioactive compounds and antioxidant capacities of 18 non-traditional tropical fruits from Brazil. Food Chemistry [Internet]. 2010;**121**(4):996-1002. Available from: http://linkinghub.elsevier.com/retrieve/pii/S0308814610001172

[22] de Almeida BC, Araújo BQ, Carvalho AA, Freitas SDL, Maciel D da SA, Ferreira AJS, et al. Antiprotozoal activity of extracts and isolated triterpenoids of 'carnauba' (*Copernicia prunifera*) wax from Brazil. Pharmaceutical Biology [Internet]. 2016;**54**(12):3280-3284. Available from: https://www.tandfonline.com/doi/full/10.1080/13880209.2016.1224257

[23] Jo W-S, Song H-Y, Song N-B, Lee J-H, Min SC, Bin SK. Quality and microbial safety of 'Fuji' apples coated with carnauba-shellac wax containing lemongrass oil. LWT - Food Science and Technology [Internet].

2014;**55**(2):490-497. Available from: http://linkinghub.elsevier.com/retrieve/pii/S0023643813003897

[24] Gonçalves FP, Martins MC, Junior GJS, Lourenço SA, Amorim L. Postharvest control of brown rot and Rhizopus rot in plums and nectarines using carnauba wax. Postharvest Biology and Technology [Internet]. 2010;**58**(3):211-217. Available from: http://linkinghub.elsevier.com/retrieve/pii/S092552141000178X

[25] Filho ACVA, Rodrigues PAS, Benjamin SR, Paim RTT, Holanda MO, Silva JYG, et al. Hypolipidemic activity of P-methoxycinnamic diester (PCO-C) isolated from *Copernicia prunífera* against triton WR-1339 and hyperlipidemic diet in mice. Environmental Toxicology and Pharmacology. 2017;**56**:198-203

[26] Alves Salmito RP, Florindo GI, Marques MMM, Da Silva ING, Vieira ÍGP. Hypoglycemic activity of *Copernicia cerifera* Mart. leaf powder extract in the treatment of alloxan-induced diabetic mice. International Journal of Pharmacy and Pharmaceutical Sciences. 2014;**6**(10):115-118

[27] Plotto A, Narciso JA. Guidelines and acceptable postharvest practices for organically grown produce. HortScience [Internet]. 2006;**41**(2):287-291. Available from: https://journals.ashs.org/view/journals/hortsci/41/2/article-p287.xml

[28] Hagenmaier RD. Wax microemulsion formulations used as fruit coatings. Proceedings of the Florida State Horticultural Society. 1998;**111**:251-258

[29] Baldwin EA, Hagenmaier RD, Bai J. Edible Coatings and Films to Improve Food Quality. 2nd ed. Boca Raton: CRC Press; 2011. pp. 1-417

[30] Nart V, Beringhs AO, França MT, de Espíndola B, Pezzini BR, Stulzer HK. Carnauba wax as a promising excipient in melt granulation targeting the preparation of mini-tablets for sustained release of highly soluble drugs. Materials Science and Engineering: C [Internet]. 2017;**70**:250-257. Available from: https://linkinghub.elsevier.com/retrieve/pii/S0928493116307445

[31] de Neto JPC, Bezerra LR, da Silva AL, de Moura JFP, Filho JMP, da Filho ECS, et al. Methionine microencapsulated with a carnauba (*Copernicia prunifera*) wax matrix for protection from degradation in the rumen. Livestock Science [Internet]. 2019. Available from: https://linkinghub.elsevier.com/retrieve/pii/S1871141318306279

[32] Dang KTH, Singh Z, Swinny EE. Edible coatings influence fruit ripening, quality, and aroma biosynthesis in mango fruit. Journal of Agricultural and Food Chemistry [Internet]. 2008;**56**(4):1361-1370. Available from: https://pubs.acs.org/doi/10.1021/jf072208a

[33] Feygenberg O, Hershkovitz V, Ben-Arie R, Jacob S, Pesis E, Nikitenko T. Postharvest use of organic coating for maintaining bio-organic avocado and mango quality. Acta Horticulturae [Internet]. 2005;**682**:1057-1062. Available from: https://www.actahort.org/books/682/682_138.htm

[34] Ergun M, Sargent SA, Fox AJ, Crane JH, Huber DJ. Ripening and quality responses of mamey sapote fruit to postharvest wax and 1-methylcyclopropene treatments. Postharvest Biology and Technology [Internet]. 2005;**36**(2):127-134. Available from: https://linkinghub.elsevier.com/retrieve/pii/S0925521404002856

[35] Barman K, Asrey R, Pal RK. Putrescine and carnauba wax pretreatments alleviate chilling injury, enhance shelf life and preserve pomegranate fruit quality during cold storage. Scientia Horticulturae [Internet]. 2011;**130**(4):795-800. Available from: https://linkinghub.elsevier.com/retrieve/pii/S0304423811004651

[36] Germano TA, Aguiar RP, Bastos MSR, Moreira RA, Ayala-Zavala JF, de Miranda MRA. Galactomannan-carnauba wax coating improves the antioxidant status and reduces chilling injury of 'Paluma' guava. Postharvest Biology and Technology [Internet]. 2019;**149**:9-17. Available from: https://linkinghub.elsevier.com/retrieve/pii/S0925521418305180

Section 2

Coumarins: Bioactivity, Synthesis and Labelling

Chapter 5

From Rat Poison to Medicine: Medical Applications of Coumarin Derivatives

Robbert Bipat

Abstract

Historical reports mention the application of medicinal plants containing coumarins against various ailments. Current research suggests that at least some of the actions described may be attributable to the action of these coumarins. Warfarin and its derivatives are coumarins used today in medical practice. Their mechanism of action lies in the competitive antagonism of vitamin K, through which they inhibit coagulation in the body by preventing the production of prothrombin. Due to this action, these coumarins are a major group of drugs with anticoagulant activity. Anticoagulants reduce the risks of undesirable blood clots leading to myocardial infarction, pulmonary embolism, and ischemic stroke among others. The anticoagulant activity can also lead to undesired bleeding. Extreme caution is warranted when given to menstruating women, patients suffering from disorders prone to bleeding like gastric ulcer and rheumatoid arthritis, and to persons with a high likelihood of blunt and sharp trauma. In addition, there is a significant augmentation of the anticoagulant activity when used in combination with non-steroidal anti-inflammatory agents and agents interfering with the metabolism of the coumarins. Recent findings propose additional uses like anti-tumor and antibiotic actions for coumarins. The clinical application of these actions has yet to be demonstrated.

Keywords: warfarin, coumarin, anticoagulant, prothrombin, coagulation cascade, adverse reaction, rodenticide, embryopathy, vitamin K

1. Introduction

Coumarins are members of the benzopyrone class of organic compounds that are found in many plants [1] and possess a variety of pharmacological properties such as antimicrobial, anti-inflammatory, antidiabetic, and antioxidant activity, as well as a significant influence on physiological processes like enzyme inhibitory activity [2]. Despite the wide availability of coumarins and their lead compounds and metabolites in natural products [3], their application up till now has been mostly limited to the anticoagulant activity of warfarin derived from dicoumarol and its analogues [4]. The mechanism of action of these anticoagulants lies in the competitive antagonism of vitamin K, through which they inhibit coagulation of blood in the body by preventing the production of prothrombin and several other coagulation factors [5]. Due to this action, these coumarins are a major group of oral drugs with anticoagulant activity. Anticoagulants reduce the risks of undesirable blood clots leading to myocardial infarction, pulmonary embolism, and ischemic stroke among others. This chapter gives an overview of medical applications of coumarins, in particular the history and evolution of warfarin and related compounds as important anticoagulant agents.

2. History

The medical application of plants containing coumarins probably started long before the isolation of this chemical compound from the Tonka bean in 1820 by Nicholas Jean Baptiste Gaston Guibourt [6]. Ancient Romans produced a cough syrup from the marshmallow (*Althea officinalis*) [7], which contains the coumarin scopoletin [8]. This coumarin demonstrated inhibition of leucocyte migration in mice [9], a process that can be linked to the alleged antitussive effect. Cough is a result of the reaction of the airways to leukotrienes and other factors secreted by leucocytes [10–12]. Inhibition of the migration of these to the affected region consequently reduces the availability of these paracrine factors. In addition, at least one of the herbs mentioned by the famous Roman General Pliny in his pharmacopeia [13] contain coumarins with proven action. For example, the extracts of the common rue or herb-of-grace *Ruta graveolens* contain xanthotoxin [14], a coumarin that reduces the mobility of human spermatozoa possibly through inhibition of membranaeous potassium channels [15].

The application of coumarin and its derivatives in current western medicine dates to the fifties of the past century with the clinical recognition of coumarins as anticoagulant agents. This event was the result of observations of poisoning of animals with coumarin derivatives that led to massive internal organ bleeding [16]. Soon it became clear that the substance that was responsible for the deadly internal bleeding of cattle was dicoumarol [17]. Shortly after this, the proposal was made to develop a coumarin derivative with rodent killing ability and gradually warfarin found its application as a potent rodenticide [18]. When it became clear that this substance also led to bleeding disorders after poisoning in human beings [19], its application as a therapeutic anticoagulant found its way in medicine [20, 21]. Seven decades later, warfarin is still in use as an anticoagulant [22]. In the meantime, several other coumarins with anticoagulant properties like acenocoumarol, phenprocoumon, and fluindione have been developed, and they are used in a variety of clinical settings [23–25].

3. Physiology of hemostasis

Hemostasis in mammals and humans is the result of three sequential processes. The first of these is the acute vasoconstriction within seconds after damaged arteries and veins, by local activity of the potent vasoconstrictor thromboxane among others [26]. The second step is the formation of a blood clot through the entrapment of platelets by fibrin within hours and finally followed by the organization of the fibrin mesh into an adhesive structure on the vessel wall [27]. In one of the last steps of the coagulation, prothrombin converts to thrombin, an enzyme that converts the plasma protein fibrinogen to fibrin monomers and activates factor XIII of the coagulation cascade. Activated factor XIII synthesizes fibrin from these monomers. The acquired fibrin molecules then trap the platelets and eventually form the blood clot [27, 28]. The coagulation process has both an intrinsic and an extrinsic pathway. The difference is that the intrinsic pathway only requires ionized calcium to be activated while the extrinsic pathway requires both calcium and tissue factor that

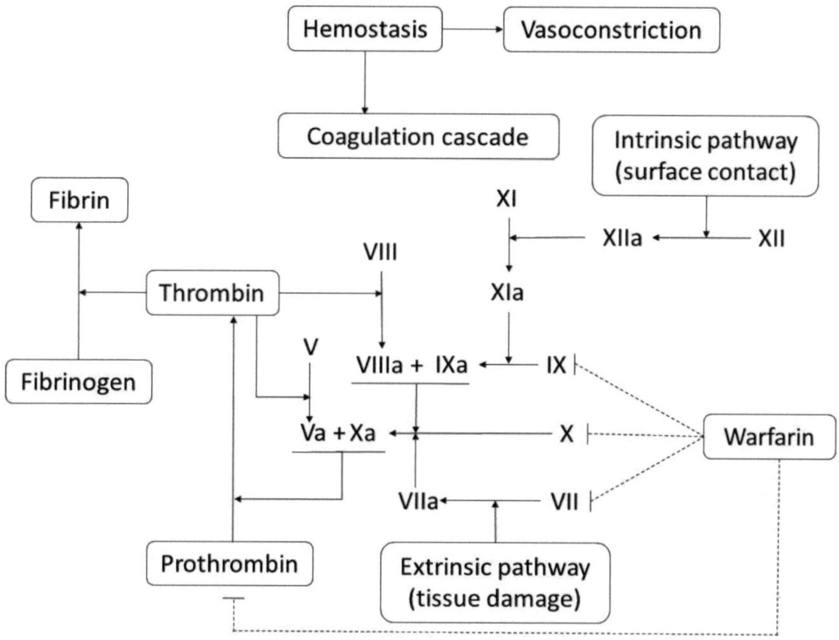

Figure 1.
Overview of the coagulation process along with the interference sites of warfarin.

is released with trauma [28]. **Figure 1** gives an overview of the coagulation process along with the interference sites of warfarin.

4. Pharmacology

4.1 Pharmacokinetics

Warfarin is highly lipid soluble [29–31]. Between 70 and 100% of the oral intake is rapidly absorbed by the intestines with a maximum plasma concentration within 2 h after oral intake [30]. The half-life of the drug is generally more than 20 h, with a large individual variety [32]. Due to the overt lipid solubility, the major part of the drug is protein bound with less than 3% being biologically available [29]. Consequently, the agent has a slow onset of action and a long duration of activity [33]. In fact, the optimal effect is delayed for a few days, until all remaining activated factors II, VII, IX, and X are depleted from the liver and the circulation [33]. Warfarin accumulates in the liver where it exerts its effect and is inactivated through oxidative metabolism by cytochromes P450 to several isomers of water-soluble hydroxywarfarin with negligible anticoagulant activity [34, 35]. These metabolites are almost completely cleared by the kidneys [36]. The hepatic accumulation and relative easy absorption in the intestines result in an enterohepatic circulation of the drug [37]. Enterohepatic circulation is a process in which substances are secreted by the liver with bile to the intestines and subsequently absorbed again by the latter [28]. This results in recycling of the product with very little elimination.

4.2 Mechanism of action

Warfarin inhibits the enzyme vitamin K epoxide reductase that recycles oxidized vitamin K [38]. Vitamin K activates the coagulating factors prothrombin (factor II) and the structurally related serine proteases known as factors VII, IX, and X in the liver cells [27]. Decreasing the biological availability of vitamin K inhibits the synthesis of these essential factors and eventually leads to inhibition of the coagulation process. This means that this compound affects both the intrinsic as well as the extrinsic cascade of coagulation since prothrombin plays a central role in both of these pathways [28] and renders it a highly effective anticoagulant drug.

5. Indications and contraindications

Hemostasis is an essential process to prevent significant external as well as internal blood loss after injury. However, under certain circumstances, it is not desirable to activate or continue this homeostatic process like in disorders with spontaneous thrombosis such as deep venous thrombosis in the legs often resulting in pulmonary embolism [39]. In addition, there are conditions that are prone to a reasonable chance of forming a blood clot during stasis of the blood circulation like in atrial fibrillation and in the limbs of patients with prolonged immobility after surgery [39]. Moreover, conditions like myocardial infarction or ischemic stroke form a preventable group of disorders with inhibition of the thrombotic process [5]. Based on its anticoagulant properties, warfarin is thus an ideal compound for the treatment and prevention of these thromboembolic conditions [5].

Based on the pharmacokinetic properties and the challenges they present, dosing of warfarin is not simple, and a careful approach is necessary. On one hand, a low plasma concentration will not achieve the effect of sufficient anticoagulation and, on the other side of the spectrum, there is the constant chance of overdosing with potential lethal internal or external bleeding. Another problem is the great variety of absorption, body distribution, and metabolism of the agent with individual patients based on the pharmacokinetic properties of warfarin [30]. Frequent monitoring of therapeutic efficiency with adequate laboratory tools like prothrombin time (PT) or international normalized ratio (INR) is absolutely necessary [40, 41] and a fixed or constant dose is close to impossible. Nevertheless, warfarin is highly effective in anticoagulation regiments when carefully dosing and assessing the potential bleeding sites as well as other potential side effects. Warfarin is initially dosed at 5–10 mg daily [42]. Subsequent doses depend on the international normalized ratio, with a therapeutic value between 2 and 4. Concomitant administration of heparins like fraxiparine is necessary when fast anticoagulant activity is desirable [43].

Warfarin is not an ideal agent in conditions when immediate treatment of thromboembolism is imminent due to the long time of onset. In cases of pulmonary embolism and acute ischemic stroke, it is desirable to start with both the oral anticoagulant and fast-acting agent like heparins [44]. The long duration of action harbors another challenge. When acute termination of anticoagulation is necessary with unwanted bleeding like in menstruating women and after blunt and sharp trauma leading to hemorrhage, it could take days before the process of coagulation completely restores after quitting oral administration [27, 30] due to the depletion of coagulant factors in liver and blood. In these cases, intravenous administration of prothrombin complex, fresh frozen plasma with coagulation factors, and high doses of vitamin K may be helpful [45].

Warfarin readily passes the placenta and may result in spontaneous abortion due to retroplacental bleeding [46], as well as prematurity [47], fetal deformity [48], stillbirth [48], and fetal cranial bleeding [49]. Administration during the first trimester of pregnancy has a high risk of embryopathy [50]. This is accompanied by deformities of bone and cartilage [51], blindness, mental retardation, and other neurologic abnormalities [52]. The occurrence of these complications and defects seem to be dose dependent [47, 53] and are most probably the result of the interference with vitamin K-dependent coagulation [38] and bone formation [54]. The effects of the central nervous system and the blindness are probably the result of microhemorrhages in the developing brain as a result of the anticoagulant activity [55]. Clotting factors are easily depleted in the fetus due to the immature liver and small circulating volume [46]. Warfarin does not enter breastmilk and is thus completely safe during lactation [56].

In conclusion, warfarin must be administered with great caution to women in their child-bearing age [57]. Therapy with this agent must be ceased immediately when it becomes clear that the patient is pregnant. Low-molecular weight heparins are a good alternative, since they do not cross the placenta and have been proven to be safe for mother, embryo, and fetus [58].

6. Drawbacks and side-effects

To say that anticoagulant coumarins have only a few side effects is an absolute understatement. Warfarin is one of the leading drugs with adverse effects requiring hospital admission [59]. Most of all, there is the constant chance of severe bleeding [60]. This can include internal hemorrhagic conditions in the head, gastrointestinal tract, female genitalia, the bladder and urethra or skeletal joints and muscles [40, 61]. They generally present as severe headache, stomach pain, and black or bloody stool, heavier than normal menstrual bleeding, discoloration of urine, and pain and swelling of the joints or muscles. Prolonged bleeding from external sharp or blunt wounds is always present [61]. All these conditions are the result of inability of the affected tissues to initiate and continue the process of hemostasis after damage to the epithelial barrier [62].

Patients suffering from hypertension, disorders of the liver, bleeding lesions, and the elderly and patients using drugs and substances that affect coagulation are at higher risk to suffer from bleeding when using warfarin [63]. Hypertension poses mechanical defects in the blood vessels, especially the arteries. Disorders of the liver reduce the ability of the body to eliminate the warfarin and thus make it more biologically available. In bleeding lesions, warfarin inhibits hemostasis. Among substances that can lead to bleeding when used with warfarin are steroidal and non-steroidal anti-inflammatory drugs, antibiotics, and alcohol. These potentiate the activity, interfere with the protein binding, and reduce the metabolism of warfarin, respectively [63]. Other side effects include injury to the kidneys with potential nephritis [64–66], inflammation of the skin [67] and blood vessels [66], and potentiation of rhabdomyolysis by simvastatin [64].

Due to resistance of rodents against warfarin, superwarfarins have been created [68]. These have a much longer time of activity and hence need only to be consumed once by the rodents, contrary to warfarin. The result however is that their effect persists much longer when deliberately or accidently consumed by humans [69] and treatment of this intoxication is a more challenging enterprise.

7. Interaction with drugs and foods

The efficacy of the anticoagulant treatment with warfarin highly depends on its bioavailability, since inhibition of the target (epoxy reductase) enzyme depends on direct binding of the drug to this protein [38]. In addition, vitamin K from external sources does not rely on recycling through this enzyme [38]. Hence, the absorption,

	Agent	Category	Possible mechanism	Effect	Reference
Allopathic medications	Amiodarone	Antiarrhythmic	Inhibition of hepatic metabolism		[78]
	Ciprofloxacin	Antibiotic	Reduction of vitamin K synthesis by intestinal bacteria	Increased bleeding	[79]
	Paroxetine	Antidepressant	Inhibition of hepatic metabolism	Potentiation	[80]
	Citalopram	Antidepressant	Inhibition of hepatic metabolism	Potentiation	[80]
	Clopidogrel	Antiplatelet medication	Inhibition of coagulation cascade	Potentiation	[81]
	Dipyridamole	Antiplatelet medication	Inhibition of coagulation cascade	Potentiation	[81]
	Diclofenac	NSAID	Inhibition of coagulation cascade	Potentiation	[82]
	Naproxen	NSAID	Inhibition of coagulation cascade	Potentiation	[82]
	Acetaminophen	Analgesic	Interference with hepatic metabolism	Increased bleeding	[83]
Food supplements	Fish oil	Lipid profile improvement	Inhibition of coagulation cascade?	Potentiation	[84]
	Pomegranate juice	Antioxidant	Interference with hepatic metabolism	Potentiation	[85]
	Glucosamine	Cartilage improvement	Unknown	Potentiation	[86]
Traditional medications	Chamomile	Medicinal herbal tea	Unknown	Increased bleeding	[87]
	Ginseng	Improving cognitive functions	Unknown	Inhibition	[88]
	St John's wort	Against depression	Induction of metabolism	Inhibition	[89]

Table 1.
Brief overview of possible interactions with warfarin.

transport, delivery, and elimination of warfarin as well as the external availability of vitamin K are potential sites of interaction with other drugs and with food and dietary supplements.

Drugs and food that influence the enterohepatic circulation can all affect the absorption of warfarin. Examples of these are the drug cholestyramine [37] and the avocado fruit [70], which prevent the reabsorption of warfarin in the intestines Concomitant administration of other protein-bound drugs may lead to greater amounts of circulating warfarin and increased risks of bleeding. Valproate sodium increases the bioavailability of warfarin through dislocation of its protein-binding sites [71]. Interference with the metabolism of warfarin is a potential of most drugs that are eliminated by hepatic metabolism. Among these are aspirin [72], non-steroidal anti-inflammatory drugs [72], serotonin reuptake inhibitors [49], anti-platelet agents and some antibiotics [72]. It can go both ways with the metabolism. Induction of the cytochromes will increase the elimination, while occupation of the binding sites by the drugs will increase the availability of warfarin.

Since warfarin acts through elimination of available bioactive vitamin K, variations of the net intake of this vitamin will certainly interfere with the drug action. A high intake of the vitamin will keep the coagulant factors at a higher level and thus inhibit the anticoagulant activity. Likewise, a lower intake will potentiate the effect of warfarin. The vitamin occurs in food in the form of phylloquinone and menaquinone. Phylloquinone is the form mostly found in plants and is also the most abundant form in food [73]. Menaquinones are mainly the product of bacterial production or conversion [74]. Consequently, simple multivitamin and other supplements, food with high vitamin K content [74] as well as antibiotics are sources of fluctuation in vitamin K intake since intestinal bacteria significantly contribute to the production of menaquinones [75].

Recently, another source of interference came into focus. In addition to the previous mentioned parameters, genetic variation in the expression of cytochrome P450 seems to play a role in the metabolism of warfarin [76], thus influencing the availability of the drug [77]. All these considerations make it clear that close monitoring of the individual coagulation ability is necessary for a successful therapy with this agent.

The abovementioned interactions are just a few of the many that are possible. **Table 1** gives examples of a variety of interactions with drugs, food, natural products, and supplements. This is only to underscore the cautious approach patients should practice when taking warfarin.

8. Future prospects and conclusion

Today, coumarins find their application predominantly as anticoagulants in medicine. The narrow therapeutic index of warfarin and related compounds sometimes limit their applicability and consequently there is a constant search for more safe agents in this drug class [90]. Unfortunately, the development of these will probably limit the use of these oral anticoagulants.

Aside from these developments, coumarins with several applications in medical practice are progressively being introduced. Investigators found that coumarin-3-carboxylic acid could be utilized as a dosimeter for radiotherapy. This substance converts to the highly fluorescent 7-hydroxy-coumarin-3-carboxylic acid, with a near perfect linear correlation upon irradiation [91].

The coumarin 2-hydroxycinnamic acid demonstrated inhibitive properties on the enzyme carbonic anhydrase [92]. Inhibition of this enzyme leads to diuresis [93] and decreases intraocular pressure in glaucoma patients [93] with clear therapeutic potential and clinical perspective.

Furano(pyrano)coumarins found in the roots of the Korean angelica (*Angelica gigas*) showed antibacterial activity in hay bacillus (*Bacillus subtilis*) cultures [94]. The coumarine derivative cloricromene reduced the inflammatory parameters in rats subjected to collagen-induced arthritis [95]. In addition, several studies found that coumarins may be useful as anti-tumor agents [4, 96].

Probably since ancient times, coumarins found their application in medicine. Currently, however, coumarins with predominantly anticoagulant properties are applied in daily medical practice. These have been developed from the initial discovery of a cattle killing weed more than six decades ago. Initially applied as a rodenticide, soon a therapeutic usable oral anticoagulant was developed, and slowly other agents entered the market. They have a small therapeutic index, rendering them toxic in a number of circumstances. The search for more safe agents with anticoagulant effects is ongoing and this may result in a decline of the use of coumarins in this field. Nevertheless, coumarins gradually find their way in other fields of medicine. Nevertheless, all these developments promise a bright future for coumarins in medical applications.

Conflict of interest

The author declares no conflict of interest.

Thanks

My gratitude goes out to my colleague Prof. Dr Dennis Mans, who encouraged me to write this chapter. I also want to thank the Anton de Kom University of Suriname for allowing me to do this on top of my regular job.

I want to thank my lovely wife and children who understand why I spend so much time in front of my PC. You are all such a supporting force for me.

Author details

Robbert Bipat
Department of Physiology, Faculty of Medical Science, Anton de Kom University of Suriname, Paramaribo, Suriname

*Address all correspondence to: robbert.bipat@uvs.edu; robbert.bipat@gmail.com

References

[1] Sarker SD, Nahar L. Progress in the chemistry of naturally occurring coumarins. In: Progress in the Chemistry of Organic Natural Products. New York: Springer; 2017. pp. 241-304

[2] Wu L, Wang X, Xu W, Farzaneh F, Xu R. The structure and pharmacological functions of coumarins and their derivatives. Current Medicinal Chemistry. 2009;**16**:4236-4260

[3] Venugopala KN, Rashmi V, Odhav B. Review on natural coumarin lead compounds for their pharmacological activity. BioMed Research International. 2013;**2013**:1-14

[4] Rohini K, Srikumar PS. Therapeutic role of coumarins and coumarin-related compounds. Journal of Thermodynamics & Catalysis. 2014;**05**:1-3

[5] Tadros R, Shakib S. Warfarin—Indications, risks and drug interactions. Australian Family Physician. 2010;**39**:476-479

[6] Guibourt N-J-B-G. Histoire naturelle des drogues simples, ou Cours d'histoire naturelle professé à l'École de pharmacie de Paris. Libraire de L'Académie Nationale de Médicine. 3rd ed. Paris. Available from: https://gallica.bnf.fr/ark:/12148/bpt6k5810074h/f378.image.langEN; 1869

[7] RedRampant.com: Roman Medicinal Herbs. Available from: http://www.redrampant.com/2009/07/roman-medicinal-herbs.html [Accessed: 8 July 2019]

[8] Al-Snafi AE. The pharmaceutical importance of *Althaea officinalis* and *Althaea rosea*: A review. International Journal of PharmTech Research. 2013;5:1378-1385

[9] Ding Z, Dai Y, Hao H, Pan R, Yao X, Wang Z. Anti-inflammatory effects of scopoletin and underlying mechanisms. Pharmaceutical Biology. 2008;**46**:854-860

[10] Kita T, Fujimura M, Ogawa H, Nakatsumi Y, Nomura S, Ishiura Y, et al. Antitussive effects of the leukotriene receptor antagonist montelukast in patients with cough variant asthma and atopic cough. Allergology International. New York: Springer; 2010;**59**:185-192

[11] Smith LJ, Kern R, Patterson R, Krell RD, Bernstein PR. Mechanism of leukotriene D4-induced bronchoconstriction in normal subjects. The Journal of Allergy and Clinical Immunology. 1987;**80**:340-347

[12] Aehringhaus U, Wölbling RH, König W, Patrono C, Peskar BM, Peskar BA. Release of leukotriene C4 from human polymorphonuclear leucocytes as determined by radioimmunoassay. FEBS Letters. 1982;**146**:111-114

[13] van Tellingen C. Pliny's pharmacopoeia or the Roman treat. Netherlands Heart Journal. 2007;**15**:118-120

[14] Naghibi Harat Z, Lakpour N, Sadeghipoor HR, Kamalinejad M, Eshraghian MR, Naghibi B, et al. Immobilising effect of *Ruta graveolens* L. on human spermatozoa: Coumarin compounds are involved. Andrologia. 2015;**47**:1183-1189

[15] Strauss U, Wissel K, Jung S, Wulff H, Hänsel W, Zhu J, et al. K(+) channel-blocking alkoxypsoralens inhibit the immune response of encephalitogenic T line cells and lymphocytes from Lewis rats challenged for experimental autoimmune encephalomyelitis. Immunopharmacology. 2000;**48**:51-63

[16] Schofield FW. The cause of a new disease in cattle stimulating

hemorrhagic septicaemia and blackleg. Journal of the American Veterinary Medical Association. 1924;**64**:553-575

[17] Stahmann MA, Huebner CF, Link KP. Studies on the hemorrhagic sweet clover disease. V. Identification and synthesis of the hemorrhagic agent. The Journal of Biological Chemistry. 1941;**138**:513-527

[18] Hayes WJ. Control of Norway rats with residual rodenticide warfarin. Public Health Reports. 1950;**65**:1537-1555

[19] Green P. Haemorrhagic diathesis attributed to warfarin poisoning. Canadian Medical Association Journal. 1955;**72**:769-770

[20] Wolff JM, Barker NW, Gifford RW, Mann FD. Experience with a new intravenous coumarin anticoagulant (warfarin, sodium derivative). Proceedings of the Staff Meetings. Mayo Clinic. 1953;**28**:489-497

[21] Shapiro S. Warfarin sodium derivative: (Coumadin® sodium). Angiology. 1953;**4**:380-390

[22] Ramachandran S, Pitchai S. Story of warfarin: From rat poison to lifesaving drug. Indian Journal of Vascular and Endovascular Surgery. 2018;**5**:174-175

[23] Minary A, Michel B, Gourieux B, Vogel T. Anticoagulant and antiplatelet combined therapy in patients 75 years and over with atrial fibrillation: A prospective observational study assessing adherence to clinical guidelines. European Journal of Hospital Pharmacy. 2018;**0**:1-6

[24] Zylla MM, Pohlmeier M, Hess A, Mereles D, Kieser M, Bruckner T, et al. Prevalence of intracardiac thrombi under phenprocoumon, direct oral anticoagulants (dabigatran and rivaroxaban), and bridging therapy in patients with atrial fibrillation and flutter. The American Journal of Cardiology. 2015;**115**:635-640

[25] Hohmann C, Hohnloser SH, Jacob J, Walker J, Baldus S, Pfister R. Non-vitamin K oral anticoagulants in comparison to phenprocoumon in geriatric and non-geriatric patients with non-valvular atrial fibrillation. Thrombosis and Haemostasis. 2019;**119**:971-980

[26] Ogletree ML. Overview of physiological and pathophysiological effects of thromboxane A2. Federation Proceedings. 1987;**46**:133-138

[27] Dahlbäck B. Blood coagulation. Lancet. 2000;**355**:1627-1632

[28] Boron W, Boulpaep E. Medical Physiology. Philadelphia: Elsevier; 2016

[29] Levy G. Protein binding of warfarin. British Journal of Clinical Pharmacology. 1995;**39**:211

[30] Breckenridge A, Orme M. Kinetics of warfarin absorption in man. Clinical Pharmacology and Therapeutics. 1973;**14**:955-961

[31] Stella VJ, Mooney KG, Pipkin JD. Dissolution and ionization of warfarin. Journal of Pharmaceutical Sciences. 1984;**73**:946-948

[32] Hornton JD, Bushwick BM. Warfarin Therapy: Evolving Strategies in Anticoagulation. USA: American Academy of Family Physicians; 1999. Available from: https://www.aafp.org/afp/1999/0201/p635.html. [Accessed: 9 July 2019]

[33] Kuruvilla M, Gurk-Turner C. A review of warfarin dosing and monitoring. Proceedings (Baylor University. Medical Center). 2001;**14**:305-306

[34] Kaminsky LS, Zhang ZY. Human P450 metabolism of warfarin.

Pharmacology & Therapeutics. 1997;**73**:67-74

[35] Obaseki AO, Coker HB. The anticoagulant activity of some selected warfarin analogues. The Journal of Pharmacy and Pharmacology. 1987;**39**:142-144

[36] Chan E, McLachlan AJ, Rowland M. Renal handling of warfarin metabolites in man. European Journal of Pharmaceutical Sciences. 1994;**1**:189-193

[37] Jähnchen E, Meinertz T, Gilfrich HJ, Kersting F, Groth U. Enhanced elimination of warfarin during treatment with cholestyramine. British Journal of Clinical Pharmacology. 1978;**5**:437-440

[38] Whitlon DS, Sadowski JA, Suttie JW. Mechanism of coumarin action: Significance of vitamin K epoxide reductase inhibition. Biochemistry. 1978;**17**:1371-1377

[39] Turpie AGG, Chin BSP, Lip GYH. Venousthromboembolism:Pathophysiology, clinical features, and prevention. BMJ. 2002;**325**:887-890

[40] Hawes EM, Viera AJ. Anticoagulation: Monitoring of patients receiving anticoagulation. FP Essent. 2014;**422**:24-30

[41] Ramos-Esquivel A. Monitoring anticoagulant therapy with new oral agents. World Journal of Methodology. 2015;**5**:212-215

[42] Md Arif K, Rahman MA. A review of warfarin dosing and monitoring. Faridpur Medical College Journal. 2018;**13**:40-43

[43] Siguret V, Gouin I, Debray M, Perret-Guillaume C, Boddaert J, Mahé I, et al. Initiation of warfarin therapy in elderly medical inpatients: A safe and accurate regimen. The American Journal of Medicine. 2005;**118**:137-142

[44] Streiff MB, Agnelli G, Connors JM, Crowther M, Eichinger S, Lopes R, et al. Guidance for the treatment of deep vein thrombosis and pulmonary embolism. Journal of Thrombosis and Thrombolysis. 2016;**41**:32-67

[45] Christensen H, Cordonnier C, Kõrv J, Lal A, Ovesen C, Purrucker JC, et al. European stroke organisation guideline on reversal of oral anticoagulants in acute intracerebral haemorrhage. European Stroke Journal. 2019;**0**:1-13. DOI: 10.1177/2396987319849763

[46] Li TC, Smith ARB, Duncan SLB. Feto-maternal haemorrhage complicating warfarin therapy during pregnancy. Journal of Obstetrics and Gynaecology. 1990;**10**:401-402

[47] Soma-Pillay P, Nene Z, Mathivha TM, Macdonald AP. The effect of warfarin dosage on maternal and fetal outcomes in pregnant women with prosthetic heart valves. Obstetric Medicine. 2011;**4**:24-27

[48] Wainwright H, Beighton P. Warfarin embryopathy: Fetal manifestations. Virchows Archiv. 2010;**457**:735-739

[49] Sansone RA, Sansone LA. Warfarin and antidepressants: Happiness without hemorrhaging. Psychiatry (Edgmont). 2009;**6**:24-29

[50] Mehndiratta S, Suneja A, Gupta B, Bhatt S. Fetotoxicity of warfarin anticoagulation. Archives of Gynecology and Obstetrics. 2010;**282**:335-337

[51] Pauli RM, Lian JB, Mosher DF, Suttie JW. Association of congenital deficiency of multiple vitamin K-dependent coagulation factors and the phenotype of the warfarin embryopathy: Clues to the mechanism of teratogenicity of coumarin derivatives. American Journal of Human Genetics. 1987;**41**:566

[52] Hou J-W. Fetal warfarin syndrome. Chang Gung Medical Journal. 2004;**27**:691-695

[53] Vitale N, De Feo M, De Santo LS, Pollice A, Tedesco N, Cotrufo M. Dose-dependent fetal complications of warfarin in pregnant women with mechanical heart valves. Journal of the American College of Cardiology. 1999;**33**:1637-1641

[54] Menon RK, Gill DS, Thomas M, Kernoff PBA, Dandona P. Impaired carboxylation of osteocalcin in warfarin-treated patients. The Journal of Clinical Endocrinology and Metabolism. 1987;**64**:59-61

[55] Pati S, Helmbrecht GD. Congenital schizencephaly associated with in utero warfarin exposure. Reproductive Toxicology. 1994;**8**:115-120

[56] Schindler D, Graham TP. Warfarin overdose in a breast-feeding woman. The Western Journal of Emergency Medicine. 2011;**12**:216-217

[57] Yurdakök M. Fetal and neonatal effects of anticoagulants used in pregnancy: A review. The Turkish Journal of Pediatrics. 2012;**54**:207-215

[58] Ní Áinle F, Wong A, Appleby N, Byrne B, Regan C, Hassan T, et al. Efficacy and safety of once daily low molecular weight heparin (tinzaparin sodium) in high risk pregnancy. Blood Coagulation and Fibrinolysis. 2008;**19**(7):689-692

[59] Pirmohamed M, James S, Meakin S, Green C, Scott AK, Walley TJ, et al. Adverse drug reactions as cause of admission to hospital: Prospective analysis of 18 820 patients. BMJ. 2004;**329**:15-19

[60] Linkins L-A, Choi PT, Douketis JD. Clinical impact of bleeding in patients taking oral anticoagulant therapy for venous thromboembolism. Annals of Internal Medicine. 2003;**139**:893

[61] Wallvik J, Själander A, Johansson L, Bjuhr O, Jansson J-H. Bleeding complications during warfarin treatment in primary healthcare centres compared with anticoagulation clinics. Scandinavian Journal of Primary Health Care. 2007;**25**:123-128

[62] Zareh M, Davis A, Henderson S. Reversal of warfarin-induced hemorrhage in the emergency department. The Western Journal of Emergency Medicine. 2011;**12**:386-392

[63] Fitzmaurice DA, Blann AD, Lip GYH. Bleeding risks of antithrombotic therapy. BMJ. 2002;**325**:828-831

[64] Mogyorosi A, Bradley B, Showalter A, Schubert ML. Rhabdomyolysis and acute renal failure due to combination therapy with simvastatin and warfarin. Journal of Internal Medicine. 1999;**246**:599-602

[65] Mendonca S, Gupta D, Valsan A, Tewari R. Warfarin related acute kidney injury: A case report. Indian Journal of Nephrology. 2017;**27**:78-80

[66] Kapoor KG, Bekaii-Saab T. Warfarin-induced allergic interstitial nephritis and leucocytoclastic vasculitis. Internal Medicine Journal. 2008;**38**:281-283

[67] Kwong P, Roberts P, Prescott SM, Tikoff G. Dermatitis induced by warfarin. JAMA. 1978;**239**:1884-1885

[68] Sarin S, Mukhtar H, Mirza MA. Prolonged coagulopathy related to superwarfarin overdose. Annals of Internal Medicine. 2005;**142**:156

[69] Watt BE, Proudfoot AT, Bradberry SM, Vale JA. Anticoagulant

rodenticides. Toxicological Reviews. 2005;**24**:259-269

[70] Blickstein D, Shaklai M, Inbal A. Warfarin antagonism by avocado. Lancet. 1991;**337**:914-915

[71] Zhou C, Sui Y, Zhao W, Dong C, Ren L, Song P, et al. The critical interaction between valproate sodium and warfarin: Case report and review. BMC Pharmacology and Toxicology. 2018;**19**:60

[72] Carpenter M, Berry H, Pelletier AL. Clinically relevant drug-drug interactions in primary care. American Family Physician. 2019;**99**:558-564

[73] Hayes A, Hennessy Á, Walton J, McNulty BA, Lucey AJ, Kiely M, et al. Phylloquinone intakes and food sources and vitamin K status in a nationally representative sample of Irish adults. The Journal of Nutrition. 2016;**146**:2274-2280

[74] Booth SL. Vitamin K: Food composition and dietary intakes. Food & Nutrition Research. 2012;**56**:1-5. DOI: 10.3402/fnr.v56i0.5505

[75] Conly JM, Stein K. The production of menaquinones (vitamin K2) by intestinal bacteria and their role in maintaining coagulation homeostasis. Progress in Food & Nutrition Science. 1992;**16**:307-343

[76] Li J, Wang S, Barone J, Malone B. Warfarin pharmacogenomics. P T. 2009;**34**:422-427

[77] Mak M, Lam C, Pineda SJ, Lou M, Xu LY, Meeks C, et al. Pharmacogenetics of warfarin in a diverse patient population. Journal of Cardiovascular Pharmacology and Therapeutics. 2019;**24**:521-533. DOI: 10.1177/1074248419843530

[78] Sanoski CA, Bauman JL. Clinical observations with the amiodarone/warfarin interaction: Dosing relationships with long-term therapy. Chest. 2002;**121**:19-23

[79] Linville D II, Emory C, Graves L III. Ciprofloxacin and warfarin interaction. The American Journal of Medicine. 1991;**90**:765

[80] Hemeryck A, De Vriendt C, Belpaire FM. Inhibition of CYP2C9 by selective serotonin reuptake inhibitors: In vitro studies with tolbutamide and (S)-warfarin using human liver microsomes. European Journal of Clinical Pharmacology. 1999;**54**:947-951

[81] Nutescu E, Chuatrisorn I, Hellenbart E. Drug and dietary interactions of warfarin and novel oral anticoagulants: An update. Journal of Thrombosis and Thrombolysis. 2011;**31**:326-343

[82] Choi KH, Kim AJ, Son IJ, Kim K-H, Kim K-B, Ahn H, et al. Risk factors of drug interaction between warfarin and nonsteroidal anti-inflammatory drugs in practical setting. Journal of Korean Medical Science. 2010;**25**:337-341

[83] Thijssen HH, Soute BA, Vervoort LM, Claessens JG. Paracetamol (acetaminophen) warfarin interaction: NAPQI, the toxic metabolite of paracetamol, is an inhibitor of enzymes in the vitamin K cycle. Thrombosis and Haemostasis. 2004;**92**:797-802

[84] Buckley MS, Goff AD, Knapp WE. Fish oil interaction with warfarin. The Annals of Pharmacotherapy. 2004;**38**:50-53

[85] Jarvis S, Li C, Bogle RG. Possible interaction between pomegranate juice and warfarin. Emergency Medicine Journal. 2010;**27**:74-75

[86] Knudsen JF, Sokol GH. Potential glucosamine-warfarin interaction

resulting in increased international normalized ratio: Case report and review of the literature and MedWatch database. Pharmacotherapy. 2008;**28**:540-548

[87] Segal R, Pilote L. Warfarin interaction with *Matricaria chamomilla*. CMAJ. 2006;**174**:1281-1282

[88] Janetzky K, Morreale AP. Probable interaction between warfarin and ginseng. American Journal of Health-system Pharmacy. 1997;**54**:692-693

[89] Henderson L, Yue QY, Bergquist C, Gerden B, Arlett P. St John's wort (*Hypericum perforatum*): Drug interactions and clinical outcomes. British Journal of Clinical Pharmacology. 2002;**54**:349-356

[90] Little JW. New oral anticoagulants: Will they replace warfarin? Oral Surgery, Oral Medicine, Oral Pathology, Oral Radiology. 2012;**113**:575-580

[91] Collins AK, Makrigiorgos GM, Svensson GK. Coumarin chemical dosimeter for radiation therapy. Medical Physics. 1994;**21**:1741-1747

[92] Maresca A, Temperini C, Pochet L, Masereel B,Scozzafava A, Supuran CT. Deciphering the mechanism of carbonic anhydrase inhibition with coumarins and thiocoumarins. Journal of Medicinal Chemistry. 2010;**53**:335-344

[93] Carta F, Supuran CT. Diuretics with carbonic anhydrase inhibitory action: A patent and literature review (2005-2013). Expert Opinion on Therapeutic Patents. 2013;**23**:681-691

[94] Lee S, Shin D-S, Ju SK, Oh K-B, Sam SK. Antibacterial coumarins from *Angelica gigas* roots. Archives of Pharmacal Research. 2003;**26**:449-452

[95] Cuzzocrea S, Mazzon E, Bevilaqua C, Costantino G, Britti D, Mazzullo G, et al. Cloricromene, a coumarine derivative, protects against collagen-induced arthritis in Lewis rats. British Journal of Pharmacology. 2000;**131**:1399-1407

[96] Klenkar J, Molnar M. Natural and synthetic coumarins as potential anticancer agents. Journal of Chemical and Pharmaceutical Research. 2015;**7**:1223-1238

Chapter 6

One-Pot Synthesis of Coumarin Derivatives

Inul Ansary and Abu Taher

Abstract

Coumarin derivatives have a myriad of applications in medical science, biomedical research, and many industrial branches. For this reason, many efforts are being dedicated to the development of novel and more practical methods for synthesizing these compounds. This chapter describes several methods of one-pot synthesis of coumarin derivatives, including von Pechmann condensation, Knoevenagel condensation, Baylis-Hillman reaction, Michael addition, Kostanecki reaction, vinyl phosphonium salt-mediated electrophilic reaction, and Heck-lactonization reaction. The methods are compared with each other, and the advantages and disadvantages of each of them are addressed.

Keywords: coumarin derivatives, one-pot synthesis, methods and procedures, advantages and disadvantages

1. Introduction

Coumarin (2H-chromen-2-one) derivatives have spawn great interest over the years because of their significant biological importance [1]. They are associated with various biological activities viz. antiviral [2, 3], antibacterial [4, 5], antimicrobial [6], anticoagulant [7], anti-inflammatory [8, 9], anticancer [10, 11], anticonvulsant [12], antioxidant [13], antifungal [14, 15], and anti-HIV [16]. They also possess the properties like inhibition of platelet aggregation [17] and inhibition of steroid 5α-reductase [18]. Besides, they are attracting considerable attention of chemists due to their wide range of applications such as optical brighteners [19], photosensitizers [20], fluorescent and laser dyes [21], and additives [22] in food, perfumes, cosmetics, and pharmaceuticals. The novel compounds are also utilized in drug and pesticidal preparations [23]. Considering these multifarious activities of coumarins, synthetic chemists are actively engaged in developing new and superior methods for the isolation of coumarin derivatives. The most widely used method for their synthesis is Pechmann reaction [24–27], which involves the condensation between phenols and β-keto esters, in the presence of an acid catalyst. This method employs both homogeneous catalysts such as concentrated H_2SO_4 [24, 25], trifluoroacetic acid (TFA) [28], and Lewis acids (LA) such as $AlCl_3$ [29], $ZnCl_2$ [30], $ZrCl_4$ [31], $TiCl_4$ [32], etc. and heterogeneous catalysts such as cation-exchange resins [33], Nafion resin/silica composites [34], zeolite H-BEA (H-beta, $SiO_2/Al_2O_3 = 14$) [35], and other solid acids.

2. Methods to synthesize coumarin derivatives

2.1 Pechmann condensation reaction

The general reaction sequence of Pechmann reaction and its mechanism, shown in **Figure 1**, involves an esterification/transesterification between the phenol **1** and β-keto ester **2** in the presence of protonic acid or Lewis acid (LA) catalyst to produce species **4** followed by an attack to the activated carbonyl carbon by the aromatic ring at ortho-position to yield the new ring in species **5**. Finally, dehydration of species **5** affords coumarin derivative **2**.

A series of substituted coumarins **8** have been synthesized in 25–77% yields by the reactions of substituted phenols **6** with ethyl acetoacetate **7** in the presence of zinc-iodine mixture in refluxing toluene (**Figure 2**) [36]. It is observed that phenols containing electron-donating substituent like —CH_3 group result in higher yields compared to unsubstituted phenols and phenols having electron-withdrawing group such as NO_2 group.

When 3-(*N,N*-dimethylamino)phenol **9** is subjected to react with ethyl 2-acetamide-3-oxobutyrate **10** in the presence of anhydrous $ZnCl_2$ in absolute ethanol under reflux condition, the acetamido coumarin **11** is obtained only in 12.4% yield (**Figure 3**) [30].

Substituted coumarins **14** have been achieved in moderate to good yields from substituted phenols **12** and methyl acetoacetate **13** under conventional and microwave heating, respectively, catalyzed by concentrated H_2SO_4 (**Figure 4**) [37]. It is found that the reactions using the latter method are faster coupled with product in better yields compared to former one.

Synthesis of substituted coumarins **16** in 62–98% yields has also been described by Maheswara et al. [38] via reactions of substituted phenols **1** with β-keto esters **15** in the presence of a heterogeneous catalyst, $HClO_4.SiO_2$ under solvent-free conditions (**Figure 5**, Condition A). The aforementioned method involves recoverable cheap catalyst and shorter reaction time with high product yields. However, relatively lower yields (35–55%) of substituted coumarins **16** have been isolated from the similar starting precursors catalyzed by Amberlyst-15 acidic catalyst [39] in toluene under refluxing condition (**Figure 5**, Condition B).

Pechmann condensation reactions for the synthesis of substituted coumarins using various homogeneous and heterogeneous catalysts have been reported in literature and some important ones are summarized in **Table 1**.

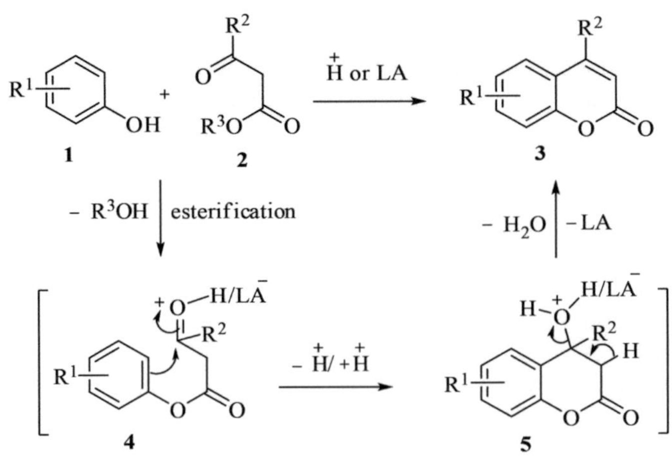

Figure 1.
Mechanism for the acid-catalyzed Pechmann condensation.

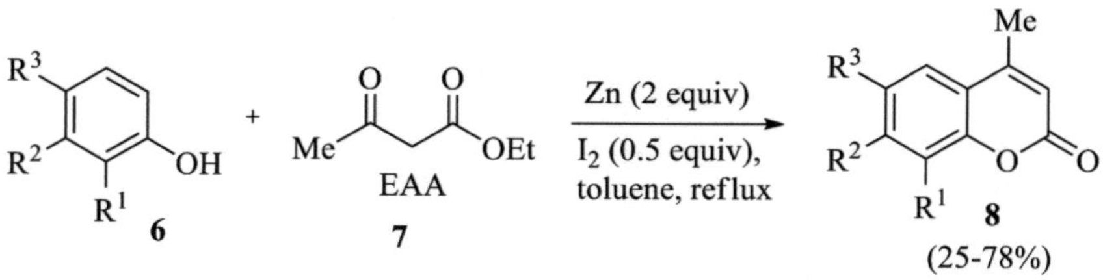

Figure 2.
Synthesis of substituted coumarins.

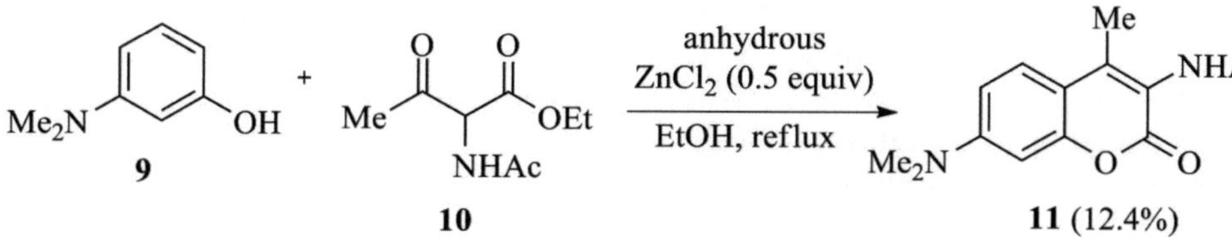

Figure 3.
Synthesis of acetamido coumarin.

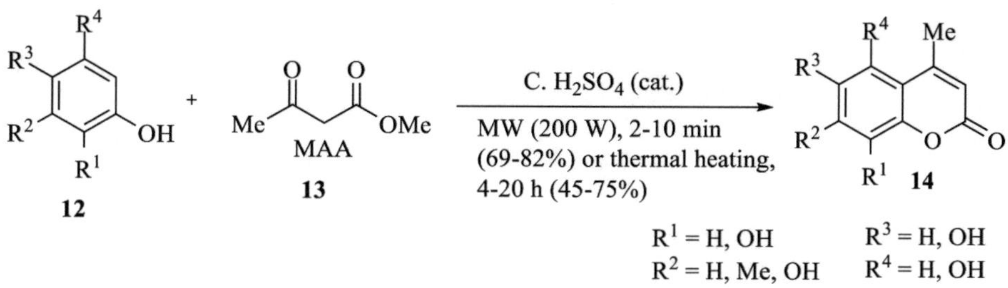

Figure 4.
Synthesis of substituted coumarins.

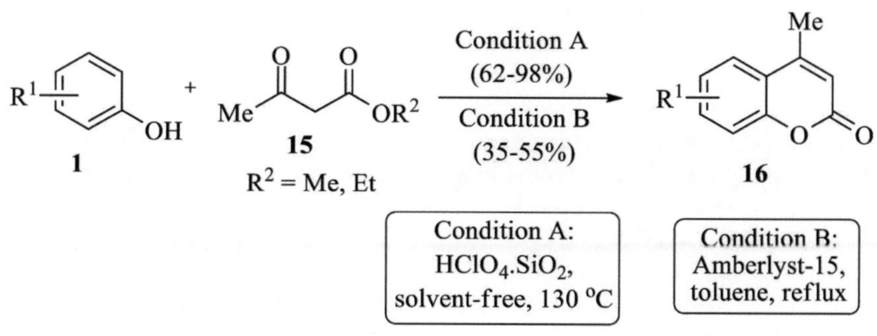

Figure 5.
Synthesis of substituted coumarins.

From **Table 1**, it is quite evident that the reactions under microwave as well as ultrasound irradiation occur at a faster rate than those of the conventional methods (entries 10, 14, 15, 16, 25, 31, 32, and 39). Unsubstituted phenol produces lower yields of corresponding coumarin derivatives and/or requires longer reaction time (entries 2–4, 7, 10, 12, 13, 24, 28, 30, and 38), higher temperature (entries 2, 3, 7, and 12), and excess amount of catalysts (entries 7 and 12) than di- and trihydric phenols. This may presumably be due to the less reactivity of unsubstituted phenol toward Pechmann condensation reaction compared to di- and trihydric phenols. In addition, the substitution of an electron-donating group such as *m/p*-Me or *p*-OMe in the phenols leads to decrease of catalytic activity and, hence, requires longer reaction time and/or gives rise to lower yields of products (entry 13). The reactivity of monohydric phenols having electron-withdrawing groups such as *m*-NH_2 and *m*-OMe is also lowered compared with simple di- and trihydric phenols (entries 19, 28, and 37). 1-Naphthol and 2-naphthol need longer reaction time (entries 13, 33, and 39) and/or furnish products with lower yields (entries 13, 37, and 40) compared to other phenols, due to the presence of another phenyl ring. However, better yield of benzocoumarin is obtained from the reaction between 1-naphthol and more reactive β-keto ester, ethyl 4-chloro-3-oxobutanoate (entry 37). It is interesting to note that β-keto ester having phenyl group at the β-position such as ethyl 3-oxo-3-phenylpropanoate is found to be less reactive in Pechmann condensation with resorcinol and 1,3-dihydroxy-5-methyl benzene due to the presence of conjugated keto center, which lengthens the reaction time than in the reactions of EAA and/or ethyl 4-chloro-3-oxobutanoate with resorcinol and 1,3-dihydroxy-5-methyl benzene (entries 21, 28, and 37). Besides, the reactivity of different types of phenols and β-keto esters, catalyst efficiency, and solvent effect of Pechmann condensation has also been studied. It is observed that $TiCl_4$ (entry 5) is the most effective catalyst as far as reaction time is considered, whereas montmorillonite K-10 (entry 1) and sulfated zirconia (SZr) (entry 9) are found to be less effective. Ionic liquids (ILs) such as 1-butyl-3-methylimidazolium hexafluorophosphate [bmim]PF_6 and 1,3-disulfonic acid imidazolium hydrogen sulfate (DSIMHS) have been used as effective and reusable catalysts and reaction media as well (entries 6 and 18).

Lewis acid–surfactant-combined catalyst (LASC) such as nano-TiO_2 on dodecyl-sulfated silica support (NTDSS) is used as a reusable and highly effective catalyst for Pechmann condensation of phenols containing different types of substituents in water led to excellent product yields (entry 20). Other recyclable solid acid catalysts have also been employed in Pechmann condensation reactions leading to coumarin derivatives in good to excellent yields under solvent-free (entries 22–24, 26–27, 29–30, and 42), microwave irradiation (entry 25) and/or ultrasound irradiation (entry 39) conditions.

More importantly, sulfonic acid-supported silica-coated magnetic nanoparticles (Fe_3O_4@SiO_2@$PrSO_3H$), $CuFe_2O_4$ nanoparticles, and zirconium(IV) complex grafted silica coated magnetic nanoparticles are found to be the most efficient catalysts toward Pechmann condensation, in which case the catalyst can be effortlessly separated by external magnet after completion of the reaction and reused for 22, 6, and 5 consecutive runs, without any significant loss in catalytic efficiency (entries 33–35).

Pechmann condensation of pyrogallol and resorcinol with ethyl acetoacetate over nanosponge MFI zeolite in comparison with conventional zeolites (MFI, BEA, and USY) and other layered MFI (lamellar, pillared, and self-pillared) have been investigated. It is important to note that the nanosponge catalysts exhibit the best catalytic performance with respect to the products' selectivity in the liquid-phase condensation reactions among all the investigated zeolites (entry 36).

On the other hand, the catalytic behavior of metal–organic frameworks such as Cu-benzene-1,3,5-tricarboxylate (CuBTC) and Fe-benzene-1,3,5-tricarboxylate (FeBTC) is investigated and compared with large-pore zeolites, beta (BEA), and ultrastable Y (USY) (entry 41). It is clear that zeolites BEA and USY are found to be more active catalysts in transformations of the most active substrates like resorcinol and pyrogallol but a low conversion of naphthol is observed. However, almost total transformation of naphthol (93–98% conversion) to the target product occurs within 23 h of the reaction time over metal–organic frameworks, CuBTC and FeBTC.

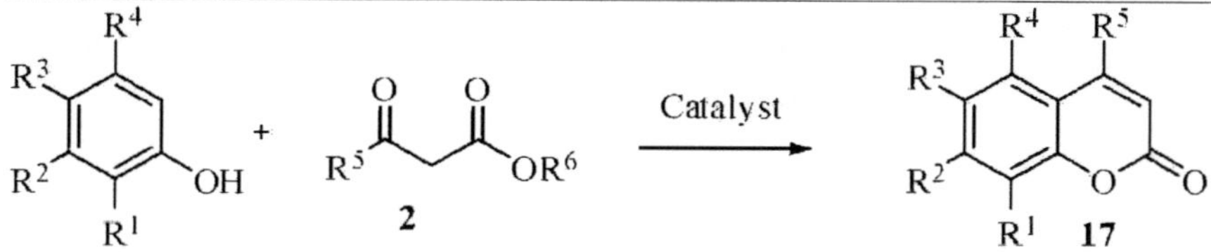

R^5 = Me, CF_3, Ph, Pr, CH_2OMe, CH_2Cl
R^6 = Me, Et
R^1 = H, OH, NO_2
R^2 = H, Me, OMe, NH_2, OEt, OCH_2Ph, OH
R^3 = H, Cl, Me, OMe, NO_2
R^4 = H, OH, Me, OMe

Entry	Catalyst	Reaction conditions	Time	Yields (%)	Reference
1	Montmorillonite K-10	K-10 (30 wt% of **12**), toluene, reflux	8–10 h	66–94	[40]
2	1-Butyl-3-methylimidazolium chloroaluminate [bmim]Cl. $2AlCl_3$	[bmim]Cl.$2AlCl_3$ (1.1 equiv. of **12**), 30–120°C	10–120 min	40–95	[41]
3	$InCl_3$	$InCl_3$ (10 mol%), 65–130°C	30–240 min	65–98	[42]
4	$ZrCl_4$	$ZrCl_4$ (2 mol%), 70°C	5–30 min	56–95	[31]
5	$TiCl_4$	$TiCl_4$ (0.5 equiv. of **12**), rt	50–70 s	56–95	[32]
6	1-Butyl-3-methylimidazolium hexafluorophosphate [bmim]PF_6	[bmim]PF_6 (4 ml), solvent-free, 100°C	45 min	90–95	[43]
7	$Bi(NO_3).5H_2O$	$Bi(NO_3).5H_2O$ (5–10 mol%), 80–130°C	15–300 min	47–94	[44]
8	SO_4^{2-}/CeO_2-ZrO_2	SO_4^{2-}/CeO_2-ZrO_2 (10 wt% of **12**), 120°C	4–143 min	80–94	[45]
9	SZr (sulfated zirconia)	SZr (1 wt% of **12**), 80°C	24 h	52–92	[46]
10	Ceric ammonium nitrate (CAN)	Condition A: CAN (10 mol%), solvent-free, 110°C Condition B: CAN (10 mol%), solvent-free, MW (300 W)	10–15 min 2–3 min	92–96 94–97	[47]
11	$ClSO_3H$	$ClSO_3H$ (0.2 ml), solvent free, 10°C	10 min	91–98	[48]
12	LiBr	LiBr (10–20 mol%), 75–125°C	15–90 min	54–92	[1]
13	Nanocrystalline-cellulose-supported sulfonic acid ionic liquid	NCC-supported sulfonic acid IL (10 wt% of **12**), solvent-free, 80°C	18 min-24 h	20–98	[49]
14	$Cu(ClO_4)_2$	$Cu(ClO_4)_2$ (20 mol%), solvent-free, US (35 kHz), 45–50°C	30–50 min	70–96	[50]
15	Selectfluor	Condition A: Selectfluor (10 mol%), solvent-free, rt. Condition B: Selectfluor (10 mol%), solvent-free, US (30 kHz, 780 W)	85–90 min 15–40 min	70–79 82–94	[51]

Entry	Catalyst	Reaction conditions	Time	Yields (%)	Reference
16	I_2	Condition A: I_2 (25 mol%), toluene, 90°C Condition B: I_2 (1 mol%), MW	18 h 1.5–5 min	42–89 80–96	[52] [53]
17	AgOTf	AgOTf (10 mol%), solvent-free, 60°C	3–12 h	60–95	[54]
18	1,3-Disulfonic acid imidazolium hydrogen sulfate (DSIMHS)	DSIMHS (7 mol%), solvent-free, 70°C	2–27 min	80–96	[55]
19	*N,N′*-dimethylaminoethanol hydrosulfate ($[N_{112}OH][HSO_4]$)	$[N_{112}OH][HSO_4]$ (5 mol%), solvent-free, 90°C	3–24 h	20–99	[56]
20	Nano-TiO_2 on dodecyl-sulfated silica support (NTDSS)	NTDSS (5 mol% TiO_2), H_2O, reflux	3–8 h	89–98	[57]
21	$ZrOCl_2.8H_2O/SiO_2$	$ZrOCl_2.8H_2O/SiO_2$ (10 mol%), solvent-free, 90°C	5–80 min	75–99	[58]
22	Polydivinylbene-bound perfluoroalkylsulfonyl imide polymers (H-PDVB-*x*-SSFAI)	H-PDVB-*x*-SSFAI (10 mol%), solvent-free, 140°C	2 h	78–94	[59]
23	Polyaniline–fluoroboric acid–dodecyl hydrogen sulfate (PANI–HBF_4–DHS)	PANI–HBF_4–DHS (20 wt.% of **12**), solvent-free, 150°C	6 h	94–98	[60]
24	Silica sulfuric acid (SSA)	SSA (15 mol%), solvent-free, 80°C	0.5–2 h	70–97	[61]
25	ZrPW (Zirconium IV Phosphotungstate) 12-TPA/ZrO_2 (12-Tungstophosphoric acid supported onto ZrO_2)	Condition A: ZrPW (0.2 g), solvent-free, 130°C Condition B: ZrPW (0.2 g), solvent-free, MW (250 W), 130°C Condition C: 12-TPA/ZrO_2 (0.2 g), solvent-free, 130°C Condition D: 12-TPA/ZrO_2 (0.2 g), solvent-free, MW (250 W), 130°C	8 h 30 min 8 h 30 min	42–65 47–66 38–63 41–65	[62]
26	12-Tungstophosphoric acid supported on SnO_2 nanoparticles (12-TPA-SnO_2)	12-TPA-SnO_2 (30 wt% of TPA), solvent-free, 120°C	2 h	78	[63]
27	Poly(4-vinylpyridine)-supported copper iodide	P_4VPy-CuI (0.1 g), solvent-free, 80°C	10–90 min	84–92	[64]
28	Polystyrene-supported $GaCl_3$ (PS–$GaCl_3$)	PS–$GaCl_3$ (10 mol%), ethanol, reflux	45–300 min	45–96	[65]
29	Silica tungstic acid (STA)	STA (5 mol%), solvent-free, 80°C	20–90 min	75–97	[66]
30	CMK-5 supported sulfonic acid (CMK-5-SO_3H)	CMK-5-SO_3H (3 mol%), solvent-free, 130°C	15–120 min	60–97	[67]
31	FeF_3	FeF_3 (0.05 g), solvent-free, MW (450 W), 110°C	6–9 min	61–98	[68]

Entry	Catalyst	Reaction conditions	Time	Yields (%)	Reference
32	$FeCl_3$	$FeCl_3$ (10 mol%), solvent-free, US (20 kHz, 130 W)	1–20 min	55–99	[69]
33	Sulfonic acid supported silica coated magnetic nanoparticles (Fe_3O_4@SiO_2@$PrSO_3H$)	Fe_3O_4@SiO_2@$PrSO_3H$ (1.6 mol%), solvent-free, 130°C	3–50 min	87–98	[70]
34	$CuFe_2O_4$ nanoparticles	$CuFe_2O_4$ (5 mol%), H_2O, rt	15–34 min	82–98	[71]
35	Zr(IV)-HMNQ@ ASMPs [Zirconium(IV)-3-hydroxy-2-methyl-1,4-naphthoquinone (HMNQ)@3-aminopropylated silica coated magnetic nanoparticles (ASMPs)]	Zr(IV)-HMNQ@ ASMPs (20 mg), solvent-free, 110°C	10 min	95–100 (selectivity)	[72]
36	MFI nanosponge zeolite (MFI-NSZ)	MFI-NSZ (0.1 g), dodecane (0.5 g, internal standard), nitrobenzene, 120–150°C	70 h	80–90 (selectivity)	[73]
37	$In(OTf)_3$	$In(OTf)_3$ (1 mol%), solvent-free, 80°C	10–87 min	68–98	[74]
38	$Mg(NTf_2)_2$	$Mg(NTf_2)_2$ (1 mol%), solvent-free, 80°C	25–60 min	85–98	[75]
39	Poly(4-vinylpyridinium) hydrogen sulfate (PVPHS)	PVPHS (2 mol%), solvent-free, US (35 kHz, 200 W)	3–18 min	62–96	[76]
40	Polyvinylpolypyrrolidone-bound boron trifluoride (PVPP-BF_3)	PVPP-BF_3 (33 mol%), ethanol, reflux	2–3 h	76–96	[77]
41	Zeolites e.g., beta (BEA) and ultrastable Y (USY) Metal–organic frameworks (MOFs) such as Cu-benzene-1,3,5-tricarboxylate (CuBTC) and Fe-benzene-1,3,5-tricarboxylate (FeBTC)	Condition A: Zeolite (0.2 g), nitrobenzene, 130°C Condition B: MOF (0.2 g), nitrobenzene, 130°C	23 h 23 h	23–91 (conversion) 2–98 (conversion)	[78]
42	$Zn_{0.925}Ti_{0.075}O$ NPs	$Zn_{0.925}Ti_{0.075}O$ (10 mol%), solvent-free, 110°C	3–5 h	51–89	[79]

Table 1.
Synthesis of substituted coumarins via Pechmann condensation reactions.

Catalytic activity of many other catalysts under different reaction conditions is delineated in the recently published review [80].

2.2 Knoevenagel condensation reaction

An efficient green one-pot synthetic method for the synthesis of 3-substituted coumarin derivatives **21/22** has been observed by Knoevenagel condensation of various *o*-hydroxybenzaldehydes **18/19** with 1,3-dicarbonyl compounds **20** using nano-ZnO catalyst under microwave or thermal conditions, which affords moderate to good yield of the products (**Figure 6**) [81]. Reactions under microwave-irradiation conditions are found to be more convenient than thermal conditions.

Various coumarin-3-carboxylic acid derivatives **25/26** have been synthesized in good yields using catalytic amounts of $SnCl_2.2H_2O$ under solvent-free condition (**Figure** 7) [82].

Ultrasound irradiation technique is also useful to synthesize 3-aryl coumarin derivatives. Treatment of *o*-hydroxybenzaldehydes **18** with aryl substituted acetyl chloride **27** in the presence of K_2CO_3 as a catalyst in tetrahydrofuran (THF) using ultrasound irradiation leads to the formation of 3-aryl coumarin derivatives **28** in moderate to high yields (**Figure 8**) [83]. This green method appears to be a convenient and simple pathway than that of conventional heating.

Coumarin-substituted benzimidazole or benzoxazole derivatives **32** that are known as coumarin dyes have been synthesized in good yields from 4-diethylamino-2-hydroxybenzaldehyde **29**, ethyl cyanoacetate **30**, and ortho-phenylenediamine/phenylenehydroxyamine derivatives **31** in the presence of reusable green solid acid like HZSM-5 zeolite, heteropoly acids, e.g., tungstophosphoric acid ($H_3PW_{12}O_{40}$), and/or tungstosilicic acid ($H_4O_{40}SiW_{12}$) in *n*-pentanol or water and even solvent-free conditions (**Figure 9**) [84].

Cellulose sulfonic acid (CSA) is an efficient catalyst for the synthesis of 3-substituted coumarin via Knoevenagel condensation reaction. Thus, 3-acetyl coumarin **34** is obtained in 88% yield in the reaction between salicylaldehyde **33** and ethyl acetoacetate **7** in the presence of CSA under solvent-free conditions (**Figure 10**) [85].

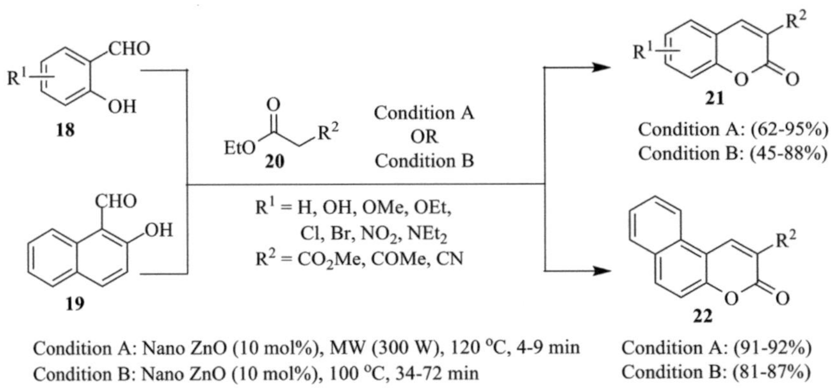

Figure 6.
Synthesis of 3-substituted coumarins.

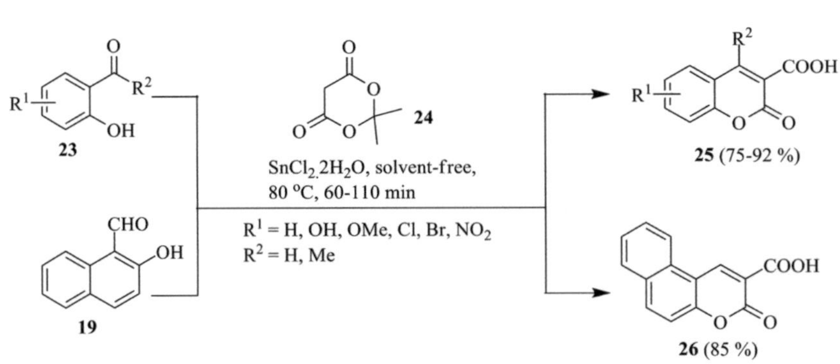

Figure 7.
Synthesis of coumarin 3-carboxylic acid derivatives.

Shaabani et al. [86] have described the synthesis of 3-substituted coumarins **21** in good yields via Knoevenagel condensation of 2-hydroxybenzaldehydes **18** with β-dicarbonyl compounds **35** in the presence of a recyclable ionic liquid 1,1,3,3-*N*,*N*,*N′*,*N′*-tetramethylguanidinium trifluoroacetate (TMGT) under thermal heating (**Figure 11**, Condition A) and/or microwave irradiation conditions (**Figure 11**, Condition B). 3-Substituted coumarins **21** are also synthesized from similar starting precursors using the 1,3-dimethylimidazolium methyl sulfate [MMIm][MSO_4] ionic liquid in the presence of L-proline as an additional promoter under heating condition (**Figure 11**, Condition C) [87].

Imidazolium based phosphinite ionic liquid (IL-$OPPh_2$) catalyzed synthesis of 3-substituted coumarin derivatives has been reported in literature; when *o*-hydroxy benzaldehydes **18** are treated with active methylene containing compounds **35** in the presence of IL-$OPPh_2$ catalyst at 60°C, 3-substituted coumarin derivatives are obtained in moderate to good yields (**Figure 12**) [88]. TSIL plays both the reaction media and catalyst as well.

Reactions of *o*-hydroxybenzaldehydes **18** with activated methylene compounds **35** catalyzed by Bronsted acid ionic liquid (BAIL) and 1-(4-sulfonic acid)butyl-3-methylimidazolium hydrogen sulfate [$(CH_2)_4SO_3HMIM$][HSO_4] in water lead to 3-substituted coumarin derivatives in good yields (**Figure 13**) [89].

Synthesis of substituted coumarins via Knoevenagel condensation using various organic catalysts such as piperidine, ammonia, L-lysine, L-proline, benzoic acid, etc. has been reported in literature and some are summarized in **Table 2**.

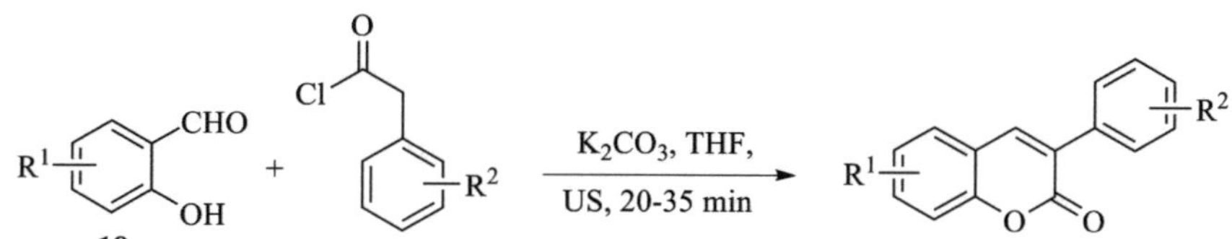

18 R^1 = H, Br, NO_2 **27** R^2 = H, NO_2 **28** (7-98%)

Figure 8.
Synthesis of 3-aryl coumarin derivatives.

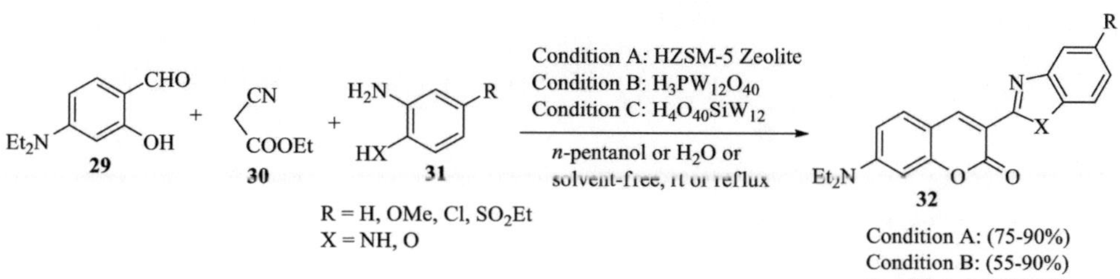

29 **30** **31** **32**

R = H, OMe, Cl, SO_2Et
X = NH, O

Condition A: (75-90%)
Condition B: (55-90%)
Condition C: (70-98%)

Figure 9.
Synthesis of coumarin-substituted benzimidazoles/benzoxazoles.

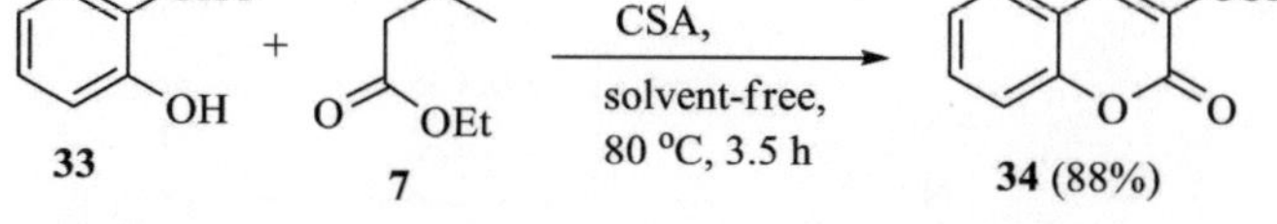

33 **7** **34** (88%)

Figure 10.
Synthesis of 3-acetyl coumarin.

It is quite evident that in **Table 2** several methodologies for the synthesis of substituted coumarins using different organic catalysts are established. Among these, L-proline-catalyzed reactions offer high yields (entry 3), which explains synthesis of 3-substituted coumarins by the condensation of *o*-hydroxybenzaldehydes with a variety of active methylene compounds catalyzed by 1,3-dimethyl imidazolium methyl sulfate [MMIm][MSO_4] and L-proline. Another L-proline-catalyzed synthesis of coumarins is known, but in that case, the yield is very poor (entry 4). Similar result is also observed under L-lysine-catalyzed synthesis of coumarins (entry 5).

A series of 3-phenyl substituted coumarin analogues have been achieved via a two-step process involving esterification using 1,1-carbonyldiimidazole (CDI) followed by condensation reaction in the presence of 1,8-diazabicyclo[5.4.0]undec-7-ene (DBU) under mild conditions (entry 1).

Microwave-assisted synthesis of coumarins is also known, which not only reduces the reaction time but also increases the yields of the products (entries 2, 6, and 7).

Benzocoumarin derivatives have been synthesized from 1-hydroxy-4-methyl-naphthalene-2-carbaldehyde and compounds containing active methylene group via piperidine-catalyzed Knoevenagel condensation reaction (entry 8). Moreover, benzothiazolyl coumarins with isothiocyanate functionality have been synthesized from commercially available 2-hydroxy-4-nitro benzoic acid in the presence of piperidine in ethanol (entry 9).

Application of sonochemistry for the synthesis of different coumarin derivatives is also useful due to better yield and shorter reaction time compared with the classical procedures (entry 10).

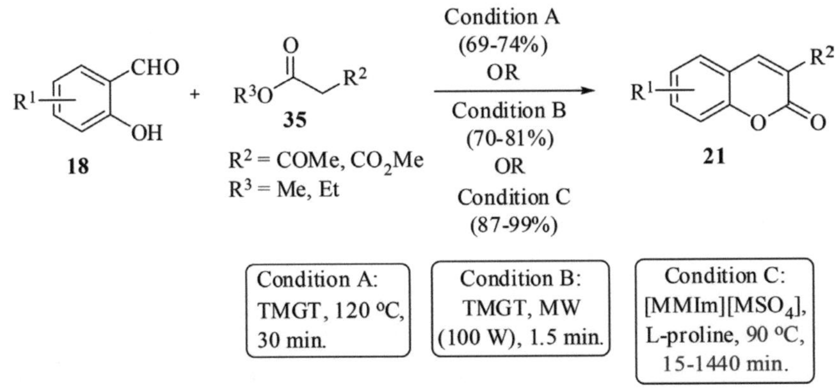

Figure 11.
Synthesis of 3-substituted coumarins.

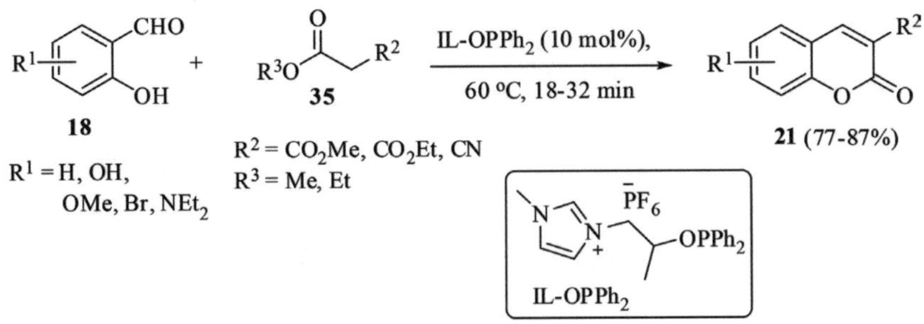

Figure 12.
Synthesis of 3-substituted coumarins.

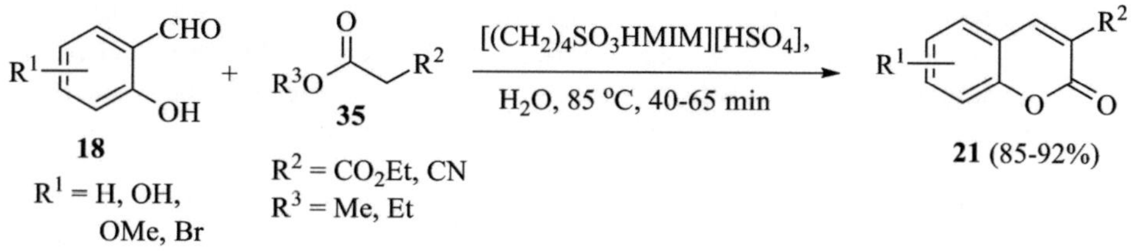

Figure 13.
Synthesis of 3-substituted coumarins.

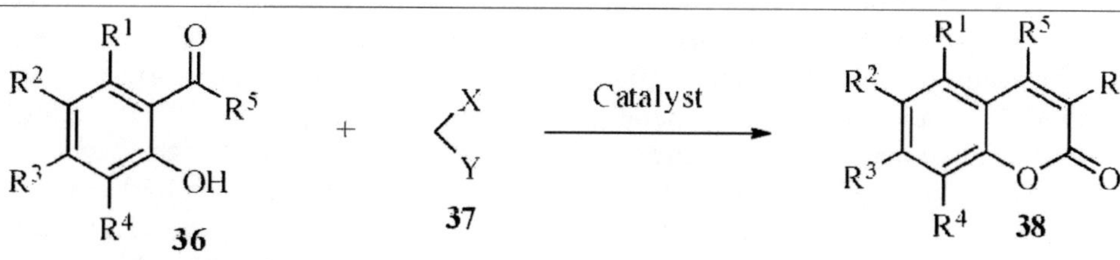

R^1 = H, OH, OMe, Cl, Br, NO_2,
R^2 = H, Me, OMe, CO_2Et, CN, Br, I
R^3 = H, Cl, Br, NEt_2, OH, OMe, NO_2
R^4 = Me, OMe, OEt, I
R^5 = H, Me

X = H, Me, CN, CO_2H, CO_2Me, CO_2Et, COMe, COPh, SO_2Me, SO_2Ph, $SO_2C_4F_9$, $CONH_2$
Y = CN, CO_2Me, CO_2Et, aryl
R' = CN, COMe, CO_2H, CO_2Et, aryl

Entry	Catalyst	Reaction conditions	Time	Yield (%)	Reference
1	CDI-DBU	(i) CDI (1.2 equiv.), DCM, rt. (ii) DBU (1.0 equiv.), DCM, rt	30 min 1–2 h	42–59	[90]
2	PhCOOH	Condition A: Polyphosphoric acid, MW (900 W), 100°C Condition B: H_2SO_4, Benzoic acid, MW (900 W), 90°C Condition C: benzoic acid, *n*-pentanol, MW (900°C), 110°C	4–6 min 3–4 min 3 min	60–75 58–75 85–95	[91]
3	L-proline	1,3-dimethyl imidazolium methyl sulfate, [MMIm] [MSO4], L-proline (1 equiv.), 90°C	15–1440 min	87–99	[87]
4	L-proline	L-proline (20 mmol%), EtOH, rt	15–20 h	54–76	[92]
5	L-lysine	L-lysine (20 mol%), H_2O, rt. –80°C	6 24 h	50–90	[93]
6	Piperidine	Piperidine (catalytic), rt.	20 min	84	[94]
7	Piperidine	Piperidine (2.0 mol%), solvent-free, MW (400 W)	1 min	50–97	[95]
8	Piperidine	Piperidine (1.48 equiv.), EtOH, reflux	30 min	85–92	[96]
9	Piperidine	Piperidine (catalytic), EtOH, reflux	2 h	82	[97]
10	Piperidine	Piperidine (1.0 equiv.), AcOH (2.5 mol%), EtOH, US, rt	5–30 min	49–90	[98]
11	Piperidine	Piperidine, EtOH, rt-reflux	1–2 h	82–92	[99]
12	Piperidine	Piperidine (7.4 equiv.), EtOH, reflux	2 h	92	[100]

Table 2.
Synthesis of substituted coumarins via Knoevenagel condensation reactions.

6,8-Diiodocoumarin derivatives have also been synthesized in good yields by Knoevenagel condensation using piperidine as catalyst (entry 11). The reaction of 3-ethoxysalicylaldehyde with ethyl acetoacetate in the presence of piperidine leads to 3-acetyl-8-ethoxycoumarin (entry 12).

2.3 Baylis-Hillman reaction

Baylis-Hillman strategy has been employed to the synthesis of substituted coumarins as shown in **Figure 14**. When 2-hydroxybenzaldehydes **18** are subjected to react with methyl acrylate **39a** (R^2 = Me) in the presence of DABCO (1,4-Diazabicyclo[2.2.2] octane), a mixture of chromenes **40** and coumarins **41** are formed [101, 102]. However, similar reactions of 2-hydroxybenzaldehydes **18** with tert-butyl acrylate **39b** (R^2 = ^{t}Bu) under classical method [103] and/or microwave irradiation [104] afford corresponding Baylis-Hillman adducts **42**, which undergo cyclization under reflux in AcOH yielding a mixture of 3-substituted chromene **43** and coumarin **44**. Treatment of the Baylis-Hillman adducts **42** with concentrated HCl in refluxing AcOH produces 3-(chloromethyl) coumarins **45** in excellent yields. Moreover, the reaction of **42** with HI under reflux in a mixture of Ac_2O and AcOH furnishes 3-methyl coumarins **46**, which upon further reaction with SeO_2 affords the corresponding 3-formyl coumarins **47**.

The suggested mechanism for the formation of the coumarin derivatives **44/45/46** is shown in **Figure 15**.

Kaye et al. have also demonstrated the synthesis of substituted coumarins employing Baylis-Hillman strategy in different ways as shown in **Figure 16** [105, 106].

2.4 Kostanecki reaction

4-Arylcoumarins **59** have been synthesized in good yields employing Kostanecki reaction between 2-hydroxybenzophenones **57** and acetic anhydride **58** in the presence of DBU under mild condition (**Figure 17**) [107].

The mechanism of the Kostanecki reaction is outlined in **Figure 18**.

Similarly, 3,4-disubstituted coumarins **65** are isolated from readily available 2-acyloxybenzophenones **64** under Kostanecki reaction conditions (**Figure 19**) [107].

2.5 Michael addition reaction

Michael addition could be applied [108] to the synthesis of 3-aroylcoumarins **68** in good yields from easily available 2-hydroxybenzaldehydes **66** and α-aroylketene dithioacetals (AKDTAs) **67** in the presence of a catalytic amount of piperidine in refluxing THF (**Figure 20**).

The reaction proceeds via initial Michael addition followed by intramolecular aldol condensation reaction as depicted in **Figure 21**.

2.6 Wittig reaction

Kumar and coworkers [109] have reported the synthesis of substituted coumarins **3** from phenolic compounds **23** containing ortho-carbonyl group and triphenyl (α-carboxymethylene)phosphorane imidazole ylide **73** via intramolecular Wittig cyclization in good yields (**Figure 22**). All the reactions proceed via formation of the phosphorane intermediates **74** as established by spectroscopic results.

2.7 Vinyl phosphonium salt-mediated electrophilic substitution reaction

A series of 4-carboxy(ethyl/methyl) coumarins **76** have been synthesized in good yields from substituted phenols **1** and di(ethyl/methyl)

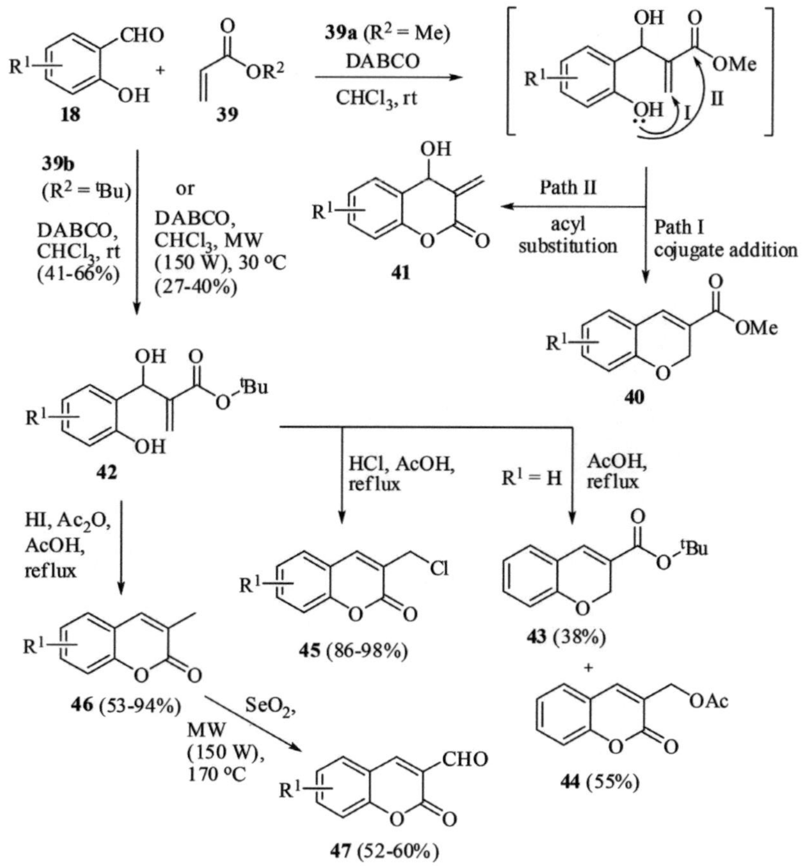

Figure 14.
Synthesis of 3-substituted coumarins.

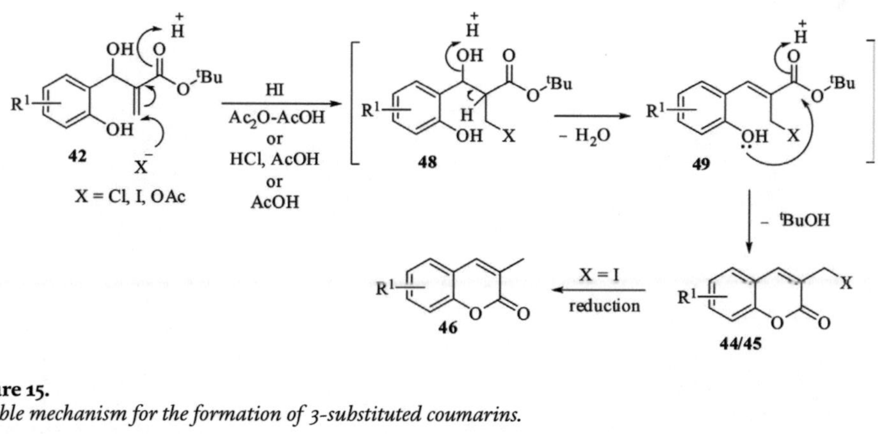

Figure 15.
Possible mechanism for the formation of 3-substituted coumarins.

acetylene-dicarboxylate **75** in the presence of phosphinite ionic liquid (IL-$OPPh_2$) under solvent-free microwave irradiation conditions (**Figure 23**) [110]. It is noticed that the diphenylphosphine group in ionic liquid accelerates the reaction.

The proposed mechanism for the formation of coumarins **76** via vinyl phosphonium salt-mediated electrophilic substitution is shown in **Figure 24**.

4-Carboxymethyl coumarins **82** have been synthesized by Yavari et al. [111] in moderate to excellent yields from the reactions of substituted phenols **1** and dimethyl acetylenedicarboxylate (DMAD) **81** in the presence of triphenylphosphine (**Figure 25**) via vinyl triphenylphosphonium salt-mediated aromatic electrophilic

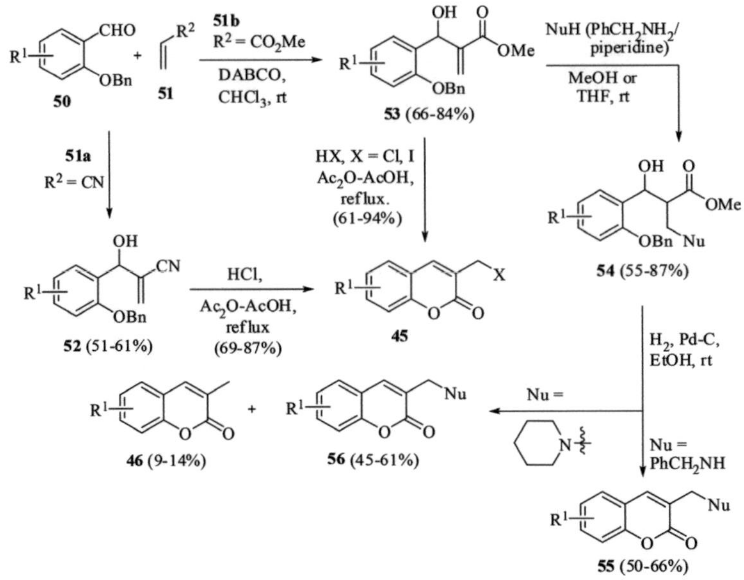

Figure 16.
Synthesis of 3-substituted coumarins.

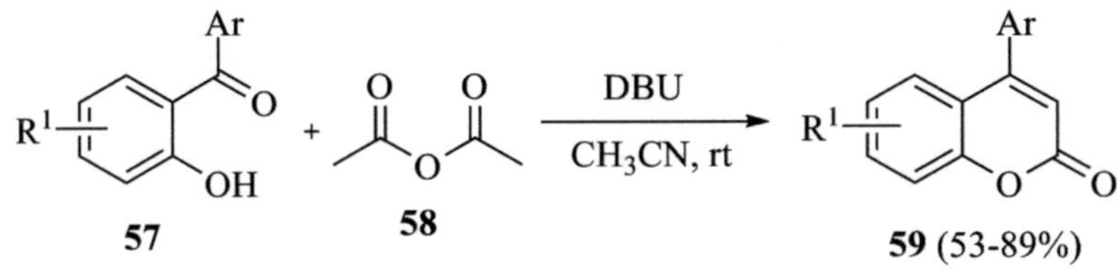

Figure 17.
Synthesis of 4-arylcoumarins.

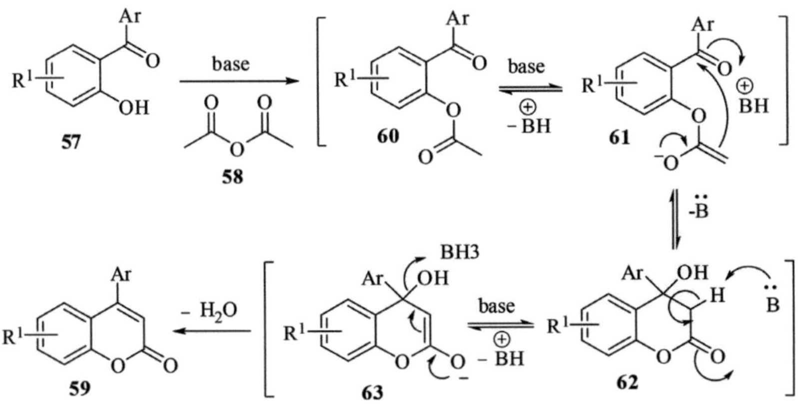

Figure 18.
Mechanism for Kostanecki reaction.

substitution reaction as mentioned in **Figure 24**. Similar results are found from the given starting materials under microwave irradiation in shorter reaction time [112].

However, reactions of di- and trihydric phenols with dimethyl acetylenedicarboxylate (DMAD) in the presence of triphenylphosphine in toluene under reflux afford polyfunctionalized coumarin analogues along with unwanted by-products in appreciable amount (**Figure 26**) [113].

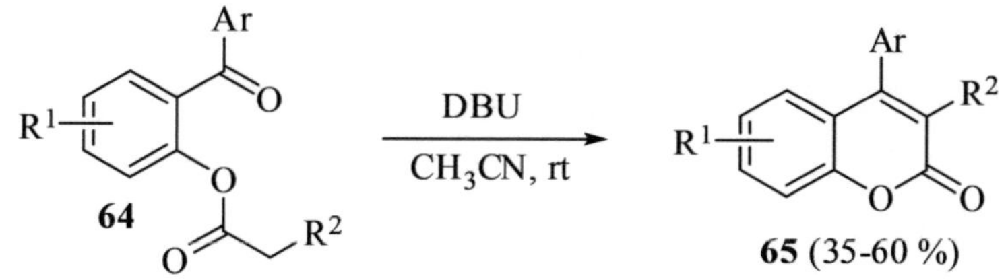

Figure 19.
Synthesis of 3,4-disubstituted coumarins.

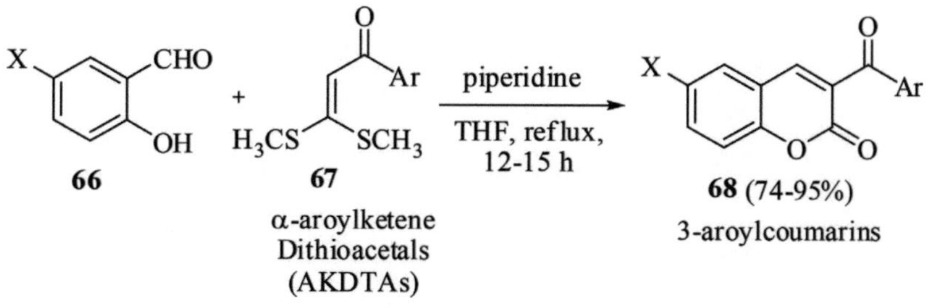

Figure 20.
Synthesis of 3-aroylcoumarins.

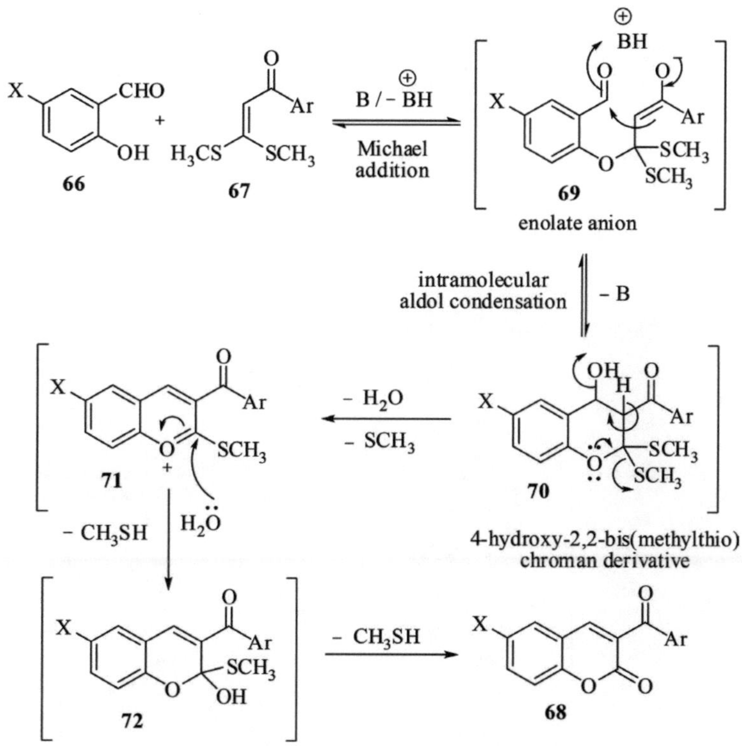

Figure 21.
Probable mechanism for the formation of 3-aroylcoumarins.

Similar reactions of 2-hydroxybenzaldehydes **18** with di(ethyl/methyl)acetylenedicarboxylates **75** leads to the corresponding 4-carboxy(ethyl/methyl)-8-formyl coumarins **93** in moderate to good yields (**Figure 27**) [114].

The methodology has also been employed to the synthesis of angular pyridocoumarins **97/98** and benzo-fused 6-azacoumarin **100** as shown in **Figure 28** [115].

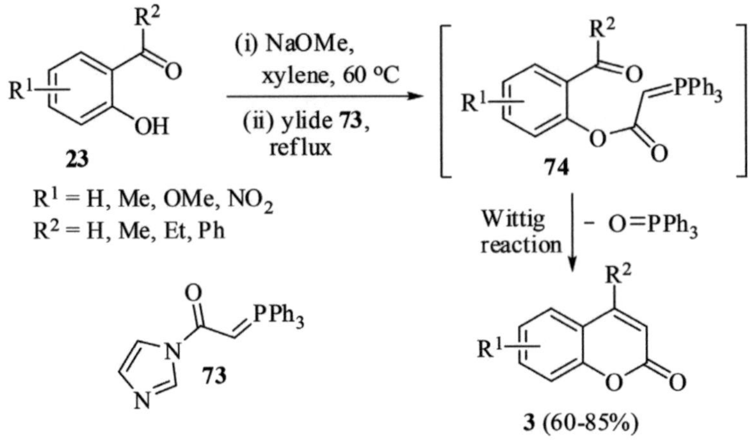

Figure 22.
Synthesis of substituted coumarins.

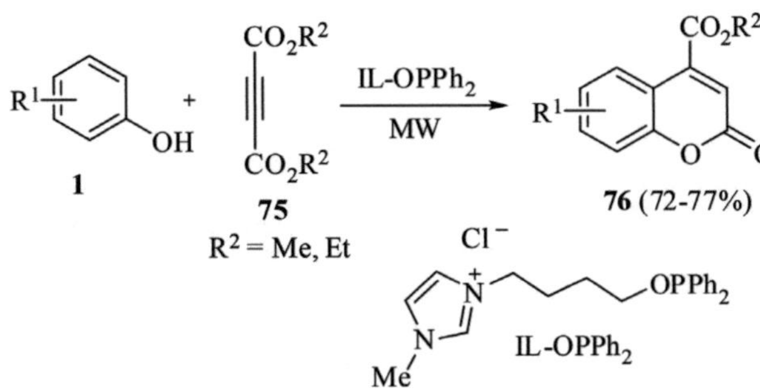

Figure 23.
Synthesis of 4-carboxy(ethyl/methyl) coumarins.

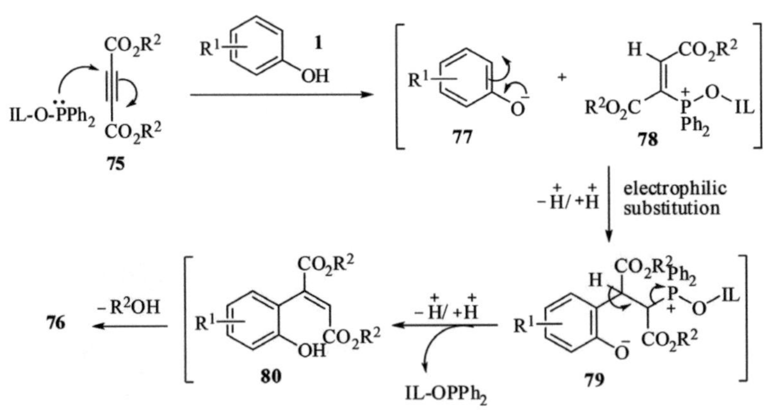

Figure 24.
Proposed mechanism for the synthesis of substituted coumarins via vinyl phosphonium salt-mediated electrophilic substitution.

2.8 Palladium-catalyzed reactions

Palladium-catalyzed reactions between substituted phenols **101** and ethyl propiolates **102** lead to substituted coumarins **103/104** (**Figure 29**) [116, 117].

Unsymmetrical monohydric phenols having *m*-OMe or *m*-Me substituent as respectively in 3-methoxyphenol and *m*-cresol show regioselectivity toward the

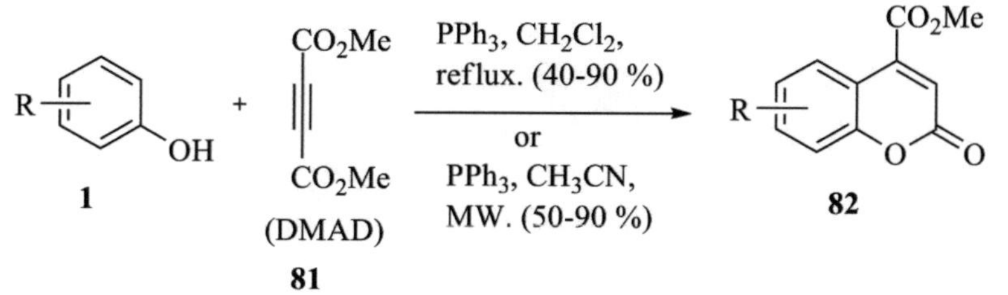

Figure 25.
Synthesis of 4-carboxymethyl coumarins.

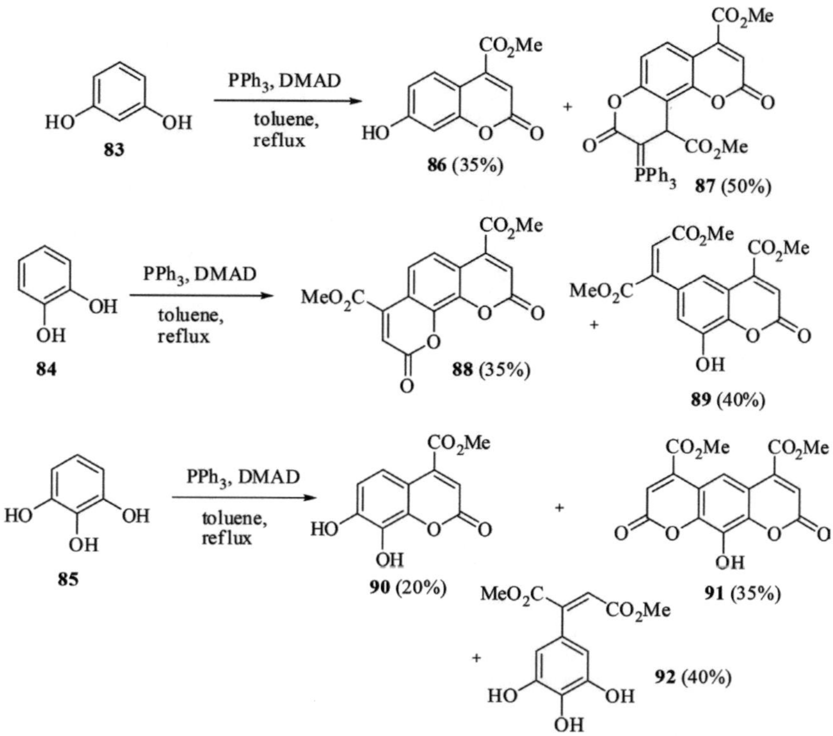

Figure 26.
Synthesis of polyfunctionalized coumarin analogues.

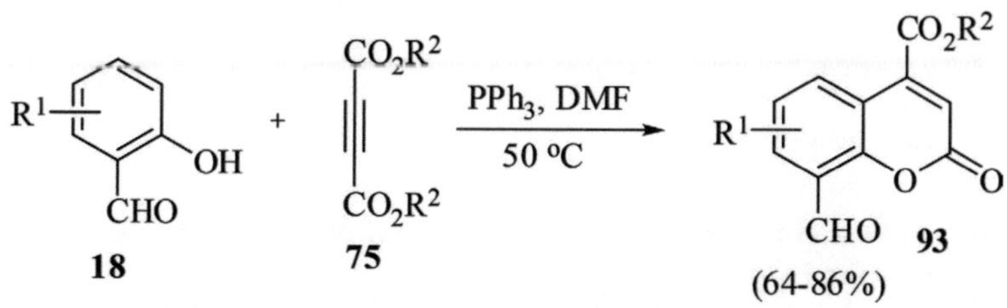

Figure 27.
Synthesis of 4-carboxy(ethyl/methyl)-8-formyl coumarins.

formation of a new bond in coumarins, which occurs at the *para* position to the methoxy group, and therefore, the regioisomers **103** are found to be formed predominantly over **104**. However, symmetrical dihydric phenol with OMe substituent like that in 5-methoxybenzene-1,3-diol affords the regioisomer **104** predominantly over **103** under the reaction condition applied. This may be due to the steric effects

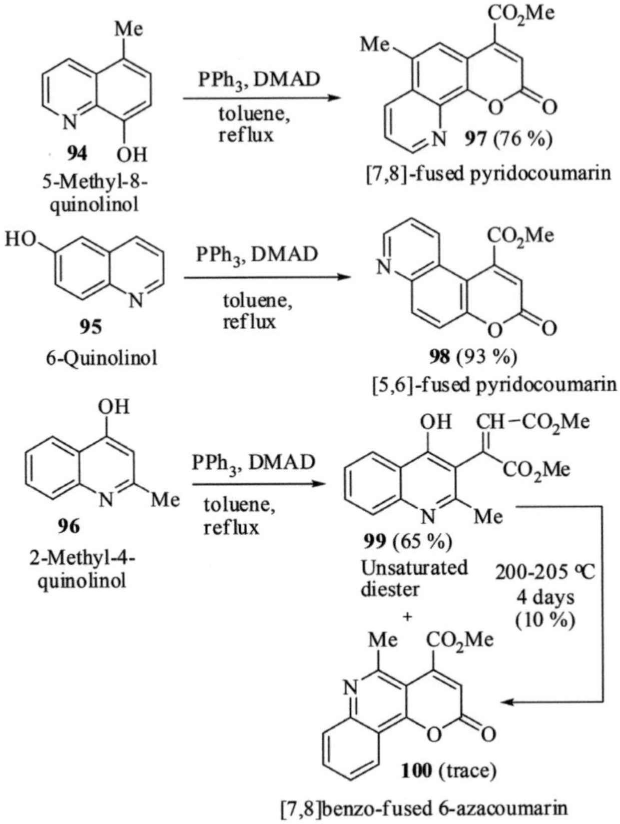

Figure 28.
Synthesis of pyridocoumarins and benzo-fused azacoumarin.

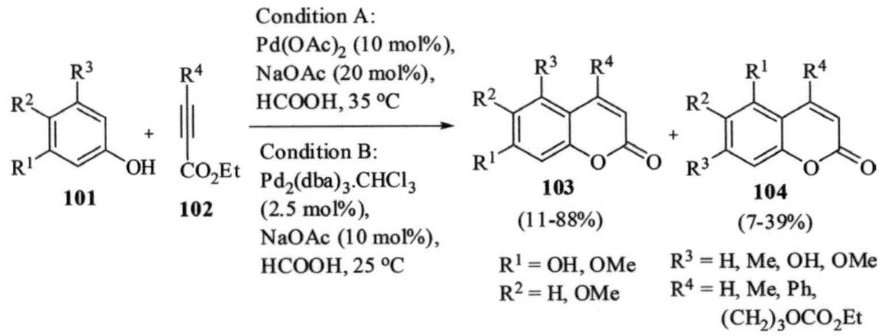

Figure 29.
Synthesis of substituted coumarins.

of the R^4 group of ethyl propiolate **102**, which dominates over the electronic effect of the methoxy group of the phenol.

A proposed mechanism for the formation of coumarins **103/104** is shown in **Figure 30**.

Substituted coumarins **3** have been synthesized in moderate yields (42–69%) via $Pd(OAc)_2$-catalyzed reaction of substituted phenols **1** with substituted propiolic acid **110** (R^3 = CO_2H) in TFA under mild conditions (**Figure 31**, Condition A) [118]. However, a mixture of catalysts $FeCl_3$ and AgOTf showed better catalytic efficiency toward yields (60–93%) of coumarin derivatives **3** (**Figure 31**, Condition B). Propiolic acid ester **110** (R^3 = CO_2Et) also furnishes the desired products **3** upon

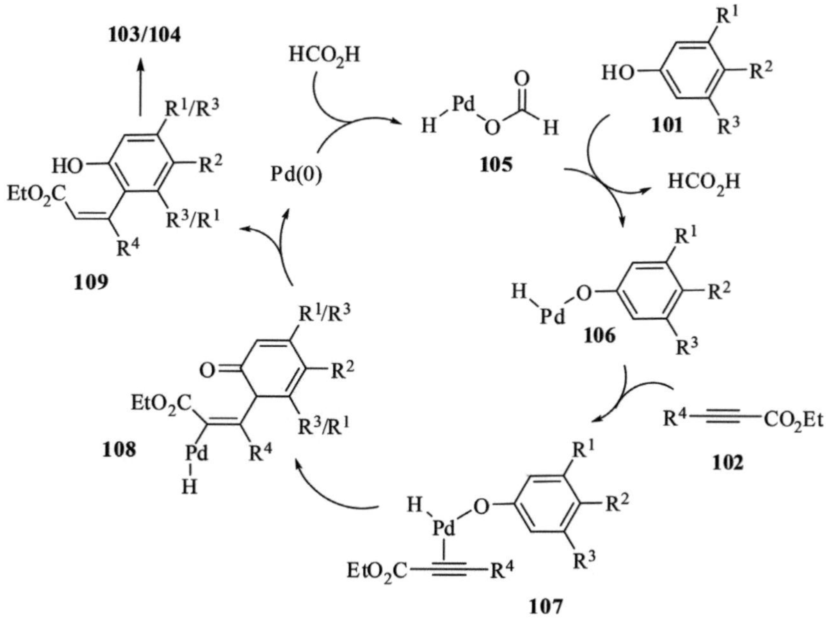

Figure 30.
Possible mechanism for Pd-catalyzed synthesis of coumarins.

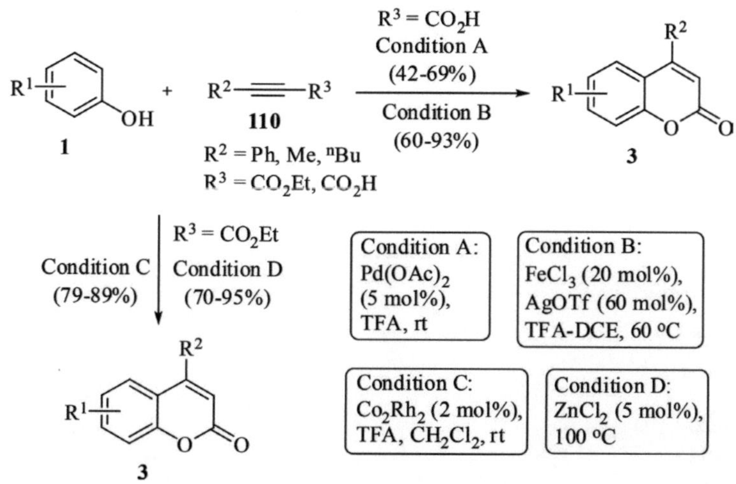

Figure 31.
Synthesis of substituted coumarins.

reactions with substituted phenols **1** under specified conditions as provided in **Figure 31** (Conditions C and D) [119–121].

4,6-Disubstituted coumarins **113** have been achieved employing palladium-catalyzed tandem Heck-lactonization of the *Z*- or *E*-enoates **112** with *o*-iodophenols **111** (**Figure 32**, Conditions A, B, and C) [122, 123].

For Heck-lactonization, the enoate *Z*-**112a** is found to be more reactive than its *E*-isomer, leading to the corresponding coumarin **113** in good yields (68–84%) under all reaction conditions studied. The enoate *Z*-**112b** leads to coumarin derivative **113** in relatively lower yields (42–56%), which may be due to the presence of the bulky ^{t}Bu ester group that hampers the lactonization step. Moreover, the reactivity of *E*-enoates depends on the β-substituent. *E*-enoates **112c** (R^2 = CH_2CHMe_2,

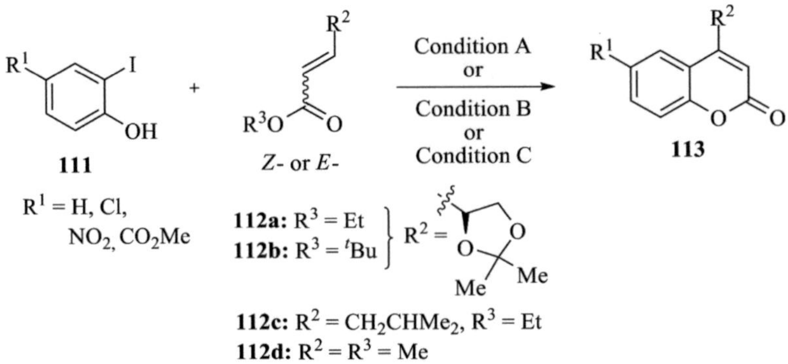

Condition A: 10 mol% of $Pd(OAc)_2$, 3 equiv Et_3N, H_2O, 80 °C, 40 h.
Condition B: 10 mol% of $PdCl_2$, 3 equiv Et_3N, H_2O, 80 °C, 40 h.
Condition C: 10 mol% of $Pd(OAc)_2$, in the presence or absence of 20 mol% of PPh_3, 3 equiv $AgCO_3$, acetone, reflux, 40 h.

Figure 32.
Synthesis of 4,6-disubstituted coumarins.

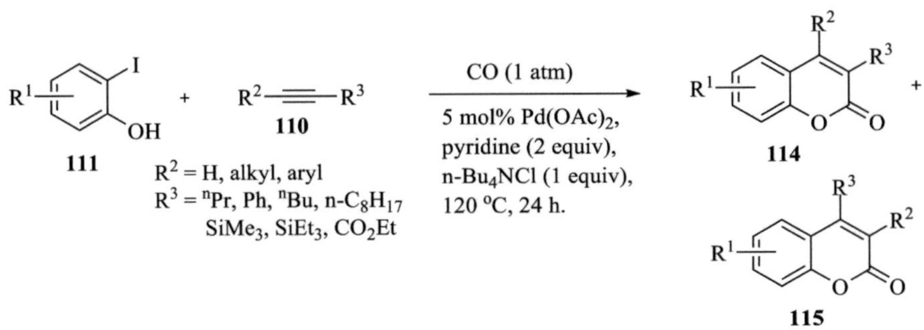

Figure 33.
Synthesis of 3, and 4-substituted and 3,4-disubstituted coumarins.

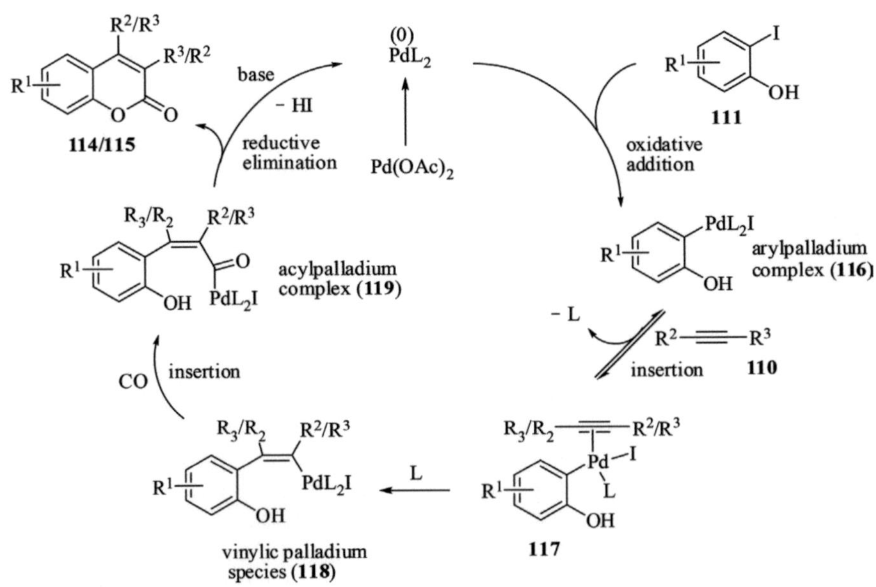

Figure 34.
Possible mechanism for the synthesis of coumarins via carbonylative annulation.

R^3 = CH_3) and **112d** (R^2 = R^3 = CH_3) having CH_2CHMe_2 and CH_3 group, respectively, at the β-carbon, and their double bonds are therefore less sterically hindered than that in *E*-enoate **112a.** This reduced hindering is a major factor for the higher reactivity of *E*-enoates **112c** and **112d** than *E*-enoate **112a**.

Palladium-catalyzed carbonylative annulation of terminal alkynes **110** (R^2 = H; R^3 = ^{n}Pr, Ph, $SiMe_3$, $SiEt_3$, CO_2Et, etc.) with *o*-iodophenols **111** affords 3-substituted coumarins **114** (R^2 = H) in poor yields (18–36%) (**Figure 33**) [124]. On the other hand, both 3- and 4-substituted coumarins **114** (R^2 = H) and **115** (R^2 = H) have been synthesized from *o*-iodophenols **111** and terminal alkynes **110** (R^2 = H; R^3 = $^nC_4H_9$, $^nC_8H_{17}$) bearing long alkyl chain. In addition, a wide variety of 3,4-disubstituted coumarins **114/115** (R^2, $R^3 \neq$ H) have also been achieved in moderate to good yields (43–78%) via carbonylative annulation between *o*-iodophenols **111** and internal alkynes **110** (R^2, $R^3 \neq$ H) [125].

The suggested mechanism of the carbonylative annulation is presented in **Figure 34**. The carbonylative annulation process is believed to proceed via (a) oxidative addition of *o*-iodophenol **111** to Pd(0), (b) insertion of alkyne **110** into the aryl-palladium complex **116**, (c) CO insertion into the resulting vinylic palladium species **118**, and (d) nucleophilic attack of the phenolic oxygen on the carbonyl carbon of the acylpalladium complex **119** with simultaneous regeneration of the Pd(0) catalyst.

3,4-Disubstituted coumarins **121** are also isolated in good to excellent yields from readily available 2-(1-hydroxyprop-2-ynyl)phenols **120** via palladium-catalyzed

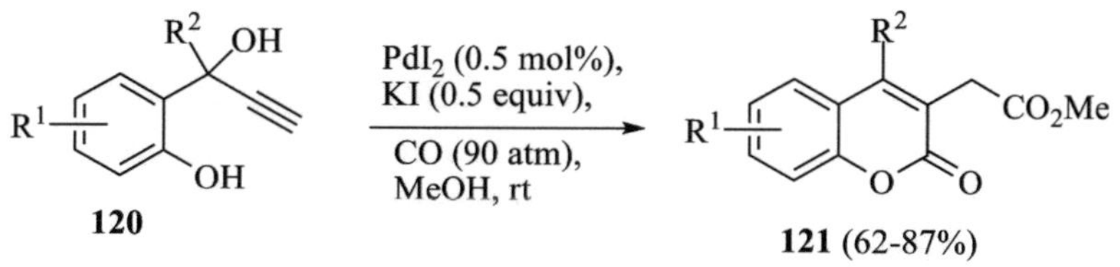

Figure 35.
Synthesis of 3,4-disubstituted coumarins.

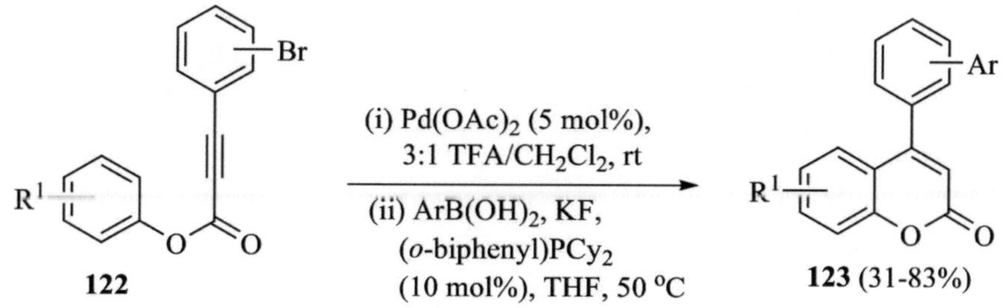

Figure 36.
Synthesis of 4-arylcoumarins.

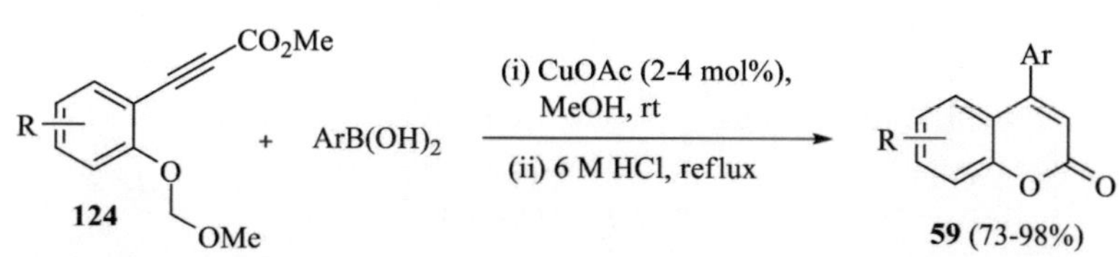

Figure 37.
Synthesis of 4-arylcoumarins.

dicarbonylation process in the presence of KI in MeOH at room temperature (**Figure 35**) [126].

Furthermore, electrophilic palladium-catalyzed cycloisomerization of brominated arylpropiolates **122** followed by Suzuki coupling with arylboronic acids furnishes 4-arylcoumarins **123** in moderate to good yields (**Figure 36**) [127]. This strongly suggests that a single loading of catalyst $Pd(OAc)_2$ could be used to conduct sequential reactions for the synthesis of substituted coumarins.

2.9 Other methods

CuOAc-catalyzed hydroarylation of methyl phenylpropiolates **124** having a methoxy methyl (MOM)-protected hydroxyl group at the ortho-position with various arylboronic acids followed by acidic workup leads to 4-arylcoumarins **59** in good to excellent yields (**Figure 37**) [128].

Substituted coumarins **126** are obtained in moderate to excellent yields by $Yb(OTf)_3$-catalyzed reactions of substituted phenols **1** with alkylidene Meldrum's acid **125** in CH_3NO_2 at 100°C (**Figure 38**) [129].

A series of 3-alkylcoumarins **128** are obtained in moderate yields from 2-hydroxybenzaldehydes **18** and α,β-unsaturated aldehydes **127** via generation of *N*-heterocyclic carbenes (NHC) in ionic liquid under conventional heating (**Figure 39**, Condition A) and/or microwave irradiation conditions (**Figure 39**, Condition B) [130].

3-Benzoylcoumarins **130/131** and coumarin-3-carbaldehydes **47** have also been isolated in moderate to good yields from the reactions of 2-hydroxybenzaldehydes **18/19** with phenylpropionyl chloride **129a** and/or propionyl chloride **129b** under esterification conditions (**Figure 40**) [131].

An electrochemical method has been developed for the synthesis of 6*H*-benzo[*c*]chromen-6-ones **133** in good to excellent yields from biphenyl-2-carboxylic acids **132** via radical arene carbon–oxygen bond formation reaction (**Figure 41**) [132].

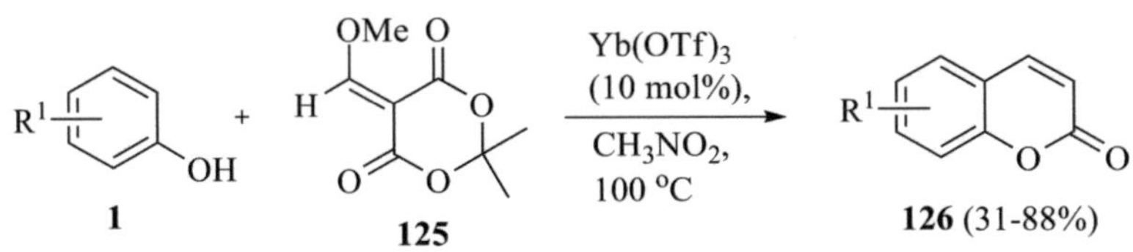

Figure 38.
Synthesis of substituted coumarins.

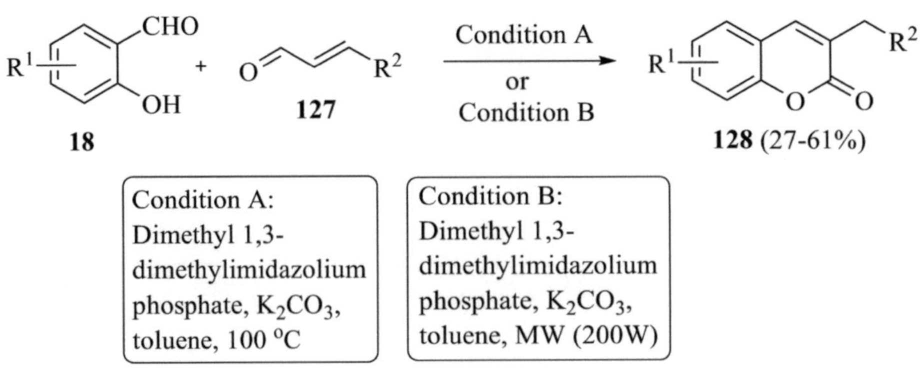

Figure 39.
Synthesis of 3-alkylcoumarins.

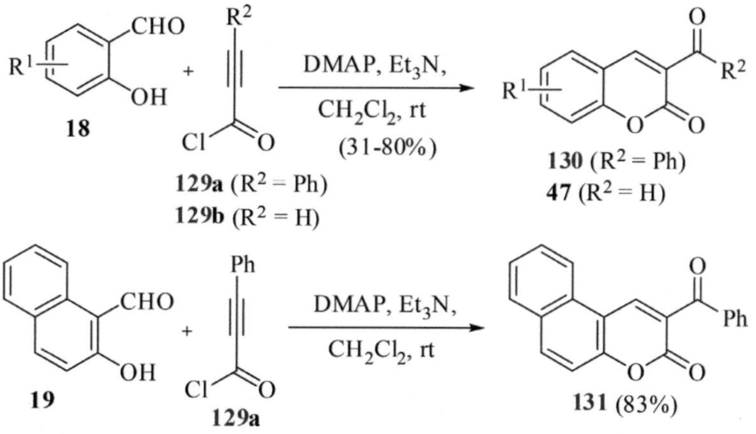

Figure 40.
Synthesis of 3-benzoyl coumarins and coumarin-3-carbaldehyde.

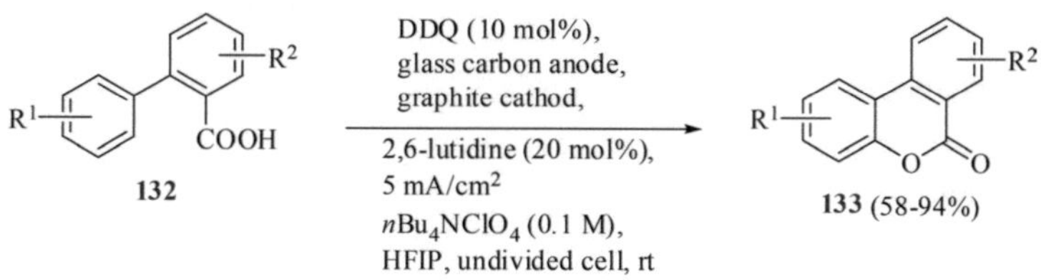

Figure 41.
Synthesis of 6H-benzo[c]chromen-6-ones.

The method involves DDQ as a redox mediator, inexpensive glassy carbon electrodes to facilitate an intramolecular lactonization of biphenyl-2-carboxylic acid derivatives, and 2,6-lutidine as an additive, in 0.1 M $^{n}Bu_4NClO_4$ electrolyte mixture of 1,1,1,3,3,3-hexafluoropropan-2-ol (HFIP).

3. Concluding remarks

In this chapter, we have discussed a plethora of methods for the one-pot synthesis of coumarin derivatives and their advantages and/or demerits compared to other methods. Both the Pechmann as well as Knoevenagel condensation reactions under microwave and/or ultrasound irradiation conditions, and catalyzed by ionic liquids and/or solid acids have several advantages including high products yields, diminutive reaction times, ease of isolation of products, recycle of catalysts, and green aspects by avoiding toxic catalysts and solvents. Chemo- and regioselective syntheses of 3-substituted coumarins have been reported via Baylis-Hillman reactions under mild conditions. On the other hand, vinyl phosphonium salt-mediated electrophilic substitution reactions of phenols afford 4-carboxyalkyl coumarin derivatives in good yields under neutral conditions. This method offers significant advantages for the synthesis of coumarins having acid sensitive functional groups. In contrast, the most widely used method von Pechmann condensation requires acidic conditions. Moreover, palladium-catalyzed Heck lactonization protocol has been employed for the regioselective synthesis of coumarin derivatives from *o*-iodophenols and enoates. It is revealed that this reaction is sensitive to steric hindrance around the double bound in the enoates. Regioselective synthesis of 3,4-disubstituted coumarins achieved from substituted 2-iodophenols and alkynes containing different substituents via palladium-catalyzed carbonylative annulative process is sensitive to the steric bulk of the alkynes, and alkynes bearing tertiary alkyl substituents generally fail to undergo annulation. Unsymmetrical alkynes produce mixtures of regioisomers with generally only modest selectivity. Kostanecki reaction protocol furnishes a notable improvement in reaction conditions for coumarin synthesis and gives rise to the advantage of its synthetic capability, especially for highly functionalized 4-arylcoumarins with structural diversity.

Acknowledgements

Dr. I. Ansary and Dr. A. Taher highly acknowledge the Department of Chemistry (Burdwan University) and Burdwan Raj College, respectively, for infrastructural facilities.

Conflict of interest

The authors declare no conflict of interest.

Author details

Inul Ansary[1*] and Abu Taher[2,3]

1 Department of Chemistry, The University of Burdwan, Burdwan, West Bengal, India

2 Burdwan Raj College, Burdwan, West Bengal, India

3 Department of Chemistry, Bankura University, Bankura, West Bengal, India

*Address all correspondence to: iansary@chem.buruniv.ac.in

References

[1] Kumar S, Saini A, Sandhu JS. LiBr-mediated, solvent free von Pechmann reaction: Facile and efficient method for the synthesis of 2H-chromen-2-ones. ARKIVOC. 2007;**2007**:18-23. DOI: 10.3998/ark.5550190.0008.f03

[2] Zembower DE, Liao S, Flavin MT, Xu ZQ, Stup TL, Buckheit RW, et al. Structural analogues of the calanolide anti-HIV agents. Modification of the trans-10,11-dimethyldihydropyran-12-ol ring (ring C). Journal of Medicinal Chemistry. 1997;**40**:1005-1017. DOI: 10.1021/jm960355m

[3] Završnik D, Muratović S, Makuc D, Plavec J, Cetina M, Nagl A, et al. Benzylidene-bis-(4-hydroxycoumarin) and benzopyrano-coumarin derivatives: Synthesis, $^{1}H/^{13}C$-NMR conformational and X-ray crystal structure studies and in vitro antiviral activity evaluations. Molecules. 2011;**16**:6023-6040. DOI: 10.3390/molecules16076023

[4] Nawrot-Modranka J, Nawrot E, Graczy K. In vivo antitumor, in vitro antibacterial activity and alkylating properties of phosphorohydrazine derivatives of coumarin and chromone. European Journal of Medicinal Chemistry. 2006;**41**:1301-1309. DOI: 10.1016/j.ejmech.2006.06.004

[5] El-Saghier AMM, Naili MB, Rammash BK, Saleh NA, Kreddan KM. Synthesis and antibacterial activity of some new fused chromenes. ARKIVOC. 2007;**2007**:83-91. DOI: 10.3998/ark.5550190.0008.g09

[6] Creaven BS, Egan DA, Kavanagh K, McCann M, Mahon M, Noble A, et al. Synthesis and antimicrobial activity of copper(II) and silver(I) complexes of hydroxynitrocoumarins: X-ray crystal structures of [Cu(hnc)2(H2O)2].2H2O and [Ag(hnc)] (hncH = 4-hydroxy-3-nitro-2H-chromen-2-one). Polyhedron. 2005;**24**:949-957. DOI: 10.1016/j.poly.2005.03.006

[7] Golfakhrabadi F, Abdollahi M, Ardakani MRS, Saeidnia S, Akbarzadeh T, Ahmadabadi AN, et al. Anticoagulant activity of isolated coumarins (suberosin and suberenol) and toxicity evaluation of *Ferulago carduchorum* in rats. Pharmaceutical Biology. 2014;**52**:1335-1340. DOI: 10.3109/13880209.2014.892140

[8] Ronad P, Dharbamalla S, Hunshal R, Maddi V. Synthesis of novel substituted 7-(benzylideneamino)-4-methyl-2H-chromen-2-one derivatives as anti-inflammatory and analgesic agents. Archiv der Pharmazie: Chemistry in Life Sciences. 2008;**341**:696-700. DOI: 10.1002/ardp.200800057

[9] Lin CM, Huang ST, Lee FW, Sawkuo H, Lin MH. 6-Acyl-4-aryl/alkyl-5,7-dihydroxycoumarins as anti-inflammatory agents. Bioorganic & Medicinal Chemistry. 2006;**14**:4402-4409. DOI: 10.1016/j.bmc.2006.02.042

[10] Wang CJ, Hsieh YJ, Chu CY, Lin YL, Tseng TH. Inhibition of cell cycle progression in human leukemia HL-60 cells by esculetin. Cancer Letters. 2002;**183**:163-168. DOI: 10.1016/S0304-3835(02)00031-9

[11] Dexeus FH, Logothetis CJ, Sella A, Fitz K, Amato R, Reuben JM, et al. Phase II study of coumarin and cimetidine in patients with metastatic renal cell carcinoma. Journal of Clinical Oncology. 1990;**8**:325-329. DOI: 10.1200/JCO.1990.8.2.325

[12] Bhat MA, Siddiqui N, Khan SA. Synthesis of novel thioureido derivatives of sulfonamides and thiosemicarbazido derivatives of coumarin as potential anticonvulsant and analgesic agents. Indian Journal of Pharmaceutical

Sciences. 2006;**68**:120-124. DOI: 10.4103/0250-474X.22984

[13] Tyagi YK, Kumar A, Raj HG, Vohra P, Gupta G, Kumari R, et al. Synthesis of novel amino and acetyl amino-4-methylcoumarins and evaluation of their antioxidant activity. European Journal of Medicinal Chemistry. 2005;**40**:413-420. DOI: 10.1016/j.ejmech.2004.09.002

[14] Sardari S, Mori Y, Horita K, Micetich RG, Nishibe S, Daneshtalab M. Synthesis and antifungal activity of coumarins and angular furanocoumarins. Bioorganic & Medicinal Chemistry. 1999;7:1933-1940. DOI: 10.1016/S0968-0896(99)00138-8

[15] Stein AC, Alvarez S, Avancini C, Zacchino S, von Poser G. Antifungal activity of some coumarins obtained from species of Pterocaulon (Asteraceae). Journal of Ethnopharmacology. 2006;**107**:95-98. DOI: 10.1016/j.jep.2006.02.009

[16] Chen Y, Zhang Q, Zhang B, Xia P, Xia Y, Yang Z-Y, et al. Anti-AIDS agents. Part 56: Synthesis and anti-HIV activity of 7-thia-di-*O*-(−)-camphanoyl-(+)-*cis*-khellactone (7-thia-DCK) analogs. Bioorganic & Medicinal Chemistry. 2004;**12**:6383-6387. DOI: 10.1016/j.bmc.2004.09.038

[17] Cravotto G, Nano GM, Palmisano G, Tagliapietra S. An asymmetric approach to coumarin anticoagulants via hetero-Diels–Alder cycloaddition. Tetrahedron: Asymmetry. 2001;**12**:707-709. DOI: 10.1016/S0957-4166(01)00124-0

[18] Fan GJ, Mar W, Park MK, Choi EW, Kim K, Kim S. A novel class of inhibitors for steroid 5α-reductase: Synthesis and evaluation of umbelliferone derivatives. Bioorganic & Medicinal Chemistry Letters. 2001;**11**:2361-2363. DOI: 10.1016/S0960-894X(01)00429-2

[19] Zahradnik M. The Production and Application of Fluorescent Brightening Agents. Chichester: John Wiley & Sons; 1992

[20] Hepworth JD, Gabbutt CD, Heron BM. In: Katritzky AR, Rees CW, Scriven EFV, editors. Comprehensive Heterocyclic Chemistry II. Vol. 5. Oxford: Pergamon Press; 1996. pp. 301-350

[21] Green GR, Evans JM, Vong AK. In: Katritzky AR, Rees CW, Scriven EFV, editors. Comprehensive Heterocyclic Chemistry II. Vol. 5. Oxford: Pergamon Press; 1996. pp. 469-500

[22] O'Kennedy R, Coumarins TRD. Biology, Applications and Mode of Action. Chichester: John Wiley & Sons; 1997

[23] Meuly WC, Kirk-Othmer. Encyclopedia of Chemical Technology. 3rd ed. Vol. 7. New York: John Wiley & Sons; 1979. p. 196

[24] von Pechmann H, Duisberg C. Ueber die Verbindungen der Phenole mit Acetessigäther. Berichte der Deutschen Chemischen Gesellschaft. 1883;**16**:2119-2128. DOI: 10.1002/cber.188301602117

[25] von Pechmann H, Duisberg C. Neue Bildungsweise der Cumarine. Synthese des Daphnetins. I. Berichte der Deutschen Chemischen Gesellschaft. 1884;**17**:929-979. DOI: 10.1002/cber.188401701248

[26] Panetta A, Rapoport H. New syntheses of coumarins. The Journal of Organic Chemistry. 1982;**47**:946-950. DOI: 10.1021/jo00345a009

[27] Miyano M, Dorn CR. Mirestrol. I. Preparation of the tricyclic intermediate. The Journal of Organic Chemistry. 1972;**37**:259-268. DOI: 10.1021/jo00967a017

[28] Woods LL, Sapp J. A new one-step synthesis of substituted coumarins. The Journal of Organic Chemistry. 1962;**27**:3703-3705. DOI: 10.1021/jo01057a519

[29] Sethna SM, Shah NM, Shah RC. Aluminium chloride, a new reagent for the condensation of β-ketonic esters with phenols. Part I. The condensations of methyl β-resorcylate, β-resorcylic acid, and resacetophenone with ethyl acetoacetate. Journal of the Chemical Society. 1938:228-232. DOI: 10.1039/JR9380000228

[30] Corrie JET. A convenient synthesis of N-(7-dimethylamino-4-methylcoumarin-3-yl)-maleimide incorporating a novel variant of the Pechmann reaction. Journal of the Chemical Society. Perkin Transactions. 1990;(7):2151-2152. DOI: 10.1039/P19900002151

[31] Smitha G, Reddy SC. $ZrCl_4$-catalyzed Pechmann reaction: Synthesis of coumarins under solvent-free conditions. Synthetic Communications. 2004;**34**:3997-4003. DOI: 10.1081/SCC-200034821

[32] Valizadeh H, Shockravi A. An efficient procedure for the synthesis of coumarin derivatives using $TiCl_4$ as catalyst under solvent-free conditions. Tetrahedron Letters. 2005;**46**:3501-3503. DOI: 10.1016/j.tetlet.2005.03.124

[33] John EVO, Israelstam SS. Notes. Use of cation exchange resins in organic reactions. I. The Von Pechmann reaction. The Journal of Organic Chemistry. 1961;**26**:240-242. DOI: 10.1021/jo01060a602

[34] Laufer MC, Hausmann H, Hölderich WF. Synthesis of 7-hydroxycoumarins by Pechmann reaction using Nafion resin/silica nanocomposites as catalysts. Journal of Catalysis. 2003;**218**:315-320. DOI: 10.1016/S0021-9517(03)00073-3

[35] Hoefnagel AJ, Gunnewegh EA, Downing RS, van Bekkum H. Synthesis of 7-hydroxycoumarins catalysed by solid acid catalysts. Journal of the Chemical Society, Chemical Communications. 1995:225-226. DOI: 10.1039/C39950000225

[36] Chavan SP, Shivasankar K, Sivappa R, Kale R. Zinc mediated transesterification of β-ketoesters and coumarin synthesis. Tetrahedron Letters. 2002;**43**:8583-8586. DOI: 10.1016/S0040-4039(02)02006-3

[37] Singh V, Singh J, Kaur KP, Kad GL. Acceleration of the Pechmann reaction by microwave irradiation: Application to the preparation of coumarins. Journal of Chemical Research, Synopses. 1997:58-59. DOI: 10.1039/A605672E

[38] Maheswara M, Siddaiah V, Damu GLV, Rao YK, Rao CV. A solvent-free synthesis of coumarins via Pechmann condensation using heterogeneous catalyst. Journal of Molecular Catalysis A: Chemical. 2006;**255**:49-52. DOI: 10.1016/j.molcata.2006.03.051

[39] Holden MS, Crouch RD. The Pechmann reaction. Journal of Chemical Education. 1998;**75**:1631. DOI: 10.1021/ed075p1631

[40] Li T-S, Zhang Z-H, Yang F, Fu C-G. Montmorillonite clay catalysis. Part 7. An environmentally friendly procedure for the synthesis of coumarins via Pechmann condensation of phenols with ethyl acetoacetate. Journal of Chemical Research, Synopses. 1998:38-39. DOI: 10.1039/a703694i

[41] Potdar MK, Mohile SS, Salunkhe MM. Coumarin syntheses via Pechmann condensation in Lewis acidic chloroaluminate ionic liquid. Tetrahedron Letters. 2001;**42**:9285-9287. DOI: 10.1016/S0040-4039(01)02041-X

[42] Bose DS, Rudradas AP, Babu MH. The indium(III) chloride-catalyzed von Pechmann reaction: A simple and effective procedure for the synthesis of 4-substituted coumarins. Tetrahedron Letters. 2002;**43**:9195-9197. DOI: 10.1016/S0040-4039(02)02266-9

[43] Potdar MK, Rasalkar MS, Mohile SS, Salunkhe MM. Convenient and efficient protocols for coumarin synthesis via Pechmann condensation in neutral ionic liquids. Journal of Molecular Catalysis A: Chemical. 2005;**235**:249-252. DOI: 10.1016/j.molcata.2005.04.007

[44] Alexander VM, Bhat RP, Samant SD. Bismuth(III) nitrate pentahydrate—A mild and inexpensive reagent for synthesis of coumarins under mild conditions. Tetrahedron Letters. 2005;**46**:6957-6959. DOI: 10.1016/j.tetlet.2005.07.117

[45] Reddy BM, Patil MK, Lakshmanan P. Sulfated $Ce_xZr_{1-x}O_2$ solid acid catalyst for solvent free synthesis of coumarins. Journal of Molecular Catalysis A: Chemical. 2006;**256**:290-294. DOI: 10.1016/j.molcata.2006.05.001

[46] Rodríguez-Domínguez JC, Kirsch G. Sulfated zirconia, a mild alternative to mineral acids in the synthesis of hydroxycoumarins. Tetrahedron Letters. 2006;**47**:3279-3281. DOI: 10.1016/j.tetlet.2006.03.030

[47] Reddy YT, Sonar VN, Crooks PA, Dasari PK, Reddy PN, Rajitha B. Ceric ammonium nitrate (CAN): An efficient catalyst for the coumarin synthesis via Pechmann condensation using conventional heating and microwave irradiation. Synthetic Communications. 2008;**38**:2082-2088. DOI: 10.1080/00397910802029091

[48] Kotharkar SA, Bahekar SS, Shinde DB. Chlorosulfonic acid-catalysed one-pot synthesis of coumarin. Mendeleev Communications. 2006;**16**:241-242. DOI: 10.1070/MC2006v016n04ABEH002256

[49] Lakouraj MM, Bagheri N, Hasantabar V. Synthesis and application of nanocrystalline-cellulose-supported acid ionic liquid catalyst in Pechmann reaction. International Journal of Carbohydrate Chemistry. 2013;**2013**:1-8. DOI: 10.1155/2013/452580

[50] Puri S, Kaur B, Parmar A, Kumar H. Ultrasound-promoted greener synthesis of 2H-chromen-2-ones catalyzed by copper perchlorate in solventless media. Ultrasonics Sonochemistry. 2009;**16**:705-707. DOI: 10.1016/j.ultsonch.2009.04.002

[51] Ranjbar-Karimi R, Hashemi-Uderji S, Mousavi M. Selectfluor promoted environmental-friendly synthesis of 2H-chromen-2-ones derivatives under various reaction conditions. Journal of the Iranian Chemical Society. 2011;**8**:193-197. DOI: 10.1007/BF03246215

[52] DeGrote J, Tyndall S, Wong KF, VanAlstine-Parris M. Synthesis of 7-alkoxy-4-trifluoromethylcoumarins via the von Pechmann reaction catalyzed by molecular iodine. Tetrahedron Letters. 2014;**55**:6715-6717. DOI: 10.1016/j.tetlet.2014.10.025

[53] Prajapati D, Gohain M. Iodine a simple, effective and inexpensive catalyst for the synthesis of substituted coumarins. Catalysis Letters. 2007;**119**:59-63. DOI: 10.1007/s10562-007-9186-6

[54] Wu J, Diao T, Sun W, Li Y. Expeditious approach to coumarins via Pechmann reaction catalyzed by molecular iodine or AgOTf. Synthetic Communications. 2006;**36**:2949-2956. DOI: 10.1080/00397910600773692

[55] Shirini F, Yahyazadeh A, Mohammadi K. A solvent-free synthesis of coumarins using 1,3-disulfonic

acid imidazolium hydrogen sulfate as a reusable and effective ionic liquid catalyst. Research on Chemical Intermediates. 2015;**41**:6207-6218. DOI: 10.1007/s11164-014-1733-3

[56] Zhang Y, Zhu A, Li Q, Li L, Zhao Y, Wang J. Cholinium ionic liquids as cheap and reusable catalysts for the synthesis of coumarins via Pechmann reaction under solvent-free conditions. RSC Advances. 2014;**4**:22946-22950. DOI: 10.1039/c4ra02227k

[57] Khalafi-Nezhad A, Haghighi SM, Panahi F. Nano-TiO_2 on dodecyl-sulfated silica: As an efficient heterogeneous Lewis acid-surfactant-combined catalyst (HLASC) for reaction in aqueous media. ACS Sustainable Chemistry & Engineering. 2013;**1**:1015-1023. DOI: 10.1021/sc4000913

[58] Karami B, Kiani M. $ZrOCl_2.8H_2O$/SiO_2: An efficient and recyclable catalyst for the preparation of coumarin derivatives by Pechmann condensation reaction. Catalysis Communications. 2011;**14**:62-67. DOI: 10.1016/j.catcom.2011.07.002

[59] Ma Z, He M, Zhang H, Ma J, Yuan H. Hydrophobic perfluoroalkylsulfonyl imide solid acid as catalyst of esterification and Pechmann condensation. Chinese Journal of Organic Chemistry. 2014;**34**:2255-2261. DOI: 10.6023/cjoc201406027

[60] Palaniappa S, John A. A novel polyaniline–fluoroboric acid–dodecylhydrogensulfate salt: Versatile reusable polymer based solid acid catalyst for organic transformations. Journal of Molecular Catalysis A: Chemical. 2005;**233**:9-15. DOI: 10.1016/j.molcata.2005.02.002

[61] Dabiri M, Salehi P, Zolfigol MA, Baghbanzadeh M. Silica sulfuric acid as an efficient and reusable catalyst for the Pechmann synthesis of coumarins under solvent-free conditions. Heterocycles. 2007;**71**:677-682. DOI: 10.3987/COM-06-10956

[62] Ghodke S, Chudasama U. Solvent free synthesis of coumarins using environment friendly solid acid catalysts. Applied Catalysis A: General. 2013;**453**:219-226. DOI: 10.1016/j.apcata.2012.12.024

[63] Ahmed AI, El-Hakam SA, Elghany MAA, El-Yazeed WSA. Synthesis and characterization of new solid acid catalysts, $H_3PW_{12}O_{40}$ supported on nanoparticle tin oxide: An efficient catalyst for the preparation of 7-hydroxy-4-methylcoumarin. Applied Catalysis A: General. 2011;**407**:40-48. DOI: 10.1016/j.apcata.2011.08.020

[64] Albadi J, Shirini F, Abasi J, Armand N, Motaharizadeh T. A green, efficient and recyclable poly(4-vinylpyridine)-supported copper iodide catalyst for the synthesis of coumarin derivatives under solvent-free conditions. C. R. Chimie. 2013;**16**:407-411. DOI: 10.1016/j.crci.2012.10.002

[65] Rahmatpour A, Mohammadian S. An environmentally friendly, chemoselective, and efficient protocol for the preparation of coumarin derivatives by Pechman condensation reaction using new and reusable heterogeneous Lewis acid catalyst polystyrene-supported $GaCl_3$. C. R. Chimie. 2013;**16**:271-278. DOI: 10.1016/j.crci.2013.01.006

[66] Karami B, Khodabakhshi S, Eskandari K. Alternative two-step route to khellactone analogues using silica tungstic acid and sodium hydrogen phosphate. Chemical Papers. 2013;**67**:474-1478. DOI: 10.2478/s11696-013-0411-z

[67] Zareyee D, Serehneh M. Recyclable CMK-5 supported sulfonic acid as an environmentally benign catalyst for solvent-free one-pot construction of coumarin through Pechmann

condensation. Journal of Molecular Catalysis A: Chemical. 2014;**391**:88-91. DOI: 10.1016/j.molcata.2014.04.013

[68] Vahabi V, Hatamjafari F. Microwave assisted convenient one-pot synthesis of coumarin derivatives via Pechmann condensation catalyzed by FeF_3 under solvent-free conditions and antimicrobial activities of the products. Molecules. 2014;**19**:13093-13103. DOI: 10.3390/molecules190913093

[69] Prousis KC, Avlonitis N, Heropoulos GA, Calogeropoulou T. $FeCl_3$-catalysed ultrasonic-assisted, solvent-free synthesis of 4-substituted coumarins. A useful complement to the Pechmann reaction. Ultrasonics Sonochemistry. 2014;**21**:937-942. DOI: 10.1016/j.ultsonch.2013.10.018

[70] Esfahani FK, Zareyee D, Yousefi R. Sulfonated core-shell mgnetic nanoparticle (Fe_3O_4@SiO_2@$PrSO_3H$) as a highly active and durable protonic acid catalyst; synthesis of coumarin derivatives through Pechmann reaction. ChemCatChem. 2014;**6**:3333-3337. DOI: 10.1002/cctc.201402547

[71] Baghbanian SM, Farhang M. $CuFe_2O_4$ nanoparticles: A magnetically recoverable and reusable catalyst for the synthesis of coumarins via Pechmann reaction in water. Synthetic Communications. 2014;**44**:697-706. DOI: 10.1080/00397911.2013.835423

[72] Sharma RK, Monga Y, Puri A. Zirconium(IV)-modified silica@magnetic nanocomposites: Fabrication, characterization and application as efficient, selective and reusable nanocatalysts for Friedel-Crafts, Knoevenagel and Pechmann condensation reactions. Catalysis Communications. 2013;**35**:110-114. DOI: 10.1016/j.catcom.2013.02.016

[73] Kim J-C, Ryoo R, Opanasenko M, Shamzhy M, Cejka J. Mesoporous MFI zeolite nanosponge as a high performance catalyst in the Pechmann condensation reaction. ACS Catalysis. 2015;**5**:2596-2604. DOI: 10.1021/cs502021a

[74] Karami B, Kiani M, Hoseini MA. $In(OTf)_3$ as a powerful and recyclable catalyst for Pechmann condensation without solvent. Chinese Journal of Catalysis. 2014;**35**:1206-1211. DOI: 10.1016/S1872-2067(14)60090-5

[75] Wang H. Magnesium bis(trifluoromethane)sulfonimide: An efficient catalyst for the synthesis of coumarins under solvent-free conditions. Monatshefte fuer Chemie. 2013;**144**:411-414. DOI: 10.1007/s00706-012-0823-4

[76] Khaligh NG. Ultrasound-assisted one-pot synthesis of substituted coumarins catalyzed by poly(4-vinylpyridinium) hydrogen sulfate as an efficient and reusable solid acid catalyst. Ultrasonics Sonochemistry. 2013;**20**:1062-1068. DOI: 10.1016/j.ultsonch.2013.01.001

[77] Mokhtary M, Najafizadeh F. Polyvinylpolypyrrolidone-bound boron trifluoride (PVPP-BF_3); a mild and efficient catalyst for synthesis of 4-metyl coumarins via the Pechmann reaction. C. R. Chimie. 2012;**15**:530-532. DOI: 10.1016/j.crci.2012.03.004

[78] Opanasenko M, Shamzhy M, Čejka J. Solid acid catalysts for coumarin synthesis by the Pechmann reaction: MOFs versus zeolites. ChemCatChem. 2013;**5**:1024-1031. DOI: 10.1002/cctc.201200232

[79] Jadhav NH, Sakate SS, Rasal NK, Shinde DR, Pawar RA. Heterogeneously catalyzed Pechmann condensation employing the tailored $Zn_{0.925}Ti_{0.075}O$ NPs: Synthesis of coumarin. ACS Omega. 2019;**4**:8522-8527. DOI: 10.1021/acsomega.9b00257

[80] Zambare AS, Khan FAK, Zambare SP, Shinde SD, Sangshetti JN. Recent advances in the synthesis of coumarin derivatives via Pechmann condensation. Current Organic Chemistry. 2016;**20**:798-828. DOI: 10.2174/1385272820666151026224227

[81] Kumar BV, Naik HSB, Girija D, Kumar BV. ZnO nanoparticle as catalyst for efficient green one-pot synthesis of coumarins through Knoevenagel condensation. Journal of Chemical Sciences. 2011;**123**:615-621. DOI: 10.1007/s12039-011-0133-0

[82] Karami B, Farahi M, Khodabakhshi S. Rapid synthesis of novel and known coumarin-3-carboxylic acids using stannous chloride dihydrate under solvent-free conditions. Helvetica Chimica Acta. 2012;**95**:455-460. DOI: 10.1002/hlca.201100342

[83] Sripathi SK, Logeeswari K. Synthesis of 3-aryl coumarin derivatives using ultrasound. International Journal of Organic Chemistry. 2013;**3**:42-47. DOI: 10.4236/ijoc.2013.31004

[84] Nourmohammadian F, Norozy S. Application of non-corrosive acids in three-component, one-pot synthesis of commercial coumarin dye. Progress in Color, Colorants and Coatings. 2010;**03**:102-109. DOI: not available

[85] Gholap SS, Deshmukh UP, Tambe MS. Synthesis and in-vitro antimicrobial screening of 3-cinnamoyl coumarin and 3-[3-(1H-indol-2-yl)-3-aryl-propanoyl]-2H-chromen-2-ones. Iranian Journal of Catalysis. 2013;**3**:171-176. DOI: not available

[86] Shaabani A, Ghadari R, Rahmati A, Rezayan AH. Coumarin synthesis via Knoevenagel condensation reaction in 1,1,3,3-N,N,N',N'-tetramethylguanidinium trifluoroacetate ionic liquid. Journal of the Iranian Chemical Society. 2009;**6**:710-714. DOI: 10.1007/BF03246160

[87] Verdía P, Santamarta F, Tojo E. Knoevenagel reaction in [MMIm][MSO_4]: Synthesis of coumarins. Molecules. 2011;**16**:4379-4388. DOI: 10.3390/molecules16064379

[88] Valizadeh H, Gholipour H. Imidazolium-based phosphinite ionic liquid (IL-$OPPh_2$) as reusable catalyst and solvent for the Knoevenagel condensation reaction. Synthetic Communications. 2010;**40**:1477-1485. DOI: 10.1080/00397910903097310

[89] Heravi MM, Ansari P, Saeedi M, Tavakoli-Hosseini N, Karimi N. Green and practical synthesis of benzopyran and 3-substituted coumarine derivatives by Bronsted acid ionic liquid [$(CH_2)_4SO_3HMIM$][HSO_4]. Bulletin of the Chemical Society of Ethiopia. 2011;**25**:315-320. DOI: 10.4314/bcse.v25i2.65915

[90] Roussaki M, Kontogiorgis CA, Hadjipavlou-Litina D, Hamilakis S, Detsi A. A novel synthesis of 3-aryl coumarins and evaluation of their antioxidant and lipoxygenase inhibitory activity. Bioorganic & Medicinal Chemistry Letters. 2010;**20**:3889-3892. DOI: 10.1016/j.bmcl.2010.05.022

[91] Nourmohammadian F, Gholami MD. Microwave-promoted one-pot syntheses of coumarin dyes. Synthetic Communications. 2010;**40**:901-909. DOI: 10.1080/00397910903026699

[92] Han J, Xin Y, Zhao J, Zhu S. L-Proline catalyzed condensation–cyclization tandem process: Facile and effective synthesis of 3-polyfluoroalkanesulfonyl coumarin. Journal of Fluorine Chemistry. 2011;**132**:409-413. DOI: 10.1016/j.jfluchem.2011.04.001

[93] You X, Yu H, Wang M, Wu J, Shang Z. A Green method for the

synthesis of 3-substituted coumarins catalyzed by L-lysine in water via Knoevenagel condensation. Letters in Organic Chemistry. 2012;**9**:19-23. DOI: 10.2174/157017812799303953

[94] Lunkad AS, Sawant RL. Conventional and microwave assisted synthesis of some new derivatives of coumarin containing pyrazoline and investigation of their antibacterial and antifungal activities. International Journal of Pharmaceutical Sciences and Research. 2018;**9**:2852-2858. DOI: 10.13040/IJPSR.0975-8232.9(7).2852-58

[95] Ajani OO, Nwinyi OC. Microwave-assisted synthesis and evaluation of antimicrobial activity of 3-{3-(s-aryl and s-heteroaromatic)acryloyl}-2H-chromen-2-one derivatives. Journal of Heterocyclic Chemistry. 2010;**47**:179-187. DOI: 10.1002/jhet.298

[96] Sashidhara KV, Kumar A, Kumar M, Sonkar R, Bhatia G, Khanna A. Novel coumarin derivatives as potential antidyslipidemic agents. Bioorganic & Medicinal Chemistry Letters. 2010;**20**:4248-4251. DOI: 10.1016/j.bmcl.2010.05.023

[97] Bhusal RP, Cho PY, Kim SA, Park H, Kim HS. Synthesis of green emitting coumarin bioconjugate for the selective determination of flu antigen. Bulletin of the Korean Chemical Society. 2011;**32**:1461-1462. DOI: 10.5012/bkcs.2011.32.5.1461

[98] De Souza M, da Silveira Pinto L. Sonochemistry as a general procedure for the synthesis of coumarins, including multigram synthesis. Synthesis. 2017;**49**:2677-2682. DOI: 10.1055/s-0036-1590201

[99] El-Wahab AHFA, Mohamed HM, El-Agrody AM, Bedair AH, Eid FA, Khafagy MM, et al. Synthesis and reactions of some new diiodocoumarin derivatives bearing side chains and some of their biological activities. American Journal of Chemistry. 2011;**1**:1-8. DOI: 10.5923/j.chemistry.20110101.01

[100] Mohamed HM, Abd El-Wahab AHF, Ahmed KA, El-Agrody AM, Bedair AH, Eid FA, et al. Synthesis, reactions and antimicrobial activities of 8-ethoxycoumarin derivatives. Molecules. 2012;**17**:971-988. DOI: 10.3390/molecules17010971

[101] Kaye PT, Robinson RS. Dabco-catalysed reactions of salicylaldehydes with acrylate derivatives. Synthetic Communications. 1996;**26**:2085-2097. DOI: 10.1080/00397919608003567

[102] Bacsa J, Kaye PT, Robinson RS. Novel products from Baylis-Hillman reactions of salicylaldehydes. South African Journal of Chemistry. 1998;**51**:47-54. Available at: https://hdl.handle.net/10520/AJA03794350_1721

[103] Kaye PT, Musa MA, Nocanda XW. Efficient and chemoselective access to 3-(chloromethyl)coumarins via direct cyclisation of unprotected Baylis-Hillman adducts. Synthesis. 2003;**2003**:531-534. DOI: 10.1055/s-2003-37655

[104] Olomola TO, Klein R, Kaye PT. Convenient synthesis of 3-methylcoumarins and coumarin-3-carbaldehydes. Synthetic Communications. 2012;**42**:251-257. DOI: 10.1080/00397911.2010.523491

[105] Kaye PT, Musa MA. A convenient and improved Baylis-Hillman synthesis of 3-substituted 2H-1-benzopyran-2-ones. Synthesis. 2002;**2002**:2701-2706. DOI: 10.1055/s-2002-35984

[106] Kaye PT, Musa MA. Application of Baylis–Hillman methodology in the synthesis of coumarin derivatives. Synthetic Communications. 2003;**33**:1755-1770. DOI: 10.1081/SCC-120018937

[107] Hwang I-T, Lee S-A, Hwang J-S, Lee K-I. A facile synthesis of highly functionalized 4-arylcoumarins via Kostanecki reactions mediated by DBU. Molecules. 2011;**16**:6313-6321. DOI: 10.3390/molecules16086313

[108] RaoHSP, SivakumarS. Condensation of r-aroylketene dithioacetals and 2-hydroxyarylaldehydes results in facile synthesis of a combinatorial library of 3-aroylcoumarins. The Journal of Organic Chemistry. 2006;**71**:8715-8723. DOI: 10.1021/jo061372e

[109] Upadhyay PK, Kumar P. A novel synthesis of coumarins employing triphenyl(a-carboxymethylene)-phosphorane imidazolide as a C-2 synthon. Tetrahedron Letters. 2009;**50**:236-238. DOI: 10.1016/j.tetlet.2008.10.133

[110] Valizadeh H, Gholipour H, Mahmoudian M. Phosphinite ionic liquid (IL-OPPh2) as a recyclable reagent for the efficient synthesis of coumarins under microwave irradiation conditions. Journal of the Iranian Chemical Society. 2011;**8**:862-871. DOI: 10.1007/BF03245917

[111] Yavari I, Hekmatshoar R, Zonouzi A. A new and efficient route to 4-carboxymethylcoumarins mediated by vinyltriphenylphosphonium salt. Tetrahedron Letters. 1998;**39**:2391-2392. DOI: 10.1016/S0040-4039(98)00206-8

[112] Hekmatshoar R, Souri S, Rahimifard M, Faridbod F. Novel synthesis of oxygenated coumarins from substituted phenols mediated by vinyl triphenylphosphonium salt under microwave irradiation. Phosphorus, Sulfur and Silicon. 2002;**177**:2827-2833. DOI: 10.1080/10426500214882

[113] Yavari I, Adib M, Hojabri L. Vinyltriphenylphosphonium salt mediated synthesis of functionalized coumarins. Tetrahedron. 2001;**57**:7537-7540. DOI: 10.1016/S0040-4020(01)00703-7

[114] Majumdar KC, Ansary I, Samanta S, Roy B. Aromatic electrophilic substitution vs. intramolecular Wittig reaction: Vinyltriphenylphosphonium salt mediated synthesis of 4-carboxyalkyl-8-formyl coumarins. Synlett. 2011;**2011**:0694-0698. DOI: 10.1055/s-0030-1259534

[115] Galariniotou E, Fragos V, Makri A, Litinas KE, Nicolaides DN. Synthesis of novel pyridocoumarins and benzo-fused 6-azacoumarins. Tetrahedron. 2007;**63**:8298-8304. DOI: 10.1016/j.tet.2007.05.102

[116] Trost BM, Toste FD. A new palladium-catalyzed addition: A mild method for the synthesis of coumarins. Journal of the American Chemical Society. 1996;**118**:6305-6306. DOI: 10.1021/ja961107i

[117] Trost BM, Toste FD, Greenman K. Atom economy. Palladium-catalyzed formation of coumarins by addition of phenols and alkynoates via a net C-H insertion. Journal of the American Chemical Society. 2003;**125**:4518-4526. DOI: 10.1021/ja0286573

[118] Kotani M, Yamamoto K, Oyamada J, Fujiwara Y, Kitamura T. A convenient synthesis of coumarins by palladium(II)-catalyzed reaction of phenols with propiolic acids. Synthesis. 2004;**2004**:1466-1470. DOI: 10.1055/s-2004-822360

[119] Kutubi Md S, Hashimoto T, Kitamura T. Improved synthesis of coumarins by iron(III)-catalyzed cascade reaction of propiolic acids and phenols. Synthesis. 2011;**2011**:1283-1289. DOI: 10.1055/s-0030-1258473

[120] Park KH, Jung IG, Chung YK. Synthesis of coumarins catalyzed by

heterobimetallic Co/Rh nanoparticles. Synlett. 2004;**2004**:2541-2544. DOI: 10.1055/s-2004-834826

[121] Leão RAC, De Moraes PF, Pedro MCBC, Costa PRR. Synthesis of coumarins and neoflavones through zinc chloride catalyzed hydroarylation of acetylenic esters with phenols. Synthesis. 2011;**2011**:3692-3696. DOI: 10.1055/s-0031-1289576

[122] Fernandes TA, Carvalho RCC, Goncalves TMD, da Silva AJM, Costa PRR. A tandem palladium-catalyzed heck-lactonization through the reaction of ortho-iodophenols with b-substituted acrylates: Synthesis of 4,6-substituted coumarins. Tetrahedron Letters. 2008;**49**:3322-3325. DOI: 10.1016/j.tetlet.2008.03.037

[123] Fernandes TA, Vaz BG, Eberlin MN, da Silva AJM, Costa PRR. Palladium-catalyzed tandem Heck-lactonization from o-iodophenols and enoates: Synthesis of coumarins and the study of the mechanism by electrospray ionization mass spectrometry. The Journal of Organic Chemistry. 2010;**75**:7085-7091. DOI: 10.1021/jo1010922

[124] Kadnikov DV, Larock RC. Palladium-catalyzed carbonylative annulation of terminal alkynes: Synthesis of coumarins and 2-quinolones. Journal of Organometallic Chemistry. 2003;**687**:425-435. DOI: 10.1016/S0022-328X(03)00786-1

[125] Kadnikov DV, Larock RC. Palladium-catalyzed carbonylative annulation of internal alkynes: Synthesis of 3,4-disubstituted coumarins. The Journal of Organic Chemistry. 2003;**68**:9423-9432. DOI: 10.1021/jo0350763

[126] Gabriele B, Mancuso R, Salerno G, Plastina P. A novel palladium-catalyzed dicarbonylation process leading to coumarins. The Journal of Organic Chemistry. 2008;**73**:756-759. DOI: 10.1021/jo702243m

[127] Li K, Zeng Y, Neuenswander B, Tunge JA. Sequential Pd(II)-Pd(0) catalysis for the rapid synthesis of coumarins. The Journal of Organic Chemistry. 2005;**70**:6515-6518. DOI: 10.1021/jo050671l

[128] Yamamoto Y, Kirai N. Synthesis of 4-arylcoumarins via Cu-catalyzed hydroarylation with arylboronic acids. Organic Letters. 2008;**10**:5513-5516. DOI: 10.1021/ol802239n

[129] Fillion E, Dumas AM, Kuropatwa BA, Malhotra NR, Sitler TC. $Yb(OTf)_3$-catalyzed reactions of 5-alkylidene meldrum's acids with phenols: One-pot assembly of 3,4-dihydrocoumarins, 4-chromanones, coumarins, and chromones. The Journal of Organic Chemistry. 2006;**71**:409-412. DOI: 10.1021/jo052000t

[130] Toräng J, Vanderheiden S, Nieger M, Bräse S. Synthesis of 3-alkylcoumarins from salicylaldehydes and α,β-unsaturated aldehydes utilizing nucleophilic carbenes: A new umpoled domino reaction. European Journal of Organic Chemistry. 2007;**2007**:943-952. DOI: 10.1002/ejoc.200600718

[131] Majumdar KC, Samanta S, Ansary I, Roy B. An unusual one-pot synthesis of 3-benzoylcoumarins and coumarin-3-carbaldehydes from 2-hydroxybenzaldehydes under esterification conditions. RSC Advances. 2012;**2**:2137-2143. DOI: 10.1039/C2RA00820C

[132] Li L, Yang Q, Jia Z, Luo S. Organocatalytic electrochemical C–H lactonization of aromatic carboxylic acids. Synthesis. 2018;**50**:2924-2929. DOI: 10.1055/s-0036-1591558

Chapter 7

Coumarin Derivatives with Antimicrobial and Antioxidant Activities

Gabriela Tataringa and Ana Maria Zbancioc

Abstract

Coumarin derivatives are structurally interesting compounds for synthesizing antimicrobial and antioxidant agents. Starting from 4-methyl-7-hydroxycoumarin, several derivatives with these properties have been obtained through different reaction steps. Their molecular structures were established by Fourier-transform infrared spectroscopy and nuclear magnetic resonance spectroscopy. The synthesized coumarin derivatives exerted meaningful activities against Gram-positive and Gram-negative bacteria as well as strains of *Candida* spp. All compounds also exhibited high and moderate antioxidant activity in assays for DPPH inhibition, total reducing power, and nitric oxide (NO) inhibition when compared to ascorbic acid.

Keywords: 4-methyl-7-hydroxycoumarin, synthesis, coumarin derivatives, antibacterial activity, antifungal activity, antioxidant activity

1. Introduction

Natural and synthetic coumarins have drawn much attention due to its broad pharmacological activities. Literature review reveals that coumarin (2-oxo-2H-chromene) and its derivatives represent one of the most active classes of heterocyclic compounds which possess a wide spectrum of biological activities [1–9]: antitumor [1, 2], antibacterial [3, 4], antifungal [5–7], anticoagulant [8], antioxidant [9], and anti-inflammatory [10].

1.1 Coumarins as antimicrobial agents

Over the past few decades, the search for newer antimicrobials remains an area of intensive investigation in the field of medicinal chemistry due to resistance developed by microorganism to conventional antibiotics. Antimicrobials are one of most significant weapons in fighting bacterial infections. Throughout history, there has been a continual battle between humans and the multitude of microorganisms that cause infection and disease [11, 12]. Coumarin derivatives have a wide range of structural modifications [13], and they can serve as molecular templates for new drugs. Coumarin derivatives are also considered as potential antimicrobial agents [14].

Medimagh-Saidana et al. reported synthesis and antimicrobial activity of some coumarin esters (**1**) (**Figure 1**). These compounds showed good activity against

Bacillus sp. and moderate activity against *Aspergillus niger*. For the data of the antibacterial activity, these compounds were found to be active against *Pseudomonas* sp. [15].

Al-Amiery et al. have synthesized some coumarin derivatives, and their antifungal activity was determined based on the growth inhibition rates of the mycelia of strains of *Aspergillus niger* and *Candida albicans* in Potato Dextrose Broth (PDB) medium against concentrations ranging from 10 to 100 μg/ml. The compound (**2**) (**Figure 1**) showed good activity as antifungals against fluconazole as standard drug [16].

Behrami et al. synthesized 8-amino-4,7-dihydroxy-chromen-2-one coumarin derivatives. The antibacterial activities of all the compounds and standard streptomycin and cefalexine at concentrations of 2, 3, and 5 mg/ml were studied against *Staphylococcus aureus*, *Bacillus subtilis*, and *Escherichia coli*. One compound (**3**) (**Figure 1**) was more active than cefalexine and lesser active than streptomycin, and it was most active among synthesized compounds [17].

Some coumarin derivatives containing thiazolidin-4-one ring were synthesized by Rama Ganesh et al. and were screened for their antibacterial activity against Gram-positive bacteria *Staphylococcus aureus* and *Bacillus subtilis* and Gram-negative bacteria *Klebsiella pneumonia* and *Escherichia coli* at the concentration of 0.001 mol/ml compared with the standard drug ciprofloxacin. Zone of inhibition of highly active compound (**4**) (**Figure 1**) was 20 mm against *Staphylococcus aureus* and *Bacillus subtilis* [18].

1.2 Coumarins as antioxidant agents

Free radicals are molecular species capable of independent existence that contain an unpaired electron in an atomic orbital; they are usually unstable and very reactive. These species are normally produced in the human body from essential metabolic processes, but they may also occur from external sources such as exposure to X-rays, ozone, cigarette smoking, air pollutants, and industrial chemicals [19].

There is an increasing interest in antioxidants, particularly in those intended to prevent the presumed deleterious effects of free radicals in the human body and to prevent the deterioration of fats and other constituents of food stuffs. In both cases, there is a preference for antioxidants from natural rather than from synthetic

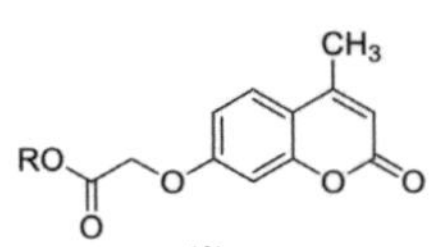

(1)

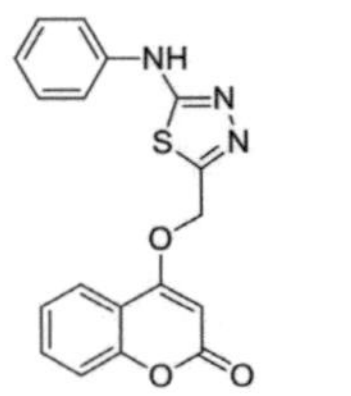

(2)

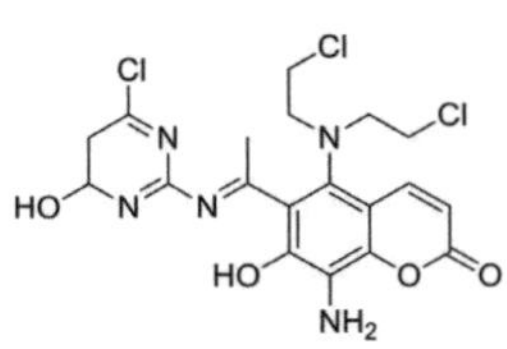

(3)

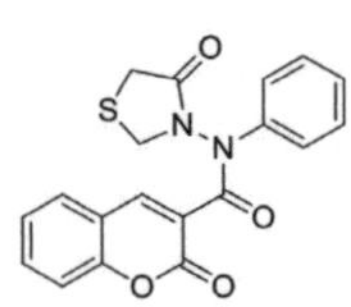

(4)

Figure 1.
Coumarin derivatives with antimicrobial activity.

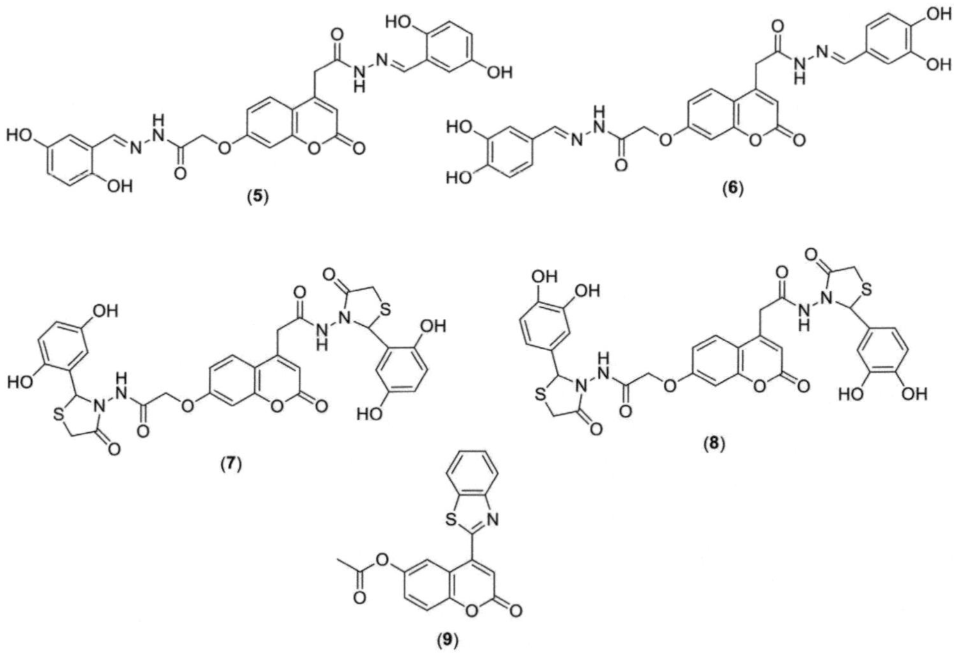

Figure 2.
Coumarin derivatives with antioxidant activity.

sources [20]. As improved antioxidant status helps to minimize the oxidative damage and thus delays or prevents pathological changes, potential antioxidant therapy should be included either as natural free-radical-scavenging antioxidant enzymes or as an agent which is capable of augmenting the activity of antioxidant enzymes [21].

The human organism possesses natural systems to annihilate these species, but when the body's ability to regulate them is overwhelmed, a condition known as oxidative stress appears, free radicals attacking important macromolecules leading to cell damage and homeostatic disruption [22].

Many coumarin derivatives have a special ability to scavenge reactive oxygen species and to influence processes involving free radical injury [23, 24].

Maja et al. synthesized a series of carbohydrazide with coumarin ring (**5**, **6**) and some coumarin derivatives with a heterocyclic ring (**7**, **8**) (**Figure 2**). All these compounds prove a good antioxidant activity [25].

Shivani et al. synthesized new coumarin-substituted derivatives of benzothiazole, and they were evaluated for antioxidant activity by DPPH radical scavenging activity. The test compound (**9**) (**Figure 2**) showed good in vitro antioxidant activity [26].

2. Synthesis of coumarin derivatives

Looking to the medicinal importance of the coumarin ring, we employed coumarin as a naturally occurring skeleton for the construction of new derivatives which might exhibit promising antimicrobial and antioxidant activities [27].

The starting materials, 4-methyl/propyl-7-hydroxycoumarin, were prepared by Pechmann synthesis which involved the condensation of resorcinol and ethylacetoacetate/ethylbutyrylacetate in the presence of H_2SO_4 concentrate [28].

2.1 Mechanism of the Pechmann condensation

The reaction is conducted with a strong Brønsted acid such as methanesulfonic acid or a Lewis acid such as $AlCl_3$. The acid catalyzes transesterification as well as keto-enol tautomerization [29]. A Michael addition leads to the formation of the coumarin skeleton. This addition is followed by rearomatization and then by elimination of water which gives the product (**Figure 3**).

The coumarin compounds have wide interest due to their diverse pharmacological properties. In particular, these biological activities make coumarin compounds more attractive and testing as novel therapeutic compounds.

As part of our aim in research of biologically active coumarin derivatives, the free hydroxyl group on the coumarin ring has allowed us to introduce some radicals that can improve the biological activity.

The fourth scheme describes the reactions of 4-methyl/propyl-7-hydroxycoumarin with ethyl bromoacetate. 2-Ethyl-((4-methy/propyl-2-oxo-2H-chromen-7-yl)oxy)acetate (**IIa–IIb**) was prepared by heating a mixture of 4-methyl/propyl-7-hydroxycoumarin and ethyl bromoacetate in the presence of K_2CO_3 anhydrous in dry acetone. After filtration, the solution was evaporated, and the solid products (**IIa** or **IIb**) were recrystallized from ethanol [30] (**Figure 4**).

The fifth scheme describes the reactions of compounds **IIa–IIb** with hydrazine hydrate. The chemistry of hydrazide and its derivatives has obtained great interest in both organic chemistry and biological science with remarkable impact. Hydrazides and hydrazones are possessing -NH-NH_2 and -NH-N=CH- groups, respectively. The availability of proton in hydrazides constitutes them as an important class of compound for new drug discovery. Therefore, researchers have shown great interest in developing these compounds as target structures for evaluating new biological activities [31].

Hydrazinolysis of compounds **IIa–IIb** gave the corresponding acetohydrazides (**IIIa–IIIb**) in good yields [30, 32]. One mole of the compound **IIa** or **IIb** in ethanol

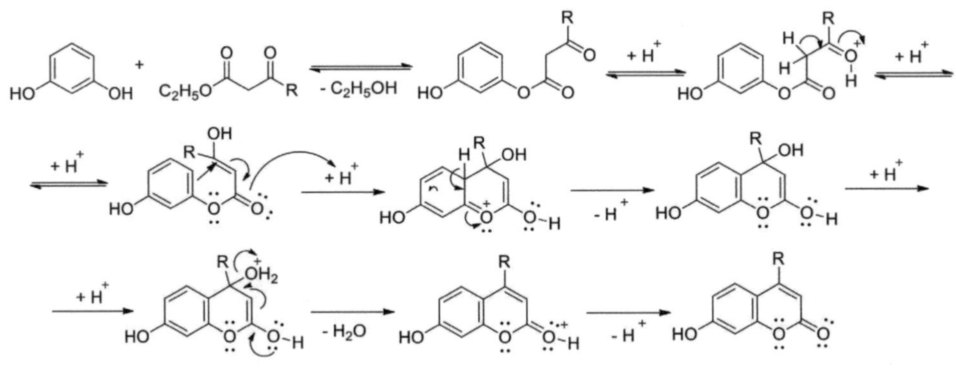

Figure 3.
Mechanism of the Pechmann condensation.

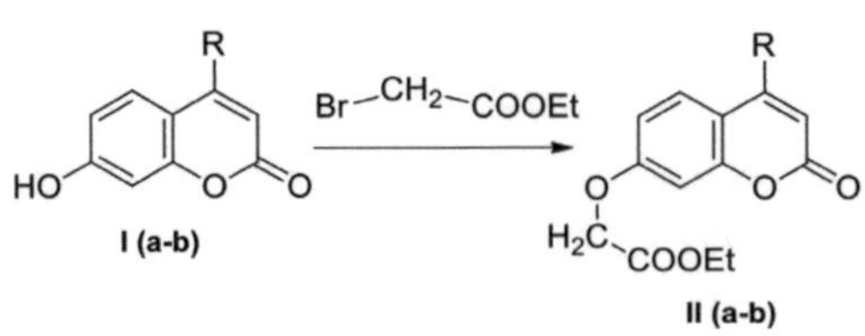

Figure 4.
*Synthesis of coumarin esters **II(a-b)**.*

was heated for 4–6 h with hydrazine hydrate (two moles). After this period of time, the mixture was cooled at room temperature, and the precipitate was filtrated and then purified by recrystallization (**Figure 5**).

The reaction of the acid hydrazides (4-methyl/propyl-2-oxo-2H-benzopyran-7-oxyacetic acid hydrazide) with CS_2 in ethanol containing KOH at room temperature has been presented in **Figure 6**. The corresponding potassium dithiocarbazate derivatives, **IVa–IVb,** are obtained [32].

The obtaining of coumarin derivatives with a thiadiazole ring has been described in **Figure** 7. Literature data show that pyrrole, pyrazole, thiadiazoles, and triazoles and their derivatives are very attractive targets due to their biological properties. In view of the above observations, we have synthesized compounds with thiadiazole ring in order to evaluate the potential antimicrobial and antioxidant activities.

These compounds were obtained following the reaction between potassium 4-methyl/propyl-2-oxo-2H-benzopyranyl-7-oxymethyldithiocarbazate and acetic acid. The reaction occurred under refluxing. The solid product was separated by filtration and then purified by recrystallization from acetic acid [33, 34].

In the synthesis of coumarin derivatives, elemental analysis and two basic spectroscopic techniques, infrared spectroscopy (IR) and nuclear magnetic resonance

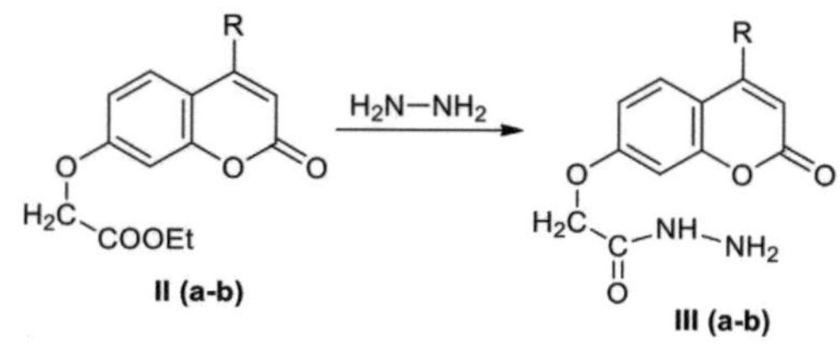

Figure 5.
*Synthesis of coumarin acetohydrazides **III(a-b)**.*

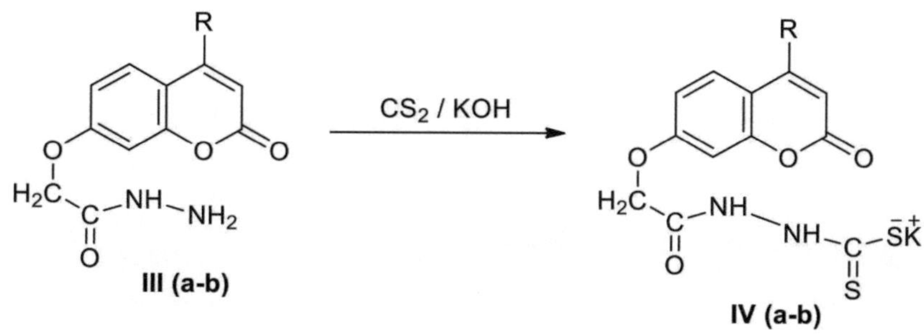

Figure 6.
*Synthesis of coumarin potassium salts **IV(a-b)**.*

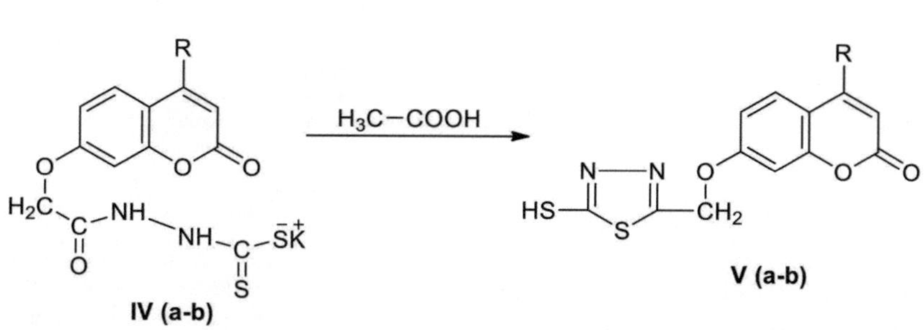

Figure 7.
*Synthesis of coumarin thiadiazoles derivatives **V(a-b)**.*

spectroscopy (NMR), were used to characterize the structures of the target compounds [35].

The physical constants and analytical data of compounds **Ia–Ib**, **IIa–IIb**, **IIIa–IIIb**, **IVa–IVb**, and **Va–Vb** have been given in **Tables 1–5**.

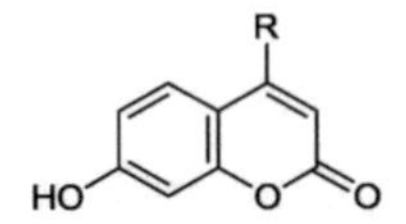

4-R-7-hydroxy-coumarin

Compd.	R	MP °C	Molecular formula	Analysis found (calculated) (%)	
				C	H
Ia	H_3C-	185	$C_{10}H_8O_3$	68.18 68.02	4.58 4.55
Ib	$H_3C-CH_2-CH_2-$	130	$C_{12}H_{12}O_3$	70.57 70.21	5.92 5.69

Table 1.
Physical constants and analytical data of compounds ***Ia–Ib***.

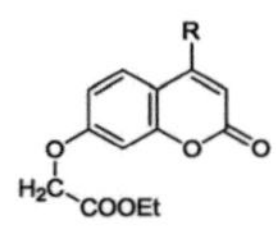

2-ethyl-((4-R-2-oxo-2H-chromen-7-yl)oxy)acetate

Compd.	R	MP °C	Molecular formula	Analysis found (calculated) (%)	
				C	H
IIa	H_3C-	100–102	$C_{14}H_{14}O_5$	64.12 64.10	5.38 5.22
IIb	$H_3C-CH_2-CH_2-$	98	$C_{16}H_{18}O_5$	66.19 65.89	6.25 6.05

Table 2.
Physical constants and analytical data of compounds ***IIa–IIb***.

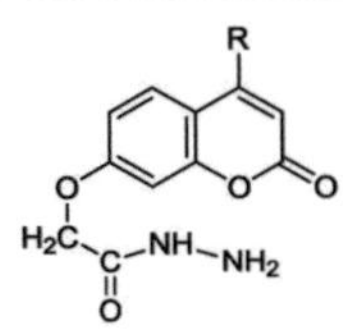

4-R-2-oxo-2H-benzopyran-7-oxyacetic acid hydrazide

Compd.	R	MP °C	Molecular formula	Analysis found (calculated) (%)	
				C	H
IIIa	H_3C-	204–205	$C_{12}H_{12}N_2O_4$	58.06 57.98	4.87 4.80
IIIb	$H_3C-CH_2-CH_2-$	147–148	$C_{14}H_{16}N_2O_4$	60.86 60.56	5.84 5.76

Table 3.
Physical constants and analytical data of compounds ***IIIa–IIIb***.

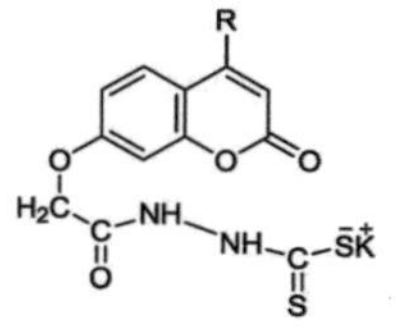

potassium 4-R-2-oxo-2H-benzopyranyl-7-oxymethyldithiocarbazate

Compd.	R	MP °C	Molecular formula	Analysis found (calculated) (%)	
				C	H
IVa	H_3C–	184	$C_{13}H_{11}N_2O_4KS_2$	43.08 42.87	3.06 3.01
IVb	H_3C–CH_2–CH_2–	176–178	$C_{15}H_{15}N_2O_4KS_2$	46.13 46.02	3.87 3.76

Table 4.
*Physical constants and analytical data of compounds **IVa–IVb**.*

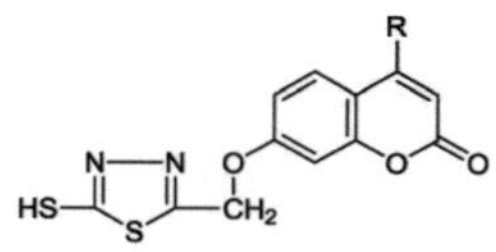

7-((5-mercapto-1,3,4-thiadiazol-2-yl)methoxy)-4-R-2H-chromen-2-one

Compd.	R	MP °C	Molecular formula	Analysis found (calculated) (%)	
				C	H
Va	H_3C–	267–268	$C_{13}H_{10}N_2O_3S_2$	50.97 50.63	3.29 3.06
Vb	H_3C–CH_2–CH_2–	172	$C_{15}H_{14}N_2O_3S_2$	53.87 53.33	4.22 4.12

Table 5.
*Physical constants and analytical data of compounds **Va–Vb**.*

The IR spectra of all synthesized compounds showed some characteristic peaks indicating the presence of particular groups (**Tables 6–10**).

^{1}H-NMR spectra of the synthesized compounds are in accordance with the assigned structures (**Table 11**). The aliphatic protons resonated in the range of 1.20–4.77 ppm. It can be seen that all the compounds exhibited the respected proton chemical shifts in the same range.

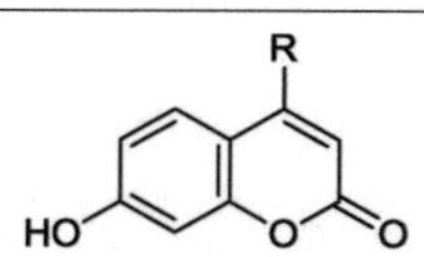

4-R-7-hydroxy-coumarin

Compd.	R	$\upsilon_{O\text{-}H}$ cm^{-1}	$\upsilon_{C\text{-}H\ aliph}$ cm^{-1}	$\upsilon_{C=O\ lactone}$ cm^{-1}	$\upsilon_{C\text{-}O}$ cm^{-1}
Ia	H_3C–	3280	2950	1680	1150
Ib	H_3C–CH_2–CH_2–	3195	2970	1695	1140

Table 6.
*IR spectral data of compounds **Ia–Ib**.*

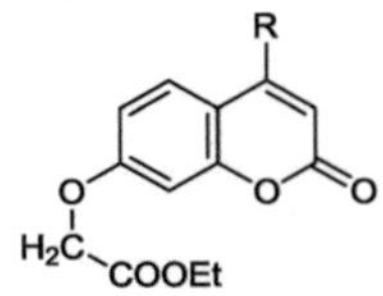

2-ethyl-((4-R-2-oxo-2H-chromen-7-yl)oxy)acetate

Compd.	R	$\upsilon_{C\text{-}H\ arom}$ cm^{-1}	$\upsilon_{C=O\ side\ chain}$ cm^{-1}	$\upsilon_{C=O\ lactone}$ cm^{-1}
IIa	$H_3C–$	3070	1750	1680
IIb	$H_3C–CH_2–CH_2–$	3080	1740	1690

Table 7.
IR spectral data of compounds ***IIa–IIb.***

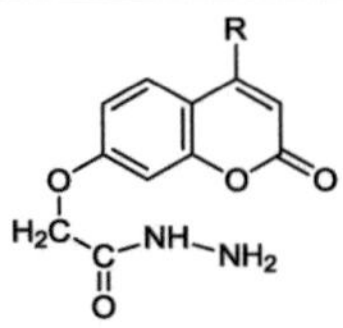

4-R-2-oxo-2H-benzopyran-7-oxyacetic acid hydrazide

Compd.	R	υ_{NH2} cm^{-1}	$\upsilon_{CO\text{-}NH}$ cm^{-1}	$\upsilon_{C\text{-}N}$ cm^{-1}
IIIa	$H_3C–$	3423, 3331	1612	1271
IIIb	$H_3C–CH_2–CH_2–$	3411, 3340	1610	1260

Table 8.
IR spectral data of compounds ***IIIa–IIIb.***

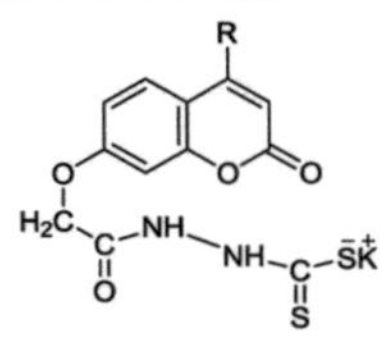

potassium 4-R-2-oxo-2H-benzopyranyl-7-oxymethyldithiocarbazate

Compd.	R	$\upsilon_{N\text{-}H}$ cm^{-1}	$\upsilon_{C=S}$ cm^{-1}
IVa	$H_3C–$	3210	1240
IVb	$H_3C–CH_2–CH_2–$	3150	1210

Table 9.
IR spectral data of compounds ***IVa–IVb.***

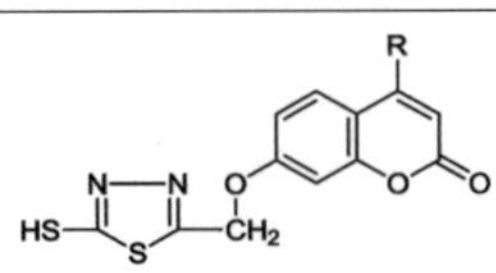

7-((5-mercapto-1,3,4-thiadiazol-2-yl)methoxy)-4-R-2H-chromen-2-one

Compd.	R	$\upsilon_{S\text{-}H}$ cm^{-1}	$\upsilon_{C=N}$ cm^{-1}	$\upsilon_{S\text{-}C}$ cm^{-1}
Va	$H_3C–$	2380	1610	621
Vb	$H_3C–CH_2–CH_2–$	2350	1620	638

Table 10.
IR spectral data of compounds ***Va–Vb.***

Compd.	R	Chemical shift (δ ppm) 500 MHz
Ia	H_3C–	2.73 ppm s, 3H: CH_3; 6.04 ppm s, H_3; 6.72 ppm s, H_8; 6.83–6.85 ppm d, H_6, $J_{H6,H5}$ = 7.5 Hz; 7.62–7.64 ppm d, H_5, $J_{H5,H6}$ = 7.5 Hz; 10.85 ppm s, 1H: OH
Ib	H_3C–CH_2–CH_2–	1.20 ppm t, 3H: CH_3 (12); 1.66 ppm m, 2H: CH_2 (11); 2.81–2.84 ppm m, 2H: CH_2 (10); 6.11 ppm s, H_3; 6.72 ppm s, H_8; 6.84–6.85 ppm d, H_6, $J_{H6,H5}$ = 7.5 Hz; 7.63–7.65 ppm d, H_5, $J_{H5,H6}$ = 7.5 Hz; 10.86 ppm s, 1H: OH
IIa	H_3C–	1.26–1.28 ppm t, 3H: CH_3 (14); 2.73 ppm s, 3H: CH_3; 4.25–4.28 ppm m, 2H: CH_2 (13); 4.61 ppm s, 2H: CH_2 (10); 6.04 ppm s, H_3; 6.84 ppm s, H_8; 7.05–7.07 ppm d H_6; 7.62–7.64 ppm d, H_5
IIb	H_3C–CH_2–CH_2–	0.95–0.98 ppm t, 3H: CH_3 (17); 1.20–1.23 ppm t, 3H: CH_3 (14); 1.60–1.65 ppm m, 2H: CH_2 (16); 2.72–2.75 ppm t, 2H: CH_2 (15); 4.15–4.20 ppm m, 2H: CH_2 (13); 4.92 ppm s, 2H: CH_2 (10); 6.17 ppm s, H_3; 6.98 ppm s, H_8; 6.96–6.97 ppm d H_6; 7.73–7.75 ppm d, H_5
IIIa	H_3C–	2.74 ppm s, 3H: CH_3; 3.86 ppm s, 2H: NH_2 (13); 4.42 ppm s, 2H: CH_2 (10); 6.03 ppm s, H_3; 6.83 ppm s, H_8; 7.05–7.07 ppm d; 7.62–7.64 ppm d, H_5; 8.02 ppm s, NH (12)
IIIb	H_3C–CH_2–CH_2–	0.95–0.98 ppm t, 3H: CH_3 (16); 1.59–1.66 ppm m, 2H: CH_2 (15); 2.72–2.75 ppm t, 2H: CH_2 (14); 4.35 ppm s, 2H: NH_2 (13); 4.61 ppm s, 2H: CH_2 (10); 6.17 ppm s, H_3; 6.98 ppm s, H_8; 6.99–7.00 ppm d H_6; 7.74–7.76 ppm d, H_5; 9.42 ppm s, NH (12)
IVa	H_3C–	2.71–2.74 ppm d, 3H: CH_3; 4.41–4.43 ppm s, 2H: CH_2 (10); 6.02–6.05 ppm q, H_3; 6.82–6.84 ppm d, H_8; 7.05–7.08 ppm q H_6; 7.62–7.64 ppm d, H_5; 9.94 ppm s, NH (12); 11.23 ppm s, NH (13)
IVb	H_3C–CH_2–CH_2–	1.20–1.22 ppm t, 3H: CH_3 (18); 1.62–1.66 ppm m, 2H: CH_2 (17); 2.79–2.82 ppm t, 2H: CH_2 (16); 4.40 ppm s, 2H: CH_2 (10); 6.10 ppm s, H_3; 6.84 ppm s, H_8; 7.04–7.08 ppm q H_6; 7.62–7.66 ppm d, H_5; 9.95 ppm s, NH (12); 11.25 ppm s, NH (13)
Va	H_3C–	2.39 ppm s, 3H: CH_3; 4.77 ppm s, 2H: CH_2 (10); 6.23 ppm s, H_3; 6.99 ppm s, H_8; 7.02–7.04 ppm d H_6; 7.70–7.72 ppm d, H_5; 10.29 ppm s, SH (16)
Vb	H_3C–CH_2–CH_2–	0.95–0.98 ppm t, 3H: CH_3 (19); 1.60–1.65 ppm m, 2H: CH_2 (18); 2.72–2.75 ppm t, 2H: CH_2 (17); 4.77 ppm s, 2H: CH_2 (10); 6.17 ppm s, H_3; 6.98 ppm s, H_8; 6.99–7.00 ppm d H_6; 7.75–7.76 ppm d, H_5; 11.27 ppm s, SH (16)

Table 11.
*^{1}H-NMR spectral data of compounds **I–V**.*

3. Pharmacological activities of coumarin derivatives

3.1 Antimicrobial activity

The compounds were screened for their antibacterial and antifungal activity according to standard protocols [36].

The antimicrobial activity was studied using Gram-positive bacteria (*Staphylococcus aureus* ATCC 25923, *Sarcina lutea* ATCC 9341, *Bacillus cereus* ATCC 14579), Gram-negative bacteria (*Escherichia coli* ATCC 25922, *Pseudomonas aeruginosa* ATCC 27853), and pathogenic yeasts (*Candida albicans* ATCC 10231, *Candida glabrata* ATCC MYA 2950, *Candida parapsilosis* ATCC 22019). All these strains were obtained from the Culture Collection of the Department of Microbiology, Faculty of Pharmacy, "Grigore T. Popa" University of Medicine and Pharmacy, Iasi, Romania.

Antimicrobial activity was evaluated by agar disc diffusion method (CLSI, 2014). A small amount of each microbial culture was diluted in sterile 0.9% NaCl until the turbidity was equivalent to McFarland standard no. 0.5 (106 CFU/ml). The suspensions were further diluted 1:10 in Mueller-Hinton agar for bacteria and Sabouraud agar for yeasts and then spread on sterile Petri plates (25 ml/Petri plate). Sterile stainless steel cylinders (5 mm internal diameter; 10 mm height) were applied on the agar surface in Petri plates. Then, 0.1 ml of each compound (10 mg/ml in DMSO) was added into the cylinders. The DMSO solvent was also tested in order to assess its intrinsic antimicrobial activity. Commercial available discs containing ampicillin (25 μg/disc), chloramphenicol (30 μg/disc), and nystatin (100 μg/disc) were also placed on the agar surface. The plates were incubated at 37°C for 24 h (bacteria) and at 24°C for 48 h (yeasts). After incubation the diameters of inhibition zones were read in triplicate. Statistical analysis of the results included the calculation of standard deviation (**Tables 12** and **13**).

The qualitative screening of the antimicrobial activity was performed in order to identify the antimicrobial spectrum of the tested compounds. The inhibitory effects of the synthetic compounds against Gram-positive and Gram-negative bacteria and fungi are given in **Tables 12** and **13** [35].

According to the results of the antibacterial studies, the efficacy of the tested compounds against Gram-positive bacteria was higher than that exhibited for Gram-negative bacteria. All the synthesized compounds were very active against *S. aureus* ATCC 25923, the most active compounds being **Ib**, **IIb**, **IIIb**, and **IVb**. The replacement of the methyl radical in the fourth position with the propyl group was correlated with an increased activity against *S. aureus* ATCC 25923.

The tested compounds exhibited excellent antibacterial activity against *S. lutea*, the most active derivatives being **IIb**, **IVb**, **Ib**, and **IIIb**.

Compd./reference	**Diameters of the growth inhibition zone (mm)**				
	S. aureus ATCC 25923	*S. lutea* ATCC 9341	*B. cereus* ATCC 14579	*E. coli* ATCC 25922	*Pseudomonas aeruginosa* ATCC 27853
Ia	14 ± 0.52	25 ± 0.79	25 ± 1.52	12 ± 0.79	8 ± 0.93
Ib	27 ± 1.29	29 ± 0.83	NA	NA	NA
IIa	14 ± 0.91	22 ± 0.79	26 ± 0.79	11 ± 0.52	8 ± 1.43
IIb	27 ± 0.52	30±	NA	NA	NA
IIIa	15 ± 0.54	20 ± 0.79	22 ± 0.79	10 ± 0.79	9 ± 0.79
IIIb	25 ± 0.52	28±	NA	NA	NA
IVa	17 ± 1.08	25 ± 0.91	24 ± 1.52	12 ± 0.93	9 ± 1.43
IVb	25 ± 1.08	30 ± 0.83	NA	NA	NA
Va	14 ± 0.52	25 ± 0.52	20 ± 1.43	10 ± 1.52	8 ± 0.52
Vb	21 ± 1.43	25 ± 0.79	NA	13 ± 0.83	NA
Ampicillin (25 μg/disc)	26 ± 0.04	36 ± 0.00	NA	21 ± 0.79	NA
Chloramphenicol (30 μg/disc)	22 ± 0.00	38 ± 0.00	24 ± 0.00	21 ± 0.52	NA

Data are mean ± SD (n = 3); NA, no activity.

Table 12.
*Antibacterial activity of compounds **I–V**.*

Compd./reference	Diameters of the growth inhibition zone (mm)		
	C. albicans ATCC 10231	*C. glabrata* ATCC MYA 2950	*C. parapsilosis* ATCC 22019
Ia	24 ± 1.83	21 ± 0.52	34 ± 1.83
Ib	10 ± 0.91	10 ± 0.79	10 ± 0.54
IIa	25 ± 0.52	27 ± 0.54	35 ± 1.83
IIb	9 ± 1.83	NA	NA
IIIa	19 ± 1.79	24 ± 0.52	24 ± 1.79
IIIb	12 ± 1.83	11 ± 0.54	11 ± 0.54
IVa	23 ± 0.91	16 ± 0.52	25 ± 1.08
IVb	10 ± 0.54	12 ± 1.08	NA
Va	16 ± 1.79	21 ± 1.83	21 ± 0.54
Vb	9 ± 0.54	9 ± 0.52	NA
Nystatin (100 μg/disc)	25 ± 0.52	25 ± 0.52	24 ± 0.00

Data are mean ± SD (n = 3); NA, no activity.

Table 13.
*Antifungal activity of compounds **I–V**.*

We found a moderate action against *B. cereus* ATCC 14579, the most active being the umbelliferone derivatives with a methyl group attached to C4: **IIa**, **Ia**, and **IVa**.

Against *Escherichia coli* ATCC 25922, the investigated compounds had a weaker action than the controls ampicillin and chloramphenicol. The most active was the compound that contains a thiadiazole ring, **Vb**.

The presence of the methyl group attached to the coumarin ring in the fourth position had a positive influence on the anti-*Pseudomonas* ATCC 27853 potential of the compounds, all the tested 4-propyl-coumarin derivatives being inactive.

We have noticed a very important action against the investigated *Candida* strains; all tested compounds were found to be very active against fungi. The compounds **IIa** and **Ia** had a greater inhibitory potential against *C. parapsilosis* ATCC 22019 than nystatin. The introduction of the sulfur atom appeared to be correlated with a good anti-*Candida* activity.

3.2 Antioxidant activity

In order to evaluate the antioxidant activity of synthesized compounds, we use three antioxidant assays: DPPH radical inhibition, total reducing power, and nitric oxide (NO) inhibition.

The DPPH assay is based on assessing the substances' ability to reduce the stable radical (diphenylpicrylhydrazyl) to diphenylpicrylhydrazine. The DPPH free radical, bearing an odd electron, gives a strong absorption maximum at $\lambda = 517$ nm (purple color). When the odd electron of the DPPH radical pairs with a hydrogen atom from an antioxidant, the reduced form DPPH-H is created, and the color turns from purple to yellow [36, 37].

A possible mechanism that can explain the antioxidant effect of the coumarin hydrazide derivatives is related to the keto-enol forms of the substances, the enol group being capable to easily donate the hydrogen (**Figure 8**) [38].

The experimental procedure for the DPPH assay was adapted from literature [27, 28, 37], only slight modifications being made. Briefly, 2.5 ml solution of DPPH (2,2-diphenyl-1-picrylhydrazyl) radical 0.1 mM in methanol was added over 0.5 ml

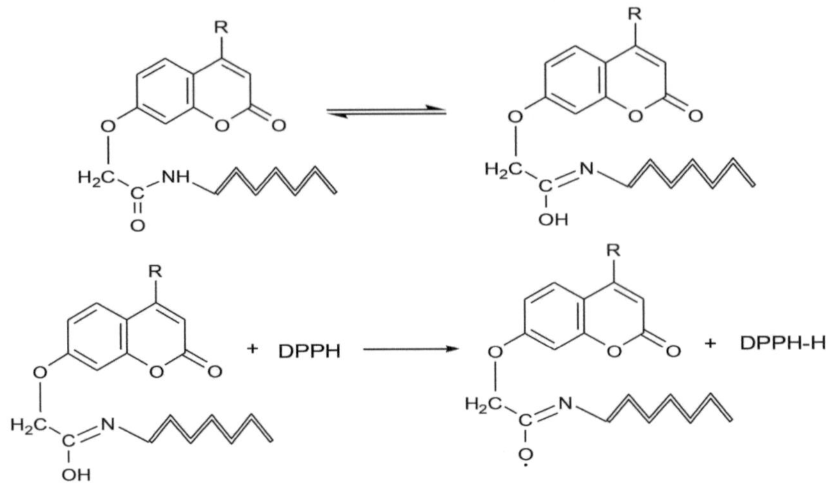

Figure 8.
Proposed mechanism for antioxidant activity of coumarin hydrazides.

of methanolic solution of the tested compound (1 mg/ml). The absorbance of the DPPH solution at 517 nm was determined spectrophotometrically before ($A_{control}$) and 15 minutes after adding the solutions of the compounds (A_{test}), and the percentage of activity was calculated. Ascorbic acid was used as a reference compound:

$$\%\ radical\ scavenging\ activity = (A_{control} - A_{test}) \times 100/A_{control}$$

where $A_{control}$ is the absorbance of the control sample (DPPH solution without test sample) and A_{test} is the absorbance of the test sample (DPPH solution + test compound).

Out of the tested compounds, the most active DPPH free radical scavengers were the coumarin hydrazide derivatives (**IIIa–IIIb, IVa–IVb**). The activities of **IVa** were similar to that of the standard, the inhibition percentage being over 90%, the introduction of sulfur atoms in the molecule having a positive influence on the scavenging potential. Compounds **IIIa** and **IVa**, containing a methyl group, were slightly more active than their analogues with propyl radical (**Table 14**) [19].

Fe(III) reduction is often used as an indicator of electron-donating activity. In the reducing power assay, antioxidants with electron-donating abilities reduce ferricyanide to ferrocyanide by donating an electron. The amount of ferrocyanide is monitored by measuring the formation of Perl's Prussian blue at 700 nm. Increasing absorbance at 700 nm indicates an increase in reducing ability [39]. Within this assay, EC_{50} values are the effective concentrations at which the absorbance is 0.5.

The solution of the test compound (0.5 ml) at different concentrations in methanol was mixed with phosphate buffer (1.25 ml, 0.2 mol/l, pH 6.6) and potassium ferricyanide 1% (1.25 ml), and the mixture was incubated at 50°C for 20 min. At the end of the incubation period, trichloroacetic acid 10% (1.25 ml) was added to the mixture and centrifuged at 3000 rpm for 10 min. The upper layer solution was collected, and 2.5 ml were mixed with distilled water (2.5 ml) and ferric chloride 0.1% (0.5 ml). The absorbance was measured after 15 min at 700 nm against a blank. The EC50 values were calculated by linear interpolation between values above and below 50% activity. Ascorbic acid was used as reference [28, 30, 36, 37].

The reducing power of the tested compounds was modest, and the results are presented in **Table 15**. The only substances that were moderately active were the hydrazide derivatives **IIIa** and **IVa**, but their activity was inferior to that exhibited by the reference substance (ascorbic acid) [19].

Compd.	Inhibition percentage (%)
Ia	21.4
Ib	21.2
IIa	19
IIb	19.83
IIIa	73
IIIb	53.56
IVa	95
IVb	87.67
Va	31.51
Vb	25.13
Ascorbic acid 1 mg/ml	96.8

Table 14.
DPPH inhibition percentages of compounds ***I–V*** *(1 mg/ml).*

Compd.		Extinction
Ia		0.0599
Ib		0.0105
IIa		0.0234
IIb		0.0010
IIIa	(1 mg/ml)	1.3432
	(0.8 mg/ml)	1.1237
	(0.6 mg/ml)	1.0821
	(0.4 mg/ml)	0.9674
	(0.2 mg/ml)	0.5499
IVa	(1 mg/ml)	0.6592
	(0.8 mg/ml)	0.5900
	(0.6 mg/ml)	0.4854
	(0.4 mg/ml)	0.4258
	(0.2 mg/ml)	0.2785
IVb		0.4887
Va		0.0472
Vb		0.3063
Ascorbic acid 1 mg/ml		2.8261

Table 15.
The reducing power of compounds ***I–V****.*

The calculated values for EC_{50} are shown in **Table 16**. This method could not be applied to compound **IIIb** due to the formation of an abundant precipitate in the process.

Nitric oxide is involved in a variety of biological functions (neurotransmission, vascular homeostasis, antimicrobial and antitumor activities). NO was primarily

Compd.	EC_{50} (mg/ml)
Ia	>>1
Ib	>>1
IIa	>>1
IIb	>>1
IIIa	0.176
IVa	0.627
IVb	1.02
Va	>>1
Vb	>1
Ascorbic acid 1 mg/ml	0.049

Table 16.
The calculated values of EC_{50}.

described as a regulator of vascular tones in the cardiovascular system. Beyond this function it can prevent platelet activation, limit leukocyte adhesion to the endothelium, and regulate myocardial contractility, and it is involved in immune system reactions.

Despite the possible beneficial effects of NO, it also contributes to oxidative damage. In general, the overwhelming production of NO contributes to the pathogenesis of both acute and chronic inflammatory processes, and NO has been recognized as one of the main signaling molecules involved in these processes [23, 40]. Therefore, compounds that act like nitric oxide inhibitors have beneficial effects.

The NO inhibition assay is based on the diazotization of sulfanilic acid at acid pH by nitric oxide. The reaction product is subsequently coupled stoichiometrically with N-(1-naphthyl)ethylenediamine, forming a colored azo compound which is measured spectrophotometrically at a peak absorbance of 548 nm [36].

0.5 ml of the tested coumarin derivative solution, as well as ascorbic acid (standard compound), was taken in separate tubes, and 2.0 ml of sodium nitroprusside

Compd.	NO inhibition (%)
Ia	15.2
Ib	19.8
IIa	12.11
IIb	14.7
IIIa	22.8
IIIb	55.5
IVa	28.45
IVb	29.44
Va	22.6
Vb	26
Ascorbic acid 50 mg/ml	77.19

Table 17.
*NO inhibition activity of compounds **I–V**.*

(10 mM) and 0.5 ml phosphate buffer saline (pH = 7.4) were added to each tube. The solutions were incubated at 25°C for 150 minutes. After the incubation, over 0.5 ml of the incubated solution 1 ml of sulfanilic acid 0.33% was added, and the mixture was left for 5 min at room temperature; after this period of time, 1 ml naphthylethylene diamine (NED) HCl reagent 0.1% was added, and the solutions were incubated for another 30 min. The absorbance was measured at 546 nm [38].

Most of the investigated compounds were moderate NO inhibitors (**Table 17**) [19].

4. Concluding remarks

We have synthesized some coumarin derivatives starting from 4-methyl-7-hydroxycoumarin with antimicrobial and antioxidant activities to different reaction steps. The IR and NMR spectra of the synthesized compounds were in accordance with the assigned structures. All the synthesized compounds were very active against *S. aureus* ATCC 25923, and they exhibited excellent antibacterial activity against *S. lutea*. The presence of the methyl group attached to the coumarin ring in the fourth position had a positive influence on the anti-*Pseudomonas* ATCC 27853 potential of the compounds, all the tested 4-propyl-coumarin derivatives being inactive. Against the investigated *Candida* strains, all tested compounds were found to be very active. The introduction of the sulfur atom appeared to be correlated with a good anti-*Candida* activity. The most active DPPH free radical scavengers were the coumarin hydrazide derivatives, the activities of these being similar to that of the standard. The reducing power of the tested compounds was modest, and only the hydrazide derivatives were moderately active. Most of the investigated compounds were moderate NO inhibitors.

The interest in the synthesis of coumarin derivatives has been gaining importance over the last decades, reflecting the importance of such compounds in both medical and chemical research. Future goals for this field of research include the discovery, synthesis, and development of compounds which display increased potency, as well as fueling structure–activity relationship studies aimed at understanding the modes of action of the most biologically active members of these classes of products.

Although coumarin is a simple molecule and many of its derivatives have been known for more than a century, it continues to maintain the interest of researchers being a plentiful source of potential drug candidate because of their significant therapeutic potential.

Conflict of interest

The authors declare no conflict of interest.

Author details

Gabriela Tataringa* and Ana Maria Zbancioc
University of Medicine and Pharmacy "Grigore T. Popa", Iasi, Romania

*Address all correspondence to: gtataringa22@yahoo.com

References

[1] Raev L, Voinov E, Ivanov I, Popov D. Antitumor activity of some coumarin derivatives. Pharmazie. 1990;**45**(9): 696-698

[2] Nofal ZM, El-Zahar M, Abd El-Karim S. Novel coumarin derivatives with expected biological activity. Molecules. 2000;**5**(2):99-113. DOI: 10.3390/50200099

[3] El-Agrody AM, Abd El-Latif MS, El-Hady NA, Fakery AH, Badair AH. Heteroaromatisation with 4-hydroxycoumarin part 2. Molecules. 2001;**6**(6):519-527. DOI: 10.3390/60600519

[4] Patibha S, Shreeya P. Synthesis, characterization and antimicrobial studies of some novel 3-arylazo-7-hydroxy-4-methylcoumarin. Indian Journal of Chemistry. 1999;**38B**: 1139-1142

[5] Patonay T, Litkei GY, Bognar R, Erdei J, Misztic C. Synthesis, antimicrobial and antifungal activity of 4-hydroxycoumarin derivatives, analogues of Novobiocin. Pharmazie. 1984;**39**:86-91

[6] Shaker RM. Synthesis and reactions of some new 4H-pyrano[3,2-c] benzopyran-5-one derivatives and their potential biological activities. Pharmazie. 1996;**51**(3):148-151

[7] Cacic M, Molnar M, Balic T, Draca N, Rajkovic V. Design and Synthesis of Some Thiazolidin-4-ones Based on (7-Hydroxy-2-oxo-2H-chromen-4-yl) Acetic Acid. Molecules. 2009;**14**: 2501-2513. DOI: 10.3390/molecules14072501

[8] Manolov I, Maichle-Moessmer C, Danchev N. Synthesis, structure, toxicological and pharmacological investigations of 4-hydroxycoumarin derivatives. European Journal of Medicinal Chemistry. 2006;**41**(7): 882-890. DOI: 10.1016/j.ejmech.2006.03.007

[9] Al-Majedy Y, Al-Amiery A, Kadhum AA, Bakar Mohamad A. Antioxidant activity of coumarins. Systematic Reviews in Pharmacy. 2017;**8**(1):24-30. DOI: 10.5530/srp.2017.1.6

[10] Emmanuel-Giota AA, Fylaktakidou KC, Hadjipavlou-Latina DJ, Litinas KE, Nicolaides DN. Synthesis and biological evaluation of several 3-(coumarin-4-yl) tetrahydroisoxazole and 3-(coumarin-4-yl)dihydropyrazole derivatives. Journal of Heterocyclic Chemistry. 2001;**38**(3): 717-722. DOI: 10.1002/jhet.5570380329

[11] Tenover FC. Mechanisms of antimicrobial resistance in bacteria. The American Journal of Medicine. 2006; **119**(6):S3-S10. DOI: 10.1016/j.amjmed.2006.03.011

[12] Rookaya M, Karenchak L, Romanowski E, Kowalski R. Fourth generation fluoroquinolones: New weapons in the arsenal of ophthalmic antibiotics. American Journal of Ophthalmology. 2002;**133**(4):463-466

[13] Veselinović J, Matejić J, Veselinović A, Soković D. Coumarin structure as a lead scaffold for antibacterial agents - molecular docking study. Biologica Nyssana. 2016;**7**(2):167-170. DOI: 10.5281/zenodo.200417

[14] Al-Majedy Y, Al-Amiery A, Kadhum AA, Bakar Mohamad A. Coumarins: The antimicrobial agents. Systematic Reviews in Pharmacy. 2017; **8**(1):62-70. DOI: 10.5530/srp.2017.1.11

[15] Medimagh-Saidana S, Romdhane A, Daami-Remadi M, Jabnoun-Khiareddine H, Touboul D, Ben Jannet H, et al. Synthesis and antimicrobial activity of novel coumarin derivatives from

4-methylumbelliferone. Medicinal Chemistry Research. 2015;**24**:3247-3257. DOI: 10.1007/s00044-015-1368-y

[16] Al-Amiery AA, Kadhum AAH, Mohamad AB. Antifungal activities of new coumarins. Molecules. 2012;**17**: 5713-5723. DOI: 10.3390/molecules17055713

[17] Behrami A, Krasniqi I. Antibacterial activity of coumarine derivatives synthesized from 8- amino-4,7-dihydroxy-chromen-2-one and comparison with standard drug. Journal of Chemical and Pharmaceutical Research. 2012;**4**(5):2495-2500

[18] RamaGanesh CK, Yadav DB, Venkatesh KB. Synthesis and biological evaluation of some innovative coumarin derivatives containing thiazolidin-4-one ring. Indian Journal of Chemistry. 2010; **49B**:1151-1154

[19] Tataringa G, Stan C, Mircea C, Jitareanu A, Zbancioc AM. Antioxidant evaluation of some coumarin derivatives. Farmácia. 2016;**64**(4): 533-538

[20] Yasameen K, Al-Majedy D, Al-Duhaidahawi K, Al-Azawi A, Kadhum A, Mohamad AB. Coumarins as potential antioxidant agents complemented with suggested mechanisms and approved by molecular modeling studies. Molecules. 2016;**21**: 135-146. DOI: 10.3390/molecules21020135

[21] Bast A, Haenen GR, Doelman CJ. Oxidants and antioxidants: State of the art. The American Journal of Medicine. 1991;**91**(3):S3-S13. DOI: 10.1016/0002-9343(91)90278-6

[22] Lobo V, Patil A, Phatak A, Chandra N. Free radicals, antioxidants and functional foods: Impact on human health. Pharmacognosy Reviews. 2010; **4**(8):118-126. DOI: 10.4103/0973-7847.70902

[23] Patel RM, Patel NJ. *In vitro* antioxidant activity of coumarin compounds by DPPH, super oxide and nitric oxide free radical scavenging methods. Journal of Advanced Pharmacy Education & Research. 2011;**1**:52-68

[24] Jamier V, Marut W, Valente S, Chereau C, Chouzenoux S, Nicco C, et al. Chalcone-Coumarin derivatives as potential anti-cancer drugs: An *in vitro* and *in vivo* investigation. Anti-Cancer Agents in Medicinal Chemistry. 2014; **14**(7):963-974. DOI: 10.2174/1871520613666131224124445

[25] Maja M, Čačić C. Antioxidant activity of some (7-hydroxy-2-oxo-2H-chromen-4yl) acetic acid derivatives. Croatian Journal of Food Science and Technology. 2012;**4**(1):54-63

[26] Gulati S, Wakode S, Kaur A, Anand K. Synthesis, biological activity and recent advancement of benzothiazoles: A Classical Review. World Journal of Pharmacy and Pharmaceutical Sciences. 2017;**6**(6):1842-1869. DOI: 10.20959/wjpps20176-9423

[27] Mohamed HM, Abd EL-Wahab HF, EL-Agrody AM, Bedair AH, Eid FA, Khafagy MM, et al. Synthesis and characterization of new diiodocoumarin derivatives with promising antimicrobial activities. Beilstein Journal of Organic Chemistry. 2011;**7**: 1688-1696. DOI: 10.3762/bjoc.7.199

[28] Hamdi N, Al-Ayed AS, Said RB, Fabienne A. Synthesis and characterization of new thiazolidinones containing coumarin moieties and their antibacterial and antioxidant activities. Molecules. 2012;**17**(8):9321-9334. DOI: 10.3390/molecules17089321

[29] Sreeja S, Arul K, Kumaran J. New Coumarin Derivatives. Design, Synthesis and Pharmacological Evaluation. LAP LAMBERT Saarbrücken, Deutschland/Germany: Academic Publishing; 2013. 5p

[30] Satyanarayana VSV, Sreevani P, Sivakumar A, Vijayakumar V. Synthesis and antimicrobial activity of new Schiff bases containing coumarin moiety and their spectral characterization. Arkivok. 2008;**xvii**:221-223. DOI: 10.3998/ark.5550190.0009.h21

[31] Khan MS, Siddiqui SP, Tarannum N. A systematic review on the synthesis and biological activity of hydrazide derivatives. Hygeia Journal for Drugs and Medicine. 2017;**9**(1):61-79. DOI: 10.15254/H.J.D.Med.9.2017.165

[32] Upmanyu N, Kumar S, Shah K, Mishra P. Synthesis and antimicrobial studies of some 4-(substituted)-ethanoylamino 3-mercapto-5-(4-substituted) phenyl-1,2,4-triazoles. Dhaka University Journal of Pharmaceutical Sciences. 2012;**11**(1): 7-18. DOI: 10.3329/dujps.v11i1.12481

[33] El-Hashash MA, Guirguis DB, Abd El-Wahed NAM, Kadhim MA. Synthesis of novel series of phthalazine derivatives with antibacterial and antifungal evaluation. Journal of Chemical Engineering & Process Technology. 2014;**5**(4):1-8. DOI: 10.4172/2157-7048.1000195

[34] Rehman A, Siddiqui SZ, Abbasi MA, Abbas N, Khan KM, Shahid M, et al. Synthesis, antibacterial screening and hemolytic activity of s-substituted derivatives of 5-benzyl-1,3,4-oxadiazole-2-thiol. International Journal of Pharmacy and Pharmaceutical Sciences. 2012;**4**(2):676-680

[35] Tataringa G, Tuchilus C, Jitareanu A, Stan CD, Zbancioc AM. Antimicrobial prospection of some coumarin derivatives. Farmácia. 2016; **66**(2):323-330

[36] Ranković B, Kosanić M, Stanojković T. *Stereocaulon paschale* lichen as antioxidant, antimicrobial and anticancer agent. Farmácia. 2014;**62**(2): 306-317

[37] Kadhum AAH, Ahmed AA, Musa AY, Mohamad AB. The antioxidant activity of new coumarin derivatives. International Journal of Molecular Sciences. 2011;**12**(9):5747-5761. DOI: 10.3390/ijms12095747

[38] Matthews EK, Seaton ED, Forsyth MJ, Humphrey PPA. Photon pharmacology of an iron-sulphur cluster nitrosyl compound acting on smooth muscle. British Journal of Pharmacology. 1994;**113**(1):87-94

[39] Kenchappa R, Bodke YD, Chandrashekar A, Sindhe MA, Peethambaret SK. Synthesis of coumarin derivatives containing pyrazole and indenone rings as potent antioxidant and antihyperglycemic agents. Arabian Journal of Chemistry. 2017;**10**(2):S3895-S3906. DOI: 10.1016/j.arabjc.2014.05.029

[40] Víteček J, Lojek A, Valacchi G, Kubala L. Arginine-based inhibitors of nitric oxide synthase: Therapeutic potential and challenges. Mediators of Inflammation. *2012*;**2012**:1-22. DOI: 10.1155/2012/318087

Chapter 8

Coumarins as Fluorescent Labels of Biomolecules

António Pereira, Sérgio Martins and Ana Teresa Caldeira

Abstract

Important areas such as environmental sciences, medicine, pharmacy, and cellular biology are dependent on very sensitive analytical techniques. One of the most common methodologies used for their bioanalytical purposes is the fluorescent labelling. The synthesis of new fluorophores and the great development of fluorescent-labelling techniques combined with the enormous technological advances in the field of fluorescence microscopy allowed to deepen the structural knowledge of biomolecules. This new organic fluorophores form covalent bonds with the sample to be analyzed, producing stable bioconjugates that show fluorescence in a wide range of wavelengths, depending on the label used. Coumarin derivatives represent one of the most important chemical classes of organic fluorescent materials being one of the most extensively investigated and commercially significant groups of organic fluorescent materials. In this chapter, it is reviewed the use of fluorescent coumarin derivatives and their application to labelling biomolecules. These fluorescent labels allow researchers to study, and understand, biomolecular assemblies that exhibit complex sensitivity and selectivity. Reactive fluorescent coumarin derivatives are actually widely used in labelling biomolecules as peptides, proteins, oligonucleotides, nucleic acids, and carbohydrates, among other biological molecules.

Keywords: coumarins, fluorophores, labelling, biomolecules, bioconjugation

1. Introduction

Important areas such as environmental sciences, medicine, medicinal chemistry, and cellular biology are dependent on very sensitive analytical techniques to detect and track biomolecules (amino acids, peptides, proteins, antibodies, oligonucleotides, nucleic acids, carbohydrates, and other biological molecules). Many of these techniques often require labelling with reporters or sensors, such as isotope labels [1], radioactive tracers [2], colorimetric biosensors [3], photoswitchable biomaterials [4], photochromic compounds [5, 6], electrochemical sensors [7], or fluorescent labels [8, 9]. The fluorescent labelling presents numerous advantages, when compared to the other techniques, due to the high sensitivity of the fluorescence technique and also due to its non-destructive nature that allows the use of small sample quantities and their fluorescent labels. The fluorescence process occurs in certain molecules called fluorophores or fluorescent dyes, and a fluorescent probe is nothing more than a fluorophore enabled to detect particular components of complex biomolecular assemblies, including live cells, with complex sensitivity and selectivity [10]. The organic fluorophores may form covalent or non-covalent linkages with the sample to be analyzed, producing the respective bioconjugates (or complexes) that can show fluorescence, from short to very long wavelengths, depending on the label used. The bioconjugation technique depends on two interrelated chemistries: the reactive functionality present on the fluorescent label and the functional groups present on the target biomolecules to be labeled. The knowledge of the basic mechanisms by which the reactive groups couple to target functionalities provides the means to intelligently design the bioconjugation strategy. Choosing the correct fluorescent label that can react with the chemical groups available on target biomolecules forms the basis for successful labelling [11].

In general, the fluorescent label should be small in size and chemically stable, with minimal interference on the structure and biological functions of the unlabeled biomolecules, producing high fluorescence quantum yield bioconjugates.

On the other hand, the labelling reaction should be extremely efficient with high yields, preferably establishing a stable covalent linkage between the fluorescent label and a specific residue in the target biomolecule. The efficiency and selectivity of several fluorescent-labeled biomolecules have been used to study and understand their dynamics, kinetics, and photophysical properties [12–18].

The amine reactive fluorescent labels are the most frequently used to prepare stable bioconjugates to a great number of biological applications since amino groups are either abundant or easily introduced into biomolecules. In contrast, to study some particular protein structures and functions, thiol-reactive reagents are chosen due to the smaller presence of thiol groups, when compared with lysine, in biomolecules [19]. In this context, cysteine is generally the amino acid chosen to label when it is desired to label selectively a protein in vitro, due to its relatively low abundance and high nucleophilicity compared to other amino acid side chains. Specific and noninterfering dual fluorescent labelling in a peptide or protein molecule allows conformational investigations in terms of intramolecular distances [20].

The expeditious development of the fluorescent-labelling techniques allowed to explore and discover several cellular functions. To study, and understand, the activity of signal transduction by visualizing protein binding or folding, the fluorescence correlation spectroscopy (FCS) and the fluorescence resonance energy transfer (FRET) are widely used [21]. Molecular tags that specifically bind to particular membrane-permeable dyes [22] allow to study protein dynamics and trafficking by fluorescence recovery after photobleaching (FRAP) as well the protein turnover [23, 24].

The great development of fluorescent-labelling techniques combined with the enormous technological advances in the field of fluorescence microscopy allowed to study, in vivo and in vitro systems, the protein distribution as well as their translocation and their interactions [25]. With specific and efficient fluorescent labelling, the proteins can be visualized in real time for the elucidation of their functions in a complex biological network, which also allows the detection of the protein-protein interactions, fundamental to understand intra- and intercellular communications [26].

Coumarins (benzopyranones or 2H-chromen-2-ones), whether natural products or synthetic ones, have also aroused a growing interest of the scientific community in the last decades due to their very significant pharmacological activity [27–37]. The nature and substitution pattern in the coumarins grant them diversified and exceptional optical properties with high fluorescence quantum yields [38]. Coumarins constitute the major class of fluorescent dyes [39–63], used as fluorescent labels and probes for physiological measurement [43–47], fluorescent whiteners [48], optical brighteners [49, 50], nonlinear optical chromophores [51–53], emission layers in organic light-emitting diodes (OLED) [54–57], and more recently, in caging [58–61], and labelling [62, 63]. Due to strong blue fluorescence of coumarin, it is easy to distinguish its light from green, yellow, and red, an enormous advantage in multicolored fluorescence investigation. Developments from the last decade show that the introduction of appropriated substituents into the coumarin ring contributes to structures with improved photophysical and spectroscopic properties [64–66]. The synthesis of new fluorophores, with absorption and emission at long wavelengths, is of extreme importance for biological purposes, and the coumarins may play a leading role in this field.

2. Chemical labelling

Of all different fluorescent-labelling techniques, the chemical labelling is actually one of the most used as it allows novel types of experiments in biomolecules using a wider range of reactive fluorescent chromophores available. The covalent attachment of the chemical probes with specific amino acid has the advantage of being an irreversible process when compared to the non-covalent binding [67]. The chemical labelling methodology produces very stable bioconjugates, easy to manipulate with high efficiency, in a great number of available fluorophores that can be coupled covalently to the target biomolecule. Chemical labelling methods produce better results in in vitro studies rather than in vivo [18]. The most used methods in chemical labelling, in the biomolecules' native functional groups, under mild aqueous conditions, and using fluorescent coumarins, are discussed below.

2.1 Amine reactive fluorescent coumarins

Presently, amine reactive fluorescent coumarins are widely used to label biomolecules, as peptides, proteins, oligonucleotides, and nucleic acids, among others. The fluorescent bioconjugates obtained are very useful in fluorescence in situ hybridization (FISH), receptor labelling immunochemistry, cell tracing, and fluorescent analog cytochemistry studies. Almost all of the techniques used in these tests implicate a robust fluorescent conjugate able to support rigorous incubation, hybridization, and washing steps, which is provided by the stability of the covalent bond between the amine reactive dye and biomolecule. Chemically, the amine labelling reaction proceeds usually through acylation pathway producing stable amide (or thiourea) bonds. The "ideal" reactions are those which require the same conditions as proteins, like functional group tolerance, compatibility, selectivity, water as solvent (or pH ~ 7), room temperature, high reaction rates, low reactant concentration, and nontoxic reagents.

A number of fluorescent amino-reactive coumarins have been developed to label various biomolecules, and the resultant conjugates are widely used in biological applications. Four major classes of amine-reactive fluorescent reagents are currently used to label biomolecules: succinimidyl esters (SE), 4-sulfotetrafluorophenyl (STP) esters, sulfonyl chlorides, and isothiocyanates [68]. **Figure 1** represents, in a general schematic diagram, the referred labelling reactions, between an amine group of a biomolecule and a fluorescent amino-reactive coumarin.

2.1.1 Fluorescent coumarin succinimidyl esters

Succinimidyl esters (SE) are proven to be very good reagents for amine modifications. These kinds of reagents are generally stable and show good reactivity and selectivity with aliphatic amines, such as the amine group of lysine side chain. Some of these kinds of reactive dyes are hydrophobic molecules and should be previously dissolved in anhydrous dimethylformamide (DMF) or dimethylsulfoxide (DMSO), but the sulfo-succinimidyl esters are water soluble. The amine labelling

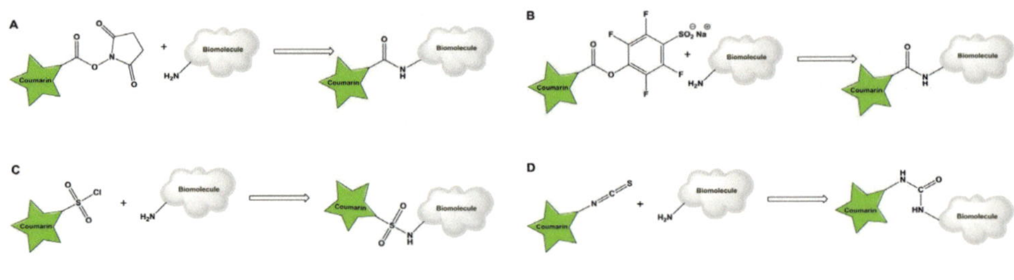

Figure 1.
Schematic diagram of amine labelling techniques using succinimidyl esters (A), 4-sulfotetrafluorophenyl esters (B), sulfonyl chlorides (C), and isothiocyanates (D).

reaction with succinimidyl esters has a handicap, due to its great pH dependence. Succinimidyl esters react with non-protonated aliphatic amine groups, and the amine acylation reaction must be carried out at pH > 7.5. In the specific case of protein labelling by succinimidyl esters, the reactions require a pH between 7.5 and 8.5. Buffers used in labelling reactions shall not contain nucleophilic compounds because they may react with the labelling reagent to form unstable intermediates that could destroy the reactive dye. Most conjugations are done at room temperature, but either high or low temperature may be required for a particular labelling reaction. Some of the fluorescent coumarin succinimidyl esters contain a seven-atom aminohexanoyl spacer between the fluorophore and the reactive group, providing better solubility and spatial separation between the fluorophore and the target molecule being labeled. This separation potentially reduces the quenching that typically occurs upon conjugation and makes the dye more available for recognition by secondary detection reagents [68]. The most important fluorescent coumarin succinimidyl esters used for labelling biomolecules are shown in **Table 1**, as the corresponding values of maximal excitation (Ex) and emission (Em) wavelengths and their physicochemical features and biological applications [19, 68].

2.1.2 Fluorescent coumarin 4-sulfotetrafluorophenyl (STP) esters

Some succinimidyl esters may not be compatible with a specific application due to their insolubility in aqueous solution. To overcome these limitations, the 4-sulfotetrafluorophenyl (STP) ester can be used. These sulfonated esters have higher water solubility than simple succinimidyl esters and sometimes eliminate the need for organic solvents in the conjugation reaction, which is a great advantage to maintain the native characteristics of biomolecules. They are, however, more polar than succinimidyl esters, which makes them less likely to react with buried amines in proteins or to penetrate cell membranes [68, 94]. **Table 2** presents the single fluorescent coumarin 4-sulfotetrafluorophenyl (STP) ester used for labelling biomolecules, as the corresponding values of maximal excitation (Ex) and emission (Em) wavelengths and their physicochemical features and biological applications [95, 96].

2.1.3 Fluorescent coumarin sulfonyl chlorides

Sulfonyl chlorides (SC) are highly reactive and are unstable in water, especially at high pH required for reaction with aliphatic amines. The labelling reactions with sulfonyl chlorides must be performed, carefully, at very low temperature in a place with local exhaust ventilation. Sulfonyl chlorides present a major reactive handicap as they can also easily react with other reactive groups present in biomolecules as phenols, thiols, aliphatic alcohols, imidazoles, and many others. Fortunately, this kind of reactions rarely occurs in proteins or in aqueous solution, allowing the use of this type of chromophores to label proteins. Sulfonyl chloride dyes are generally hydrophobic molecules and should be dissolved in anhydrous dimethylformamide (DMF), but never in dimethylsulfoxide (DMSO) due to their highly instability in this solvent.

The labelling reactions of amines with SC reagents are strongly pH dependent, and the sulfonylation-based conjugations may require a pH 9.0–10.0 for optimal conjugations, which potentiates the sulfonyl chlorides' degradation by hydrolysis reactions. In general, sulfonylation-based conjugations have much lower yields than the succinimidyl ester-based conjugations. As in the case of succinimidyl esters, the buffers used in sulfonyl chloride reactions shall not contain nucleophilic compounds, because they may react with the labelling reagent to form unstable

Coumarin	Ex/Em (nm)	Physicochemical features and biological applications	Ref.
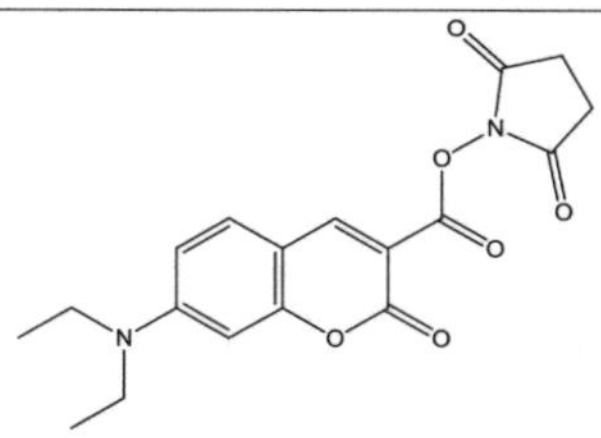 2,5-dioxopyrrolidin-1-yl-7-diethylaminocoumarin-3-carboxylate (DEAC SE)	432/472	Strong blue-fluorescent bioconjugates. Quite hydrophobic fluorescent dye, used for labelling live cells	[19, 69–72]
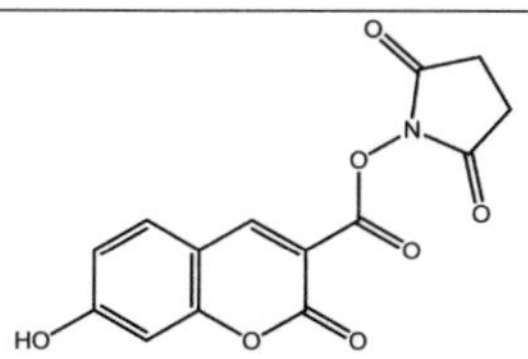 2,5-dioxopyrrolidin-1-yl 7-hydroxycoumarin-3-carboxylate	363/447	One of the most popular blue-fluorescent dyes for labelling proteins and nucleic acids and increasingly used to label peptides, nucleotides, and carbohydrates	[19, 73]
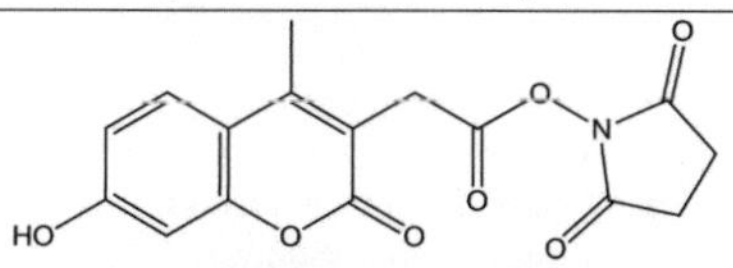 2,5-dioxopyrrolidin-1-yl 2-(7-hydroxy-4-methylcoumarin) acetate	364/458	Widely used for preparing bioconjugates of blue fluorescence but pH-dependent and environment-sensitive fluorescence	[19, 74, 75]
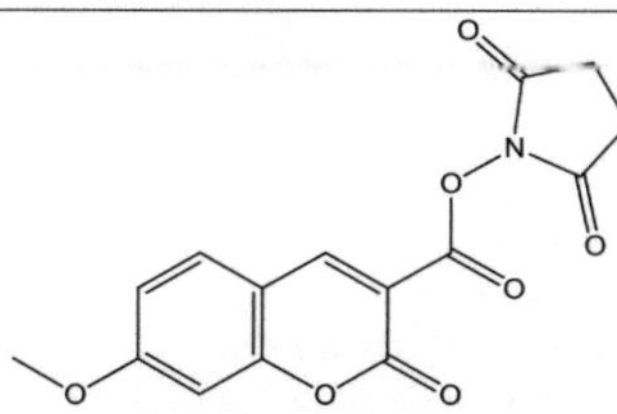 2,5-dioxopyrrolidin-1-yl 7-methoxycoumarin-3-carboxylate	358/410	Used to label peptides and nucleotides with strong blue fluorescence and also used to label cell membranes although its fluorescence is quite short	[19, 72, 76, 77]
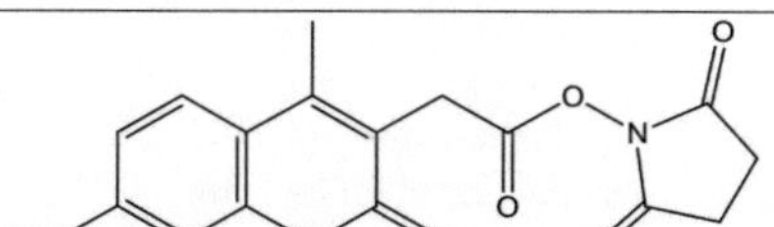 2,5-dioxopyrrolidin-1-yl 2-(7-amino-4-methylcoumarin-3-yl)acetate	350/450	Used for fluorohistochemical examination of human kidney glomeruli. Reacts under mild conditions	[78, 79]

Coumarin	Ex/Em (nm)	Physicochemical features and biological applications	Ref.
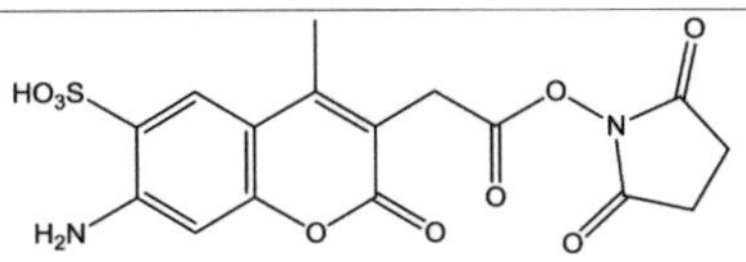 7-amino-3-(2-((2,5-dioxo-pyrrolidin-1-yl)oxy)-2-oxoethyl)-4-methylcoumarin-6-sulfonic acid (Alexa Fluor™ 350 SE)	346/442	Blue-fluorescent dye, water soluble and pH insensitive from pH 4 to pH 10, used for stable signal generation in imaging and flow cytometry	[68, 80–83]
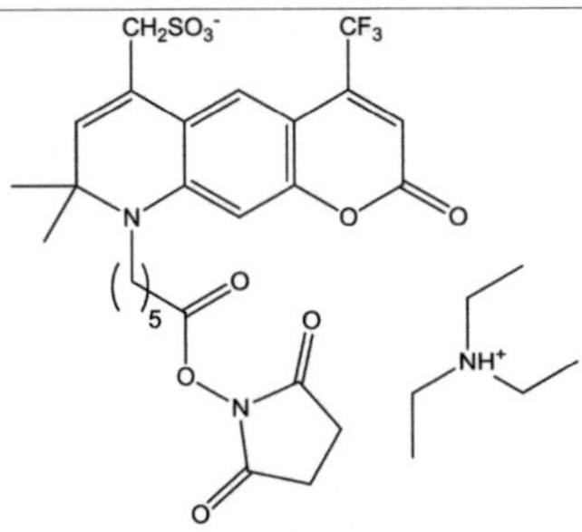 Triethylammonium (9-(6-((2,5-dioxopyrrolidin-1-yl)oxy)-6-oxohexyl)-8,8-dimethyl-2-oxo-4-(trifluoromethyl)-8,9-dihydro-2H-pyrano[3,2-g]quinolin-6-yl)methanesulfonate (Alexa Fluor™ 430 SE)	430/545	Bright green-fluorescent dye, water soluble and pH insensitive from pH 4 to pH 10. Used for stable signal generation in imaging and flow cytometry	[68, 80, 81, 84, 85]
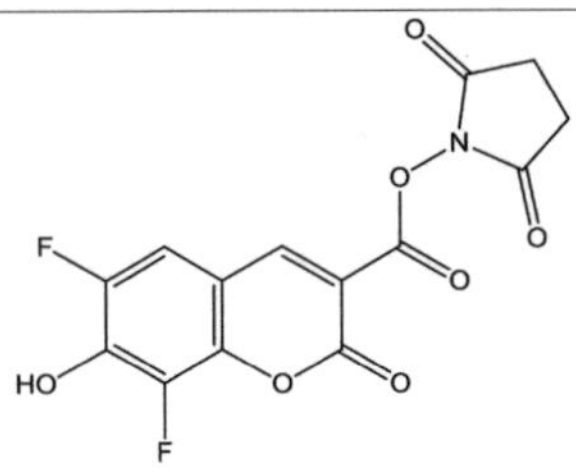 2,5-dioxopyrrolidin-1-yl 6,8-difluoro-7-hydroxycoumarin-3-carboxylate (Pacific Blue™ SE)	410/455	Conjugates of this dye are strongly fluorescent even at neutral pH. Ideally suited for 405 nm violet diode laser excitation on the Applied Biosystems® Attune™ Acoustic Focusing cytometer and similarly equipped fluorescence microscopes	[68, 86–88]
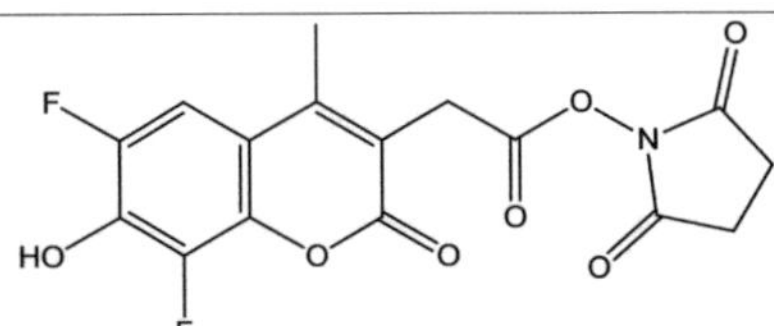 2,5-dioxopyrrolidin-1-yl 2-(6,8-difluoro-7-hydroxy-4-methylcoumarin-3-yl)acetate (Marina Blue™ SE)	365/460	Conjugates that are strongly fluorescent, even at neutral pH. Optimally detected using optical filters configured for 4′,6-diamidino-2-phenylindole (DAPI)	[68]
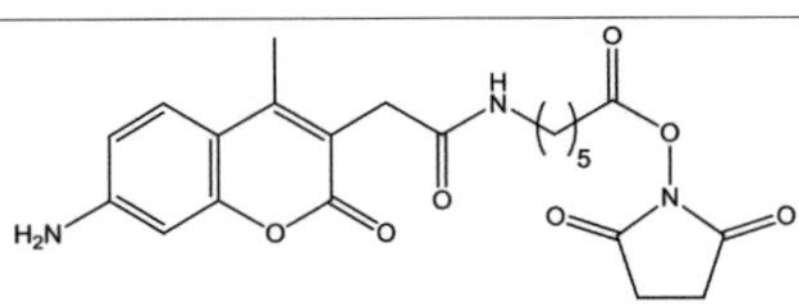 2,5-dioxopyrrolidin-1-yl 6-(2-(7-amino-4-methylcoumarin-3-yl)acetamido)hexanoate (AMCA-X SE)	353/442	Conjugates yield blue fluorescence that can be used as a contrasting color in multicolor applications. Because its fluorescence may not be as bright as that of other dyes or may be obscured by autofluorescence, it is only recommended for use with highly abundant targets	[68, 89–91]

Coumarin	Ex/Em (nm)	Physicochemical features and biological applications	Ref.
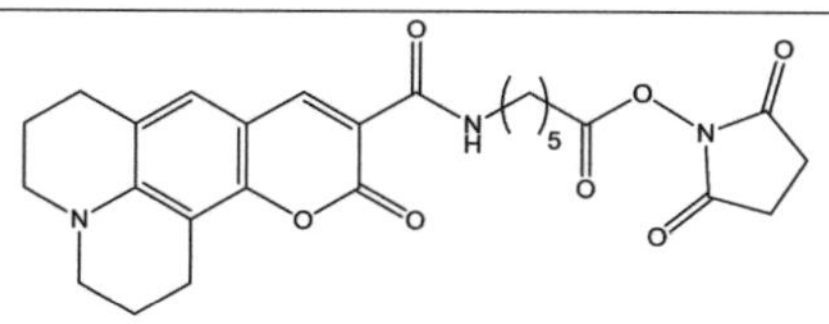 2,5-dioxopyrrolidin-1-yl 6-(11-oxo-2,3,5,6,7,11-hexahydro-1H-pyrano[2,3-f]pyrido[3,2,1-ij] quinoline-10-carboxamido)hexanoate (Coumarin 343 X SE)	437/477	Blue-emitting coumarin can be used to design fluorescence resonance energy transfer (FRET)-based assays with fluorescein amidite (FAM) as acceptor and to construct systems which harvest blue light energy	[92, 93]

Table 1.
Fluorescent coumarin succinimidyl esters used for biomolecule labelling.

intermediates that could destroy the reactive dye [19, 97–99]. **Table 3** shows fluorescent coumarin sulfonyl chlorides used for labelling biomolecules, as the corresponding values of maximal excitation (Ex) and emission (Em) wavelengths and their physicochemical features and biological applications. In addition to the coumarins presented in **Table 3**, new sulfonyl chloride coumarins have been developed, with high potential as fluorescent probes [100, 101].

2.1.4 Fluorescent coumarin isothiocyanates

Isothiocyanates form thioureas upon reaction with amines, but some thiourea products are much less stable than the conjugates that are prepared from the corresponding succinimidyl esters. Most part of isothiocyanate-reactive dyes are hydrophobic molecules and should be dissolved either in anhydrous dimethylformamide (DMF) or in dimethylsulfoxide (DMSO), and their reactions may require a pH 9.0–10.0 for optimal conjugations. As in the previous cases, the buffers used shall not contain nucleophilic compounds. The isothiocyanate conjugations are done at room temperature, but either high or low temperature may be required for a particular labelling reaction [19, 102]. The unique fluorescent coumarin isothiocyanate used for labelling biomolecules is shown in **Table 4**, but new isothiocyanate coumarins have been synthesized, with high potential as fluorescent probes [103, 104].

2.2 Thiol-reactive fluorescent coumarins

Cysteine is, in comparison with lysine, a rare amino acid present in biomolecules, and, for this reason, thiol-reactive reagents are used to label selectively a biomolecule at a defined site, probing their function, interaction, and biological structure. A great number of thiol-reactive dyes have been developed to analyze the proteins' topography in biological membranes, to measure the distances within (or between) proteins, and to observe and understand the changes in protein conformation using environmental sensitive probes.

Maleimides and iodoacetamides are the principal types of thiol-reactive coumarin dyes reported in the literature. Despite many similarities in their reactivity and selectivity toward thiol-reactive moieties, maleimides have a great advantage in relation to iodoacetamides, due to their high stability, solubility in simple solvent mixtures, and their high reactivity in the neutral pH range. Air oxidation of thiol compounds (to

Coumarin	Ex/Em (nm)	Physicochemical features and biological applications	Ref.
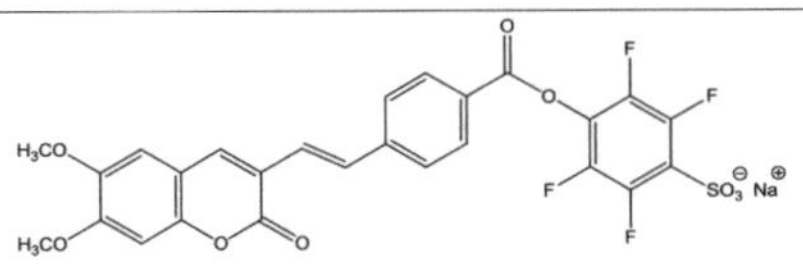 Sodium (E)-4-((4-(2-(6,7-dimethoxycoumarin-3-yl)vinyl)benzoyl)oxy)-2,3,5,6-tetrafluorobenzenesulfonate	392/490	Used to label proteins and nucleotides with strong blue fluorescence. Blue-fluorescent dye, water soluble and pH insensitive with excellent photostability	[95, 96]

Table 2.
Fluorescent coumarin 4-sulfotetrafluorophenyl (STP) ester used for biomolecule labelling.

Coumarin	Ex/Em (nm)	Physicochemical features and biological applications	Ref.
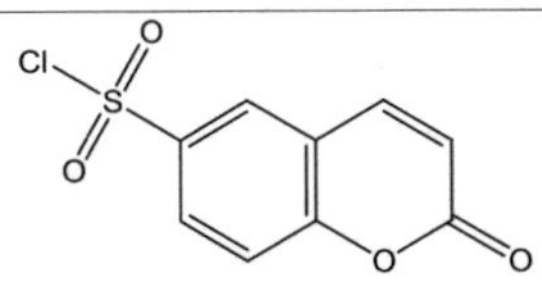 Coumarin-6-sulfonyl chloride	360/460	Used to label amines, amino acids, and phenols in mild conditions. Fluorescence produced in alkaline solution or in the presence of β-cyclodextrin	[99, 100]
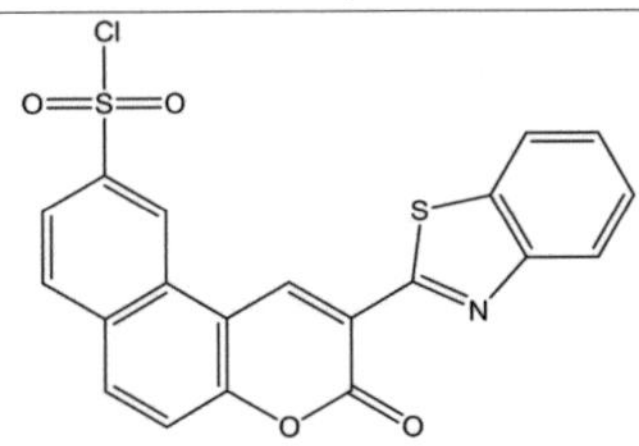 2-(benzo[d]thiazol-2-yl)-3-oxo-3H-benzo[f]chromene-9-sulfonyl chloride	405/435	Biosensor sensitive toward polarity changes in bio environments	[101]

Table 3.
Fluorescent coumarin sulfonyl chlorides used for biomolecule labelling.

Coumarin	Ex/Em (nm)	Physicochemical features and biological applications	Ref.
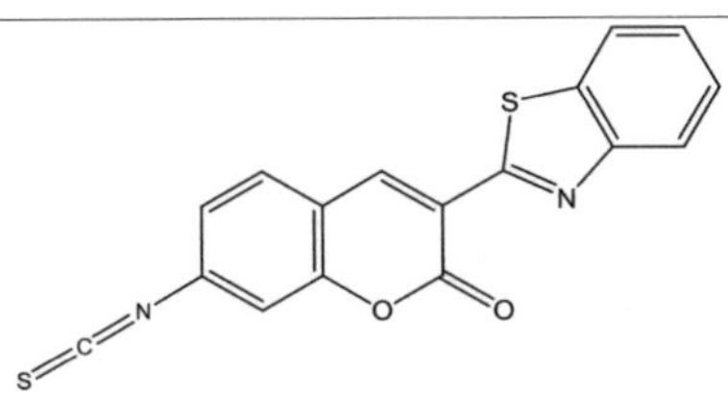 3-(benzo[d]thiazol-2-yl)-7-isothiocyanatocoumarin	485/535 (conjug.)	Selective determination of flu antigen	[102]

Table 4.
Fluorescent coumarin isothiocyanate used for biomolecule labelling.

disulfides) is a major competing reaction for the iodoacetamide modifications of thiol compounds [18, 19, 105]. Due to the disinterest on the development of new coumarin iodoacetamides, for the above reasons, only the fluorescent coumarin maleimides will be focused in this section. **Figure 2** represents, in a general schematic diagram, the thiol-labelling reaction with fluorescent coumarin maleimides.

2.2.1 Fluorescent coumarin maleimides

Maleimides readily react with thiol moieties of biomolecules to form thioether conjugates even under neutral conditions. The thioether bond formed is quite stable and is known to be responsible for the light produced, especially in the solution. Maleimides require conjugation conditions less rigorous than those of iodoacetamides and do not react with histidine and methionine under physiological conditions. Most labelling reactions can be done at room temperature at neutral pH. However, either elevated or reduced pH or temperature may be required for a particular labelling reaction [18, 19, 68]. In **Table 5**, the most important fluorescent coumarin maleimides used for labelling biomolecules are presented, as the corresponding values of maximal excitation (Ex) and emission (Em) wavelengths and their physicochemical features and biological applications.

2.3 Tyrosine-reactive fluorescent coumarins

The hydroxyl groups of the amino acids can be labeled with the same reagents used for the lysine residues, but the labelling reaction is carried out in organic solvent, like anhydrous dimethylformamide (DMF) or dimethylsulfoxide (DMSO), which absorbs the formed water molecule avoiding possible hydrolysis reactions. The amino acid

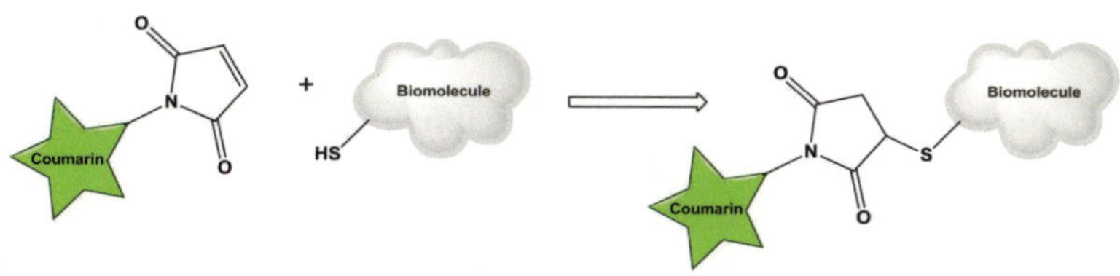

Figure 2.
Schematic diagram of thiol-labelling technique using maleimides.

Coumarin	Ex/Em (nm)	Physicochemical features and biological applications	Reference
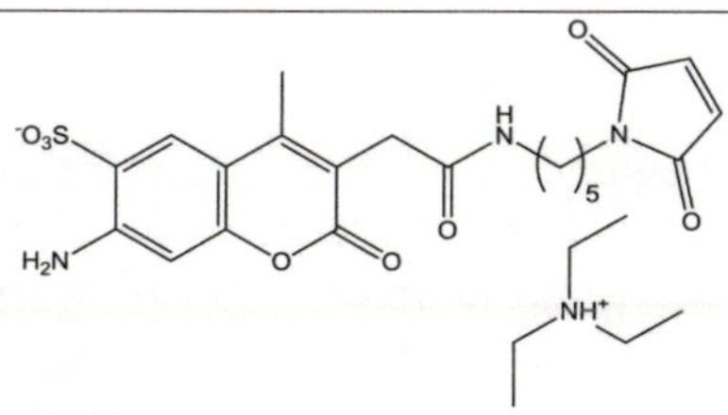 Triethylammonium 7-amino-3-(2-((5-(2,5-dioxo-2,5-dihydro-1H-pyrrol-1-yl)pentyl) amino)-2-oxoethyl)-4-methylcoumarin-6-sulfonate (Alexa Fluor™ 350 C5 Maleimide)	345/444	Blue-fluorescent dye, with moderate photostability, water soluble and pH insensitive from pH 4 to pH 10, used for stable signal generation in imaging and flow cytometry	[68, 80, 81, 106–108]
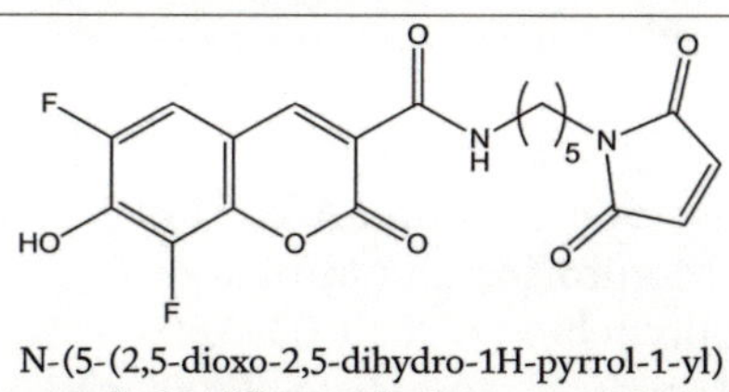 N-(5-(2,5-dioxo-2,5-dihydro-1H-pyrrol-1-yl) pentyl)-6,8-difluoro-7-hydroxycoumarin-3-carboxamide (Pacific Blue™ C5-Maleimide)	410/455	Excellent reagent for thiol-selective modification, quantitation, and analysis and usually requires a higher pH than reaction of maleimides with thiols. Does not react with methionine, histidine, or tyrosine	[68, 80, 109]

Coumarin	Ex/Em (nm)	Physicochemical features and biological applications	Reference
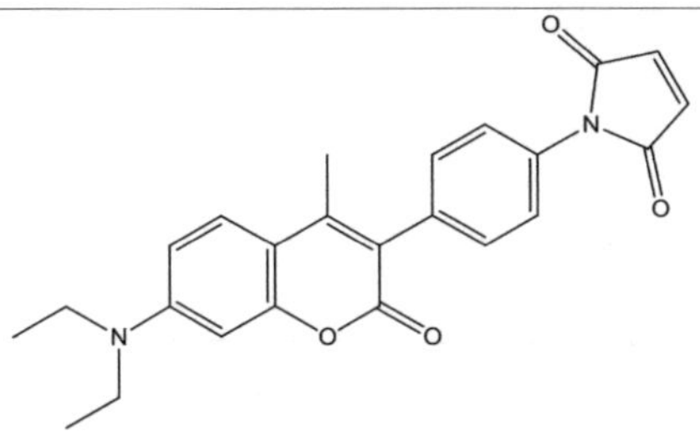 1-(4-(7-(diethylamino)-4-methylcoumarin-3-yl)phenyl)-1H-pyrrole-2,5-dione	391/472	Labelling of protein thiol groups in tissue sections. Fluorescence probe for glutathione intact cells Used to monitor release of thiols, to quantitate thiol in microplate reactions, and to distinguish proliferating cancer cells by nuclear protein staining	[68, 105]
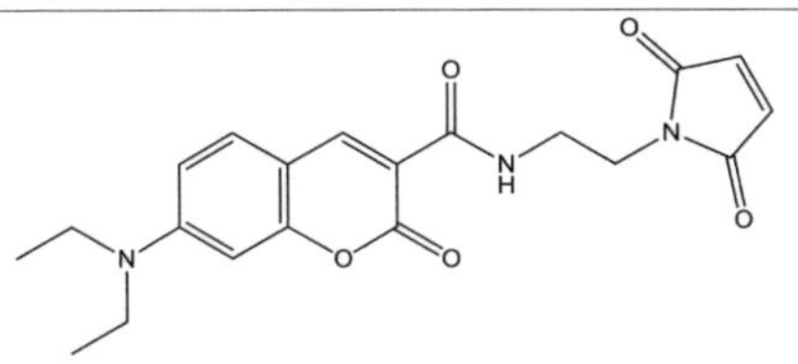 7-(diethylamino)-N-(2-(2,5-dioxo-2,5-dihydro-1H-pyrrol-1-yl)ethyl)coumarin-3-carboxamide	419/467	Used as a fluorescent biological sensing device and for real-time measurements for the release of inorganic phosphates during enzymatic reaction. Also, used for intramolecular fluorescence energy transfer (FRET) experiments	[68, 110–112]

Table 5.
Fluorescent coumarin maleimides used for biomolecule labelling.

hydroxyl groups do not allow highly specific labelling reactions due to the existence of several hydroxyl groups in biomolecules (serine, threonine, and tyrosine) [113].

One of the well-known labelling methods is the reaction with diazonium salts resulting in the formation of azo compounds, as 4-trifluoromethylcoumarin-7-diazonium chloride [114]. Although these aryl diazonium ions are promising for the desired application, their storage and delivery are challenging, and they often require in situ generation. The pH range should be between 8 and 10 for the formation of a phenolate anion [115].

3. Concluding remarks

Reactive fluorescent coumarins have been increasingly attracting special interest as fluorescent labels, with a wide range of applications in bioimaging and biolabelling, due to their extremely attractive and stable scaffold. Coumarins will allow the development of new low-cost fluorescent dyes due to its easy synthesis with high yields, large Stokes shift, pH independence of absorbance and emission, and excellent photostability, which represents a great value for the biological fluorescence imaging techniques.

Acknowledgements

The authors thank the European Regional Development Fund ALENTEJO 2020 through the project "MEDUSA-Microorganisms Monitoring and Mitigation—Developing and Unlocking novel Sustainable Approaches (ALT20-03-0145-FEDER-000015)" for financial support. Sérgio Martins acknowledge Fundação para a Ciência e a Tecnologia (FCT) for the economic support through the doctoral grant SFRH/BD/128807/2017.

Conflict of interest

There are no conflicts of interest to declare.

Author details

António Pereira[1,2]*, Sérgio Martins[1] and Ana Teresa Caldeira[1,2]

1 HERCULES Laboratory, Évora University, Évora, Portugal

2 Chemistry Department, School of Sciences and Technology, Évora University, Évora, Portugal

*Address all correspondence to: amlp@uevora.pt

References

[1] Ong S, Mann M. A practical recipe for stable isotope labeling by amino acids in cell culture (SILAC). Nature Protocols. 2006;**1**:2650-2660. DOI: 10.1038/nprot.2006.427

[2] Varki A. Radioactive tracer techniques in the sequencing of glycoprotein oligosaccharides. The FASEB Journal. 1991;**5**(2):226-235. DOI: 10.1096/fasebj.5.2.2004668

[3] Li H, Rothberg L. Colorimetric detection of DNA sequences based on electrostatic interactions with unmodified gold nanoparticles. Proceedings of the National Academy of Sciences of the United States of America. 2004;**101**:14036-14039. DOI: 10.1073/pnas.0406115101

[4] Kumita J, Smart O, Woolley G. Photo-control of helix content in a short peptide. Proceedings of the National Academy of Sciences of the United States of America. 2000;**97**:3803-3808. DOI: 10.1073/pnas.97.8.3803

[5] Soh N, Yoshida K, Nakajima H, Nakano K, Imato T, Fukaminato T, et al. A fluorescent photochromic compound for labeling biomolecules. Chemical Communications. 2007;**48**:5206-5208. DOI: 10.1039/b713663c

[6] Benedetto F, Mele E, Camposeo A, Athanassiou A, Cingolani R, Pisignano D. Photoswitchable organic nanofibers. Advanced Materials. 2008;**20**:314-318. DOI: 10.1002/adma.200700980

[7] Staveren D, Metzler-Nolte N. Bioorganometallic chemistry of ferrocene. Chemical Reviews. 2004;**104**:5931-5985. DOI: 10.1021/cr0101510

[8] Giepmans B, Adams S, Ellisman M, Tsien R. The fluorescent toolbox for assessing protein location and function. Science. 2006;**312**:217-224. DOI: 10.1126/science.1124618

[9] Zimmer M. Green fluorescent protein (GFP): Applications, structure, and related photophysical behavior. Chemical Reviews. 2002;**102**:759-781. DOI: 10.1021/cr010142r

[10] Hanson G, Hanson B. Fluorescent probes for cellular assays. Combinatorial Chemistry & High Throughput Screening. 2008;**11**(7):505-513. DOI: 10.2174/138620708785204090

[11] Hermanson T. Bioconjugate Techniques. 3rd ed. London: Elsevier; 2013. DOI: 10.1016/C2009-0-64240-9

[12] Kurien B, Scofield R. A brief review of other notable protein detection methods on blots. Methods in Molecular Biology. 2009;**536**:557-571. DOI: 10.1007/978-1-59745-542-8_56

[13] Maurel D, Comps-Agrar L, Brock C, Rives M, Bourrier E, Ayoub M, et al. Cell-surface protein–protein interaction analysis with time-resolved FRET and snap-tag technologies: Application toGPCR oligomerization. Nature Methods. 2008;**5**:561-567. DOI: 10.1038/nmeth

[14] Yin J, Straight P, McLoughlin S, Zhou Z, Lin A, Golan D, et al. Genetically encoded short peptide tag for versatile protein labeling by Sfp phosphopantetheinyl transferase. Proceedings of the National Academy of Sciences of the United States of America. 2005;**102**(44):15815-15820. DOI: 10.1073/pnas.0507705102

[15] Evans J, Yue D. New turf for CFP/YFP FRET imaging of membrane signaling molecules. Neuron. 2003;**38**:145-147. DOI: 10.1016/S0896-6273(03)00234-4

[16] Pan D, Qin H, Cooperman B. Synthesis and functional activity of

tRNAs labeled with fluorescent hydrazides in the D-loop. RNA. 2009;**15**(2):346-354. DOI: 10.1261/rna.1257509

[17] Katritzky A, Narindoshvilia T. Fluorescent amino acids: Advances in protein-extrinsic fluorophores. Organic & Biomolecular Chemistry. 2009;7(4):627-634. DOI: 10.1039/b818908k

[18] Sahoo H. Fluorescent labeling techniques in biomolecules: A flashback. RSC Advances. 2012;**2**:7017-7029. DOI: 10.1039/c2ra20389h

[19] Biomol. Classic Reactive Fluorescent Labeling Dyes & Their Applications [Internet]. 2018. Available from: https://www.biomol.com/resources/biomol-blog/classic-fluorescent-labeling-dyes [Accessed: February 8, 2019]

[20] Sahoo H. Förster resonance energy transfer–A spectroscopic nanoruler: Principle and applications. Journal of Photochemistry and Photobiology C. 2011;**12**:20-30. DOI: 10.1016/j.jphotochemrev.2011.05.001

[21] Mochizuki N, Yamashita S, Kurokawa K, Ohba Y, Nagai T, Miyawaki A, et al. Spatio-temporal images of growth-factor-induced activation of Ras and Rap1. Nature. 2001;**411**(6841):1065-1068. DOI: 10.1038/35082594

[22] Gaietta G, Deerinck T, Adams S, Bouwer J, Tour O, Laird D, et al. Multicolor and electron microscopic imaging of connexin trafficking. Science. 2002;**296**(5567):503-507

[23] Miyashita T. Confocal microscopy for intracellular co-localization of proteins. Methods in Molecular Biology. 2004;**261**:399-410. DOI: 10.1385/1-59259-762-9:399

[24] Zuleger N, Kelly D, Richardson A, Kerr A, Goldberg M, Goryachev A, et al. System analysis shows distinct mechanisms and common principles of nuclear envelope protein dynamics. The Journal of Cell Biology. 2011;**193**:109-123. DOI: 10.1083/jcb.201009068

[25] Fernández-Suárez M, Ting A. Fluorescent probes for super-resolution imaging in living cells. Nature Reviews Molecular Cell Biology. 2008;**9**(12):929-943. DOI: 10.1038/nrm2531

[26] Bustamante C, Cheng W, Mejia Y. Revisiting the central dogma one molecule at a time. Cell. 2011;**144**(4):480-497. DOI: 10.1016/j.cell.2011.01.033

[27] Riveiro M, De Kimpe N, Moglioni A, Vázquez R, Monczor F, Shayo C, et al. Coumarins: Old compounds with novel promising therapeutic perspectives. Current Medicinal Chemistry. 2010;**17**(13):1325-1338. DOI: 10.2174/092986710790936284

[28] Hoult J, Payá M. Pharmacological and biochemical actions of simple coumarins: Natural products with therapeutic potential. General Pharmacology. 1996;**27**(4):713-722. DOI: 10.1016/0306-3623(95)02112-4

[29] Pierson J, Dumètre A, Hutter S, Delmas F, Laget M, Finet J, et al. Synthesis and antiprotozoal activity of 4-arylcoumarins. European Journal of Medicinal Chemistry. 2010;**45**(3):864-869. DOI: 10.1016/j.ejmech.2009.10.022

[30] Combes S, Barbier P, Douillard S, McLeer-Florin A, Bourgarel-Rey V, Pierson J, et al. Synthesis and biological evaluation of 4-arylcoumarin analogues of combretastatins. Part 2. Journal of Medicinal Chemistry. 2011;**54**(9):3153-3162. DOI: 10.1021/jm901826e

[31] Musa M, Cooperwood J, Khan M. A review of coumarin derivatives in pharmacotherapy of breast cancer. Current Medicinal Chemistry. 2008;**15**(26):2664-2679. DOI: 10.2174/092986708786242877

[32] Kostova I. Synthetic and natural coumarins as cytotoxic agents. Current Medicinal Chemistry Anti-Cancer Agents. 2005;**5**(1):29-46. DOI: 10.2174/1568011053352550

[33] Chin Y, Huang W, Hsu F, Lin Y, Lin M. Synthesis and evaluation of antibacterial activities of 5,7-dihydroxycoumarin derivatives. Archiv Der Pharmazie. 2011;**344**(6):386-393. DOI: 10.1002/ardp.201000233

[34] Carotti A, Altomare C, Catto M, Gnerre C, Summo L, De Marco A, et al. Lipophilicity plays a major role in modulating the inhibition of monoamine oxidase B by 7-substituted coumarins. Chemistry & Biodiversity. 2006;**3**(2):134-149. DOI: 10.1002/cbdv.200690017

[35] Thuong P, Hung T, Ngoc T, Ha do T, Min B, Kwack S, et al. Antioxidant activities of coumarins from Korean medicinal plants and their structure-activity relationships. Phytotherapy Research. 2010;**24**(1):101-106. DOI: 10.1002/ptr.2890

[36] Kabeya L, de Marchi A, Kanashiro A, Lopes N, da Silva C, Pupo M, et al. Inhibition of horseradish peroxidase catalytic activity by new 3-phenylcoumarin derivatives: Synthesis and structure-activity relationships. Bioorganic & Medicinal Chemistry. 2007;**15**(3):1516-1524. DOI: 10.1016/j.bmc.2006.10.068

[37] Jung J, Park O. Synthetic approaches and biological activities of 4-hydroxycoumarin derivatives. Molecules. 2009;**14**(11):4790-4803. DOI: 10.3390/molecules14114790

[38] Kuznetsova N, Kaliya O. The photochemistry of coumarins. Russian Chemical Reviews. 1992;**61**(7):683-696. DOI: 10.1070/RC1992v061n07ABEH000992

[39] Jones G, Jackson W, Choi C, Bergmark W. Solvent effects on emission yield and lifetime for coumarin laser dyes. Requirements for a rotatory decay mechanism. The Journal of Physical Chemistry. 1985;**89**(2):294-300. DOI: 10.1021/j100248a024

[40] Jagtap A, Satam V, Rajule R, Kanetkar V. The synthesis and characterization of novel coumarin dyes derived from 1,4-diethyl-1,2,3,4-tetrahydro-7-hydroxyquinoxalin-6-carboxaldehyde. Dyes and Pigments. 2009;**82**:84-89. DOI: 10.1016/j.dyepig.2008.11.007

[41] Wagner B. The use of coumarins as environmentally-sensitive fluorescent probes of heterogeneous inclusion systems. Molecules. 2009;**14**(1):210-237. DOI: 10.3390/molecules14010210

[42] Katerinopoulos H. The coumarin moiety as chromophore of fluorescent ion indicators in biological systems. Current Pharmaceutical Design. 2004;**10**(30):3835-3852. DOI: 10.2174/1381612043382666

[43] Kim J, Kim H, Kim S, Lee J, Do J, Kim H, et al. Fluorescent coumarinyldithiane as a selective chemodosimeter for mercury(II) ion in aqueous solution. Tetrahedron Letters. 2009;**50**(43):5958-5961. DOI: 10.1016/j.tetlet.2009.08.045

[44] Sheng R, Wang P, Gao Y, Wu Y, Liu W, Ma J, et al. Colorimetric test kit for Cu2+ detection. Organic Letters. 2008;**10**(21):5015-5018. DOI: 10.1021/ol802117p

[45] Kim H, Park J, Choi M, Ahn S, Chang S. Selective chromogenic and fluorogenic signalling of Hg2+ ions using a fluorescein-coumarin conjugate. Dyes and Pigments. 2010;**84**:54-58. DOI: 10.1016/j.dyepig.2009.06.009

[46] Lin W, Yuan L, Cao X, Tan W, Feng Y. A coumarin-based chromogenic sensor for transition-metal ions showing ion-dependent bathochromic shift. European Journal of Organic Chemistry. 2008;**29**:4981-4987. DOI: 10.1002/ejoc.200800667

[47] Jung H, Kwon P, Lee J, Kim J, Hong C, Kim J, et al. Coumarin-derived Cu(2+)-selective fluorescence sensor: Synthesis, mechanisms, and applications in living cells. Journal of the American Chemical Society. 2009;**131**(5): 2008-2012. DOI: 10.1021/ja808611d

[48] Wilze K, Johnson A. Handbook of Detergents, Chemistry, Production, and Application of Fluorescent Whitening Agents, Part F. Boca Raton: Taylor & Francis. CRC Press; 2007. DOI: 10.1201/9781420014655.ch28

[49] Dorlars A, Schellhammer C, Schroeder J. Heterocycles as structural units in new optical brighteners. Angewandte Chemie International Edition. 1975;**14**:665-679. DOI: 10.1002/anie.197506651

[50] Kido J, Iizumi Y. Fabrication of highly efficient organic electroluminescent devices. Applied Physics Letters. 1998;**73**:2721-2723. DOI: 10.1063/1.122570

[51] Moylan C. Molecular hyperpolarizabilities of coumarin dyes. Physical Chemistry. 1994;**98**(51): 13513-13516. DOI: 10.1021/j100102a014

[52] Painelli A, Terenziani F. Linear and non-linear optical properties of push-pull chromophores: Vibronic and solvation effects beyond perturbation theory. Synthetic Metals. 2001;**124**(1):171-173. DOI: 10.1016/S0379-6779(01)00431-3

[53] Benight S, Johnson L, Barnes R, Olbricht B, Bale D, Reid P, et al. Reduced dimensionality in organic electro-optic materials: Theory and defined order. The Journal of Physical Chemistry B. 2010;**114**:11949-11956. DOI: 10.1021/jp1022423

[54] Lee M, Yen C, Yang W, Chen H, Liao C, Tsai C, et al. Efficient green coumarin dopants for organic light-emitting devices. Organic Letters. 2004;**6**(8):1241-1244. DOI: 10.1021/ol049903d

[55] Swanson S, Wallraff G, Chen J, Zhang W, Bozano L, Carter K, et al. Stable and efficient fluorescent red and green dyes for external and internal conversion of blue OLED emission. Chemistry of Materials. 2003;**15**(12):2305-2312. DOI: 10.1021/cm021056q

[56] Yu T, Zhang P, Zhao Y, Zhang H, Meng J, Fan D. Synthesis, characterization and high-efficiency blue electroluminescence based on coumarin derivatives of 7-diethylamino-coumarin-3-carboxamide. Organic Electronics. 2009;**10**:653-660. DOI: 10.1016/j.orgel.2009.02.026

[57] Chang C, Cheng H, Lu Y, Tien K, Lin H, Lin C, et al. Enhancing color gamut of white OLED displays by using microcavity green pixels. Organic Electronics. 2010;**11**(4):247-254. DOI: 10.1016/j.orgel.2009.11.002

[58] Mayer G, Heckel A. Biologically active molecules with a "light switch". Angewandte Chemie International Edition. 2006;**45**(30):4900-4921. DOI: 10.1002/anie.200600387

[59] Geissler D, Antonenko Y, Schmidt R, Keller S, Krylova O, Wiesner B, et al. (Coumarin-4-yl)methyl esters as highly efficient, ultrafast phototriggers for protons and their application to acidifying membrane surfaces. Angewandte Chemie International Edition. 2005;**44**(8):1195-1198. DOI: 10.1002/anie.200461567

[60] Yu H, Li J, Wu D, Qiu Z, Zhang Y. Chemistry and biological applications of photo-labile organic molecules. Chemical Society Reviews. 2010;**39**(2):464-473. DOI: 10.1039/b901255a

[61] Pinheiro A, Baptista P, Lima J. Light activation of transcription: Photocaging of nucleotides for control over RNA polymerization. Nucleic Acids Research. 2008;**36**(14):e90. DOI: 10.1093/nar/gkn415

[62] Goddard J, Reymond J. Recent advances in enzyme assays. Trends in Biotechnology. 2004;**22**(7):363-370. DOI: 10.1016/j.tibtech.2004.04.005

[63] Heiner S, Detert H, Kuhn A, Kunz H. Hydrophilic photolabelling of glycopeptides from the murine liver–intestine (LI) cadherin recognition domain. Bioorganic & Medicinal Chemistry. 2006;**14**:6149-6164. DOI: 10.1016/j.bmc.2006.06.014

[64] Mizukami S, Okada S, Kimura S, Kikuchi K. Design and synthesis of coumarin-based Zn(2+) probes for ratiometric fluorescence imaging. Inorganic Chemistry. 2009;**48**(16):7630-7638. DOI: 10.1021/ic900247r

[65] Hara K, Kurashige M, Dan-oh Y, Kasada C, Shinpo A, Suga S, et al. Design of new coumarin dyes having thiophene moieties for highly efficient organic-dye-sensitized solar cells. New Journal of Chemistry. 2003;**27**(5):783-785. DOI: 10.1039/b300694h

[66] Camur M, Bulut M, Kandaz M, Guney O. Effects of coumarin substituents on the photophysical properties of newly synthesized phthalocyanine derivatives. Supramolecular Chemistry. 2009;**21**:624-631. DOI: 10.1080/10610270802613596

[67] Gonçalves M. Fluorescent labeling of biomolecules with organic probes. Chemical Reviews. 2009;**109**:190-212. DOI: 10.1021/cr0783840

[68] Johnson I, Spence M. Molecular Probes™ Handbook: A Guide to Fluorescent Probes and Labeling Technologies. 11th ed. New York: Life Technologies. Thermo Fisher Scientific; 2010. ISBN 13:978-0982927915

[69] Webb M, Corrie J. Fluorescent coumarin-labeled nucleotides to measure ADP release from actomyosin. Biophysical Journal. 2001;**81**:1562-1569. DOI: 10.1016/S0006-3495(01)75810-9

[70] Arguello J, Kaplan J. Glutamate 779, an intramembrane carboxyl, is essential for monovalent cation binding by the Na,K-ATPase. The Journal of Biological Chemistry. 1994;**269**(9): 6892-6899

[71] Arguello J, Kaplan J. Evidence for essential carboxyls in the cation-binding domain of the Na,K-ATPase. The Journal of Biological Chemistry. 1991;**266**(22):14627-14635

[72] Berthelot T, Talbot J, Laïn G, Déleris G, Latxague L. Synthesis of nepsilon-(7-diethylaminocoumarin-3-carboxyl)- and nepsilon-(7-methoxycoumarin-3-carboxyl)-l-fmoc lysine as tools for protease cleavage detection by fluorescence. Journal of Peptide Science. 2005;**11**:153-160. DOI: 10.1002/psc.608

[73] Oda Y, Kinoshita M, Nakayama K, Ikeda S, Kakehi K. Flow injection analysis of binding reaction between fluorescent lectin and cells. Analytical Biochemistry. 1999;**269**(2):230-235. DOI: 10.1006/abio.1999.4020

[74] Gabor G, Chadha S, Walt D. Sensitivity enhancement of fluorescent pH indicators using pH-dependent energy transfer. Analytica Chimica

Acta. 1995;**313**:131-137. DOI: 10.1016/0003-2670(95)00248-X

[75] Exley D, Ekeke G. Fluoroimmunoassay of 5a-dihydrotestosterone. Journal of Steroid Biochemistry. 1981;**14**:1297-1302. DOI: 10.1016/0022-4731(81)90335-6

[76] Tisljar U, Knight C, Barrett A. An alternative quenched fluorescence substrate for Pz-peptidase. Analytical Biochemistry. 1990;**186**(1):112-115. DOI: 10.1016/0003-2697(90)90582-T

[77] Mita H, Yasueda H, Hayakawa T, Shida T. Quantitation of platelet-activating factor by high performance liquid chromatography with fluorescent detection. Analytical Biochemistry. 1989;**180**:131-135. DOI: 10.1016/0003-2697(89)90100-0

[78] Khalfan H, Abuknesha R, Rand-Weaver M, Price R, Robinson D. Aminomethyl coumarin acetic acid: A new fluorescent labelling agent for proteins. The Histochemical Journal. 1986;**18**(9):497-499. DOI: 10.1007/BF01675617

[79] Eldaw A, Khalfan H. Aminomethyl coumarin acetic acid and fluorescein isothiocyanate in detection of leishmanial antibodies: A comparative study. Transactions of the Royal Society of Tropical Medicine and Hygiene. 1988;**82**(4):561-562. DOI: 10.1016/0035-9203(88)90506-8

[80] Panchuk-Voloshina N, Haugland R, Bishop-Stewart J, Bhalgat M, Millard P, Mao F, et al. Alexa dyes, a series of new fluorescent dyes that yield exceptionally bright, photostable conjugates. The Journal of Histochemistry and Cytochemistry. 1999;**47**(9):1179-1188. DOI: 10.1177/002215549904700910

[81] Cox W, Singer V. Fluorescent DNA hybridization probe preparation using amine modification and reactive dye coupling. BioTechniques. 2004;**36**(1):114-122. DOI: 10.2144/04361RR02

[82] Li M, Reddy L, Bennett R, Silva N, Jones L, Thomas D. A fluorescence energy transfer method for analyzing protein oligomeric structure: Application to phospholamban. Biophysical Journal. 1999;**76**(5):2587-2599. DOI: 10.1016/S0006-3495(99)77411-4

[83] Nguyen T, Joshi N, Francis M. An affinity-based method for the purification of fluorescently-labeled biomolecules. Bioconjugate Chemistry. 2006;**17**(4):869-872. DOI: 10.1021/bc060130i

[84] Weerachatyanukul W, Xu H, Anupriwan A, Carmona E, Wade M, Hermo L, et al. Acquisition of arylsulfatase A onto the mouse sperm surface during epididymal transit. Biology of Reproduction. 2003;**69**(4):1183-1192. DOI: 10.1095/biolreprod.102.010231

[85] Lewis B, Rathman S, McMahon R. Detection and quantification of biotinylated proteins using the storm 840 optical scanner. The Journal of Nutritional Biochemistry. 2003;**14**(4):196-202. DOI: 10.1016/S0955-2863(02)00283-8

[86] Kumari S, Wälchli S, Fallang L, Yang W, Lund-Johansen F, Schumacher T, et al. Alloreactive cytotoxic T cells provide means to decipher the immunopeptidome and reveal a plethora of tumor-associated self-epitopes. Proceedings of the National Academy of Sciences of the United States of America. 2014;**111**(1):403-408. DOI: 10.1073/pnas.1306549111

[87] Saavedra-Lozano J, Cao Y, Callison J, Sarode R, Sodora D, Edgar J, et al. An anti-CD45RO immunotoxin kills

HIV-latently infected cells from individuals on HAART with little effect on CD8 memory. Proceedings of the National Academy of Sciences of the United States of America. 2004;**101**:2494-2499. DOI: 10.1073/pnas.0308381100

[88] Irish J, Myklebust J, Alizadeh A, Houot R, Sharman J, Czerwinski D, et al. B-cell signaling networks reveal a negative prognostic human lymphoma cell subset that emerges during tumor progression. Proceedings of the National Academy of Sciences of the United States of America. 2010;**107**(29):12747-12754. DOI: 10.1073/pnas.1002057107

[89] Karachunski P, Ostlie N, Monfardini C, Conti-Fine B. Absence of IFN-gamma or IL-12 has different effects on experimental myasthenia gravis in C57BL/6 mice. Journal of Immunology. 2000;**164**(10):5236-5244. DOI: 10.4049/jimmunol.164.10.5236

[90] Henegariu O, Bray-Ward P, Ward D. Custom fluorescent-nucleotide synthesis as an alternative method for nucleic acid labeling. Nature Biotechnology. 2000;**18**(3):345-348. DOI: 10.1038/73815

[91] Zhang P, Beck T, Tan W. Design of a molecular beacon DNA probe with two fluorophores. Angewandte Chemie International Edition. 2001;**40**(2):402-405. DOI: 10.1002/1521-3773

[92] Lumiprobe. Life science solutions [Internet]. 2019. Available from: https://www.lumiprobe.com/p/coumarin-343-x-nhs-ester [Accessed: February 15, 2019]

[93] Zheng L, Zhao H, Han Y, Qian H, Vukovic L, Mecinović J, et al. Catalytic transport of molecular cargo using diffusive binding along a polymer track. Nature Chemistry. 2019;**11**:359-366. DOI: 10.1038/s41557-018-0204-7

[94] Gee K, Archer E, Kang H. 4-Sulfotetrafluorophenyl (STP) esters: New water-soluble amine-reactive reagents for labeling biomolecules. Tetrahedron Letters. 1999;**40**(8):1471-1474. DOI: 10.1016/S0040-4039(98)02695-1

[95] González-Pérez M, Ooi S, Martins S, Ramalho J, Pereira A, Caldeira A. Gaining insight into the photophysical properties of a coumarin STP ester with potential for bioconjugation. New Journal of Chemistry. 2018;**42**:16635-16645. DOI: 10.1039/c8nj03548b

[96] Yin S, Ramalho J, Pereira A, Martins S, Salvador C, Caldeira A. A simple method for labelling and detection of proteinaceous binders in art using fluorescent coumarin derivatives. The European Physical Journal Plus. 2019;**134**:71-80. DOI: 10.1140/epjp/i2019-12478-4

[97] Al-Kindy S, Miller J. Coumarin-6-sulphonyl chloride: A novel label in fluorimetry and phosphorimetry Part 1. Synthesis and luminescence properties anal. Chimica Acta. 1989;**227**:145-153. DOI: 10.1016/S0003-2670(00)82653-7

[98] Al-Kindy S, Suliman F, Al-Hamadi A. Fluorescence enhancement of coumarin-6-sulfonyl chloride amino acid derivatives in cyclodextrin media. Analytical Sciences. 2001;**17**(4):539-543. DOI: 10.2116/analsci.17.539

[99] Signore G, Nifosì R, Albertazzi L, Bizzarri R. A novel coumarin fluorescent sensor to probe polarity around biomolecules. Journal of Biomedical Nanotechnology. 2009;**5**(6):722-729. DOI: 10.1166/jbn.2009.1089

[100] Jashari A, Hey-Hawkins E, Mikhova B, Draeger G, Popovsk E. An improved synthesis of 4-chlorocoumarin-3-sulfonyl chloride and its reactions with different bidentate nucleophiles to give pyrido[1′,2′:2,3]- and thiazino[3′,2′:2,3]-

1,2,4-thiadiazino[6,5-c]benzopyran-6-one 7,7-dioxides. Molecules. 2007;**12**(8):2017-2028. DOI: 10.3390/12082017

[101] Al-Kindy S, Al-Sharji N, Al-Harasi A, Suliman F, AL-Lawati H, Schulman S. Synthesis and spectroscopic study of 2,7-diethylamino-2-oxo-2H-chromen-3-yl benzothiazole-6-sulfonyl chlorides and its derivatives. Arabian Journal of Chemistry. 2017;**10**:S114-S120. DOI: 10.1016/j.arabjc.2012.06.015

[102] Bhusal R, Cho P, Kim S-A, Park H, Kim H. Synthesis of green emitting coumarin bioconjugate for the selective determination of flu antigen. Bulletin of the Korean Chemical Society. 2011;**32**(5):1461-1462. DOI: 10.5012/bkcs.2011.32.5.1461

[103] Mayekar S, Chaskar A, Mulwad V. Facile synthesis of coumarinyl isothiocyanate from amino coumarin. Synthetic Communications. 2010;**1**(40):46-51. DOI: 10.1080/00397910902916080

[104] Patil S, Deokar H. Synthesis and structural determination of novel heterocyclic derivatives of coumarin isothiocyanates. Der Chemica Sinica. 2014;**5**(6):74-78

[105] Sippel T. New fluorochromes for thiols: Maleimide and iodoacetamide derivatives of a 3-phenylcoumarin fluorophore. The Journal of Histochemistry and Cytochemistry. 1981;**29**(2):314-316. DOI: 10.1177/29.2.7019305

[106] Kuiper J, Pluta R, Huibers W, Fusetti F, Geertsma E, Poolman B. A method for site-specific labeling of multiple protein thiols. Protein Science. 2009;**18**(5):1033-1041. DOI: 10.1002/pro.113

[107] Zhou Z, Koglin A, Wang Y, McMahon A, Walsh C. An eight residue fragment of an acyl carrier protein suffices for post-translational introduction of fluorescent pantetheinyl arms in protein modification *in vitro* and *in vivo*. Journal of the American Chemical Society. 2008;**130**(30):9925-9930. DOI: 10.1021/ja802657n

[108] Zavala L, Pardo-López L, Cantón P, Gómez I, Soberón M, Bravo A. Domains II and III of Bacillus thuringiensis Cry1Ab toxin remain exposed to the solvent after insertion of part of domain I into the membrane. The Journal of Biological Chemistry. 2011;**286**(21):19109-19117. DOI: 10.1074/jbc.M110.202994

[109] Ramachandiran V, Grigoriev V, Lan L, Ravkov E, Mertens S, Altman J. A robust method for production of MHC tetramers with small molecule fluorophores. Journal of Immunological Methods. 2007;**319**(1-2):13-20. DOI: 10.1016/j.jim.2006.08.014

[110] Kunzelmann S, Webb M. A biosensor for fluorescent determination of ADP with high time resolution. The Journal of Biological Chemistry. 2009;**284**(48):33130-33138. DOI: 10.1074/jbc.M109.047118

[111] Brune M, Corrie J, Webb M. A fluorescent sensor of the phosphorylation state of nucleoside diphosphate kinase and its use to monitor nucleoside diphosphate concentrations in real time. Biochemistry. 2001;**40**(16):5087-5094. DOI: 10.1021/bi002484h

[112] Anderson S, Williams C, O'donnell M, Bloom L. A function for the psi subunit in loading the Escherichia coli DNA polymerase sliding clamp. The Journal of Biological Chemistry. 2007;**282**(10):7035-7045. DOI: 10.1074/jbc.M610136200

[113] Boutureira O, Bernardes G. Advances in chemical protein modification. Chemical Reviews. 2015;**115**(5):2174-2195. DOI: 10.1021/cr500399p

[114] Shadmehr M, Davis G, Mehari B, Jensen S, Jewett J. Coumarin triazabutadienes for fluorescent labeling of proteins. Chembiochem. 2018;**19**:2550-2552. DOI: 10.1002/cbic.201800599

[115] Gavrilyuk J, Ban H, Nagano M, Hakamata W, Barbas C. Formylbenzene diazonium hexafluorophosphate reagent for tyrosine-selective modification of proteins and the introduction of a bioorthogonal aldehyde. Bioconjugate Chemistry. 2012;**23**(12):2321-2328. DOI: 10.1021/bc300410p

Section 3

Medical and Human Health Applications

Chapter 9

Ocimum Phytochemicals and Their Potential Impact on Human Health

Debjoy Bhattacharjya, Sinchan Adhikari, Arijit Biswas, Anil Bhuimali, Parthadeb Ghosh and Soumen Saha

Abstract

The genus *Ocimum* (Lamiaceae) is distributed all over the world and can be found in many environments. *Ocimum* species is a rich source of various phytochemicals including tannins, phenolic acids, anthocyanins, phytosterols, and policosanols. These phytochemicals have the potential to significantly impact human health. The economic importance of *Ocimum* is also evident; *Ocimum* oil and its constituents and derivatives are used as flavoring agents throughout the world in food, pharmaceutical, herbal, perfumery, and flavoring industry. The important advantages of *Ocimum* plants in various treatments are their safety besides being less expensive, efficacy and availability throughout the world. This paper will focus on the biological effects of *Ocimum* essential oils, with particular attention on the molecular mechanism underlying their action.

Keywords: *Ocimum* sp., essential oil, composition, biological activity

1. Introduction

Living plants produce a vast quantity of chemicals required for their performance and improvement. Some of these chemicals are primary metabolites, which consist of proteins (amino acids), carbohydrates, fats, nucleic acids, etc. but, besides these primary chemicals, the plants further provide just so-called secondary metabolites, which are specific to some taxonomic groups (families, genera). About 80% of the world's population still depends on the traditional system of medicine for curing several health hazards [1]. Despite the vast scientific development in contemporary medicine, Ayurvedic system of medicine is widely practiced and accepted by people not only in India but also in many developed countries. According to the World Health Organization (WHO), about 80% of patients in India still, rely on the practitioners of the traditional system of medicines. The therapeutic use of herbal crude medicines in different rural and urban communities is most of the time regulated by their traditional beliefs, and thus a majority of the herbal drugs are used as "folk" medicines and well practiced since long past. Furthermore, increasing dependency on medicinal plants in the industrialized communities have been found to the extraction and improvement of several remedies and chemotherapeutics from these plants and from traditionally established rural herbal remedies [2]. In these communities, herbal remedies become deeply engaged in the practice of minor conditions and again on the explanation of the increasing costs of specific health maintenance. Although synthetic drugs enhanced the

IntechOpen

demand against green remedies because of their rapid-acting implements, people have been turned up to understand the benefits correlated with essential remedies. Chemically prepared drugs may act at earlier, but they side effects which influence the human body separately in the long run, because medicinal plants work in an integrated or probiotic with limited or no negative effects on the body [3].

The genus *Ocimum* includes approximately 150 species, possessing a great variation in plant morphology and biology, essential oil content, and chemical composition [4]. The economic importance and global dissemination of *Ocimum*, with its many uses in cooking and folk medicine, make it important to investigate its pharmacological and toxicological effects in order to ensure its efficacy and safety. In India, among the medicinal herbs known for their healing properties, the genus *Ocimum* (commonly known as 'Basil' or 'Tulsi') is very important for its curative potential. Basils contain a wide range of essential oils rich in phenolic compounds and a wide array of other natural products including flavonoids and anthocyanins [5] having great pharmacological importance.

Nowadays, scientists are mainly focused on exploring the potential of plant antioxidants for curing several diseases. Antioxidants are compounds that can delay or inhibit the oxidation of lipids or other molecules by inhibiting the initiation or propagation of oxidative chain reactions [6]. Antioxidants counteract the harmful reactive oxygen species (ROS) and free radicals generated in the living organism during regular metabolism, especially under stress conditions. Several in vitro and in vivo studies have already proved the antioxidative property of beta-carotene, alfa-tocopherol, ascorbic acids, phenolics, flavonoids present in different plants [7].

Presently due to our regular stressful lifestyle, we are suffering from several types of diseases like aging, diabetes, degenerative disorders, etc., which mainly develop due to the development of ROS in our body. A dynamic balance is already operating in our body to reduce the harmful effects of generated ROS that is not adequate enough. Therefore, it is obvious to enrich our diet with antioxidants for developing protection. The purpose of this paper will focus on the recent research of the major nutrients and phytochemicals of *Ocimum* and their potential health benefits related to the dietary prevention of chronic diseases.

2. Health benefits of phytochemicals

"Phyto" in the word of phytochemicals is derived from the Greek word "phyto," which means plant. Phytochemicals are a naturally occurring group of chemicals in plants and plant-derived foods, which may function in reducing the risk of chronic diseases [8]. Although it is estimated at least more than 5000 dietary phytochemicals have been discovered, it is believed that a high percentage of phytochemicals in foods still remain unknown [8]. Critical reviews of studies available in the literature support the concept that phytochemicals (polyphenols, tocopherols, tocotrienols, carotenoids, and ascorbic acid) has been associated with the maintenance of good health as well as prevention/treatment of many health conditions including cancer, cardiovascular diseases, diabetes, hypertension, stroke, metabolic syndrome, and other degenerative diseases. It is largely accepted that the additive effects of the combinations of various phytochemicals in whole plant-based foods are shown to have stronger protective actions than single, isolated phytochemical compounds [9].

3. Role of secondary metabolites

New drug model seeks to meet on bringing compounds active toward target proteins. Even though newly pharmaceutical companies and support organizations take influence in molecular design, combinatorial chemistry and synthetic chemistry, natural productions, and especially those of plant source, remain as a prerequisite cause of new remedies, current medicine leads and other new synthetic entities (NCEs) [10, 11].

Plants produce a diverse array of compounds that can broadly be categorized into primary and secondary metabolites. Primary metabolites are the ones which are required for the normal growth and biological processes and are produced in the pathways that are crucial for plant survival. The other class of metabolites, though generally termed secondary, is also very crucial for plants from the ecological perspective.

These secondary metabolites are classified on the basis of their biosynthetic pathway and the following types are frequently observed-terpenes, phenylpropanoids, alkaloids, saponins, and glucosinolates. The availability of carbon, nitrogen, and sulfur along with energy from the primary metabolism governs the biosynthesis of these compounds [12].

4. Classification of secondary metabolites

Metabolites are the mediators and amounts of metabolism. The term metabolite is commonly confined to narrow fragments. Metabolites have specific functions,

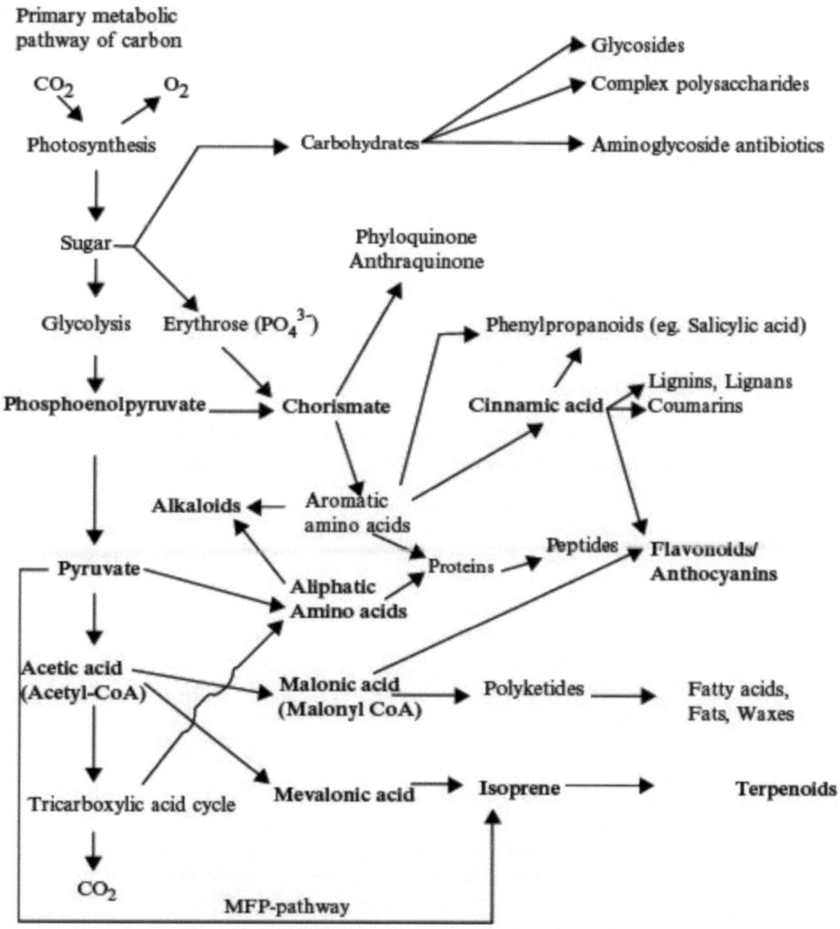

Figure 1.
Major pathways of biosynthesis of secondary metabolites [15].

consisting of fuel, structure, signaling, stimulatory and inhibitory effects on enzymes, and the catalytic activity of their holding (mostly as a cofactor to a stimulant), defense and interaction with distinct pathogens. Plant metabolites are categorized based on their biosynthetic pathways. The pathways of biosynthesis are responsible for the occurrence of both primary and secondary metabolites (**Figure 1**) [13, 14]. Plant secondary metabolites can be classified on the basis of chemical structure (for example, having rings, containing a sugar), composition (consisting of nitrogen or not), their solubility in numerous solvents, or the pathway by which they are synthesized (e.g., phenylpropanoid, which provides tannins).

5. Essential oil of *Ocimum* species

Since ancient times, essential oils are known for their medicinal use, and they are very much interesting and impressive natural plant commodities. They continue to be of paramount importance until the present day. Essential oils have been tested as perfumes, flavors for foods and beverages, or to provide both bodies and care for thousands of years [16].

Species	Class of compounds	Reference
O. basilicum L.	**Monoterpene hydrocarbons** α-Phellandrene, α-pinene, α-terpinene, α-terpinolene, α-myrcene, β-phellandrene, β-pinene, camphene, *cis*-β-ocimene, *cis*-ocimene, δ-3-carene, β-ocimene, (E)-β-ocimene, limonene, myrcene, *ρ*-cymene, sabinene, terpinolene, thujene, γ-terpinene	[18–20, 23]
	Oxygenated monoterpene α-Citral, α-fenchyl acetate, α-terpineol, borneol, bornyl acetate, camphor, carvacrol, carvone, 1,8-cineole, *cis*-linalool oxide, *cis*-rose oxide, citronellol, endo-fenchol, estragol, eugenol, exo-2-hydroxycineole-acetate, fenchone, geranial, geraniol, geranyl acetate, hotrienol, *iso*-neomenthol, *iso*-pinocamphone, *trans*-pinocamphone, L-camphor, L-carvone, lavandulol, linalool, linalool *cis*-furanoid, linalool *trans*-furenoid, linalyl acetate, menthol, menthone, methyl chavicol, myrtenal, myrtenol, neral, nerol, ocimene oxide, pinocarvone, P-menth-1,8-dien-4-ol, piperitone, pulegone, terpinen-4-ol, terpinyl formate, *trans*-linalool oxide, *trans*-myroxide, *trans*-sabinene hydrate, *trans-p*-menth-2-en-1-ol, thymol, verbenone, (Z)-sabinene hydrate	[18, 19, 21, 25]
	Sesquiterpene hydrocarbons β-acoradiene, β-bourbonene, β-caryophyllene, β-cedrene, β-copaene, β-cubebene, β-elemene, β-guaiene, β-ocimene, β-selinene, cyclohexane, 2,4 diisopropenyl-1-methyl-1-vinyl, 1,4,7-cycloundecatriene,1,5,9,9-tetramethyl, α-acoradiene, α-amorphene, α-bulnesene, α-cadinene, α-cedrene, α-copaene, α-cubebene, α-guaiene, α-gurjunene, γ-gurjunene, α-7-epi-selinene, α-humulene, aromadendrene, α-(Z)-bergamotene, α-zingiberene, epsilon-muurolene, dehydroaromadendrene, germacrene-A, bicycloelemene, bicyclogermacrene, cadinene, cadina-3,5-diene, *cis*-calamene, *cis*-muurola-4(14),5-diene, (E)-β-farnesene, (E)-caryophyllene, 1-epibicyclosesqui-phellandrene, germacrene-B, germacrene-D, guaia-1(10),11-diene, iso-caryophyllene, isoledene, longifolene, δ-selinene, St α-ylangene, *trans*-α-bisabolene, *trans*-α-bergamotene, *trans*-β-farnesene, *trans*-β-ocimene, *trans*-caryophyllene, valencene, γ-cadinene, γ-terpin, δ-cadinene, γ-muurolene, (Z)-calamenene	[19, 26, 21–24]

Species	Class of compounds	Reference
	Oxygenated sesquiterpenes α-Cadinol, α-humulene oxide, alloaromadendrene, β-basibolol, β-basibolol isomer, β-eudesmol, cubenol, caryophyllene oxide, 1,10-di-epi-cubenol, dihydroactinidiolide, isospathulenol, muurolol, spathulenol, T-cadinol, viridiflorol, (Z)-nerolidol	[19, 21]
	Triterpene Alphitolic acid, betulin, betulinic acid, 3-epimaslinic acid, euscaphic acids, oleonolic acid, pomolic acid, ursolic acid, basilol, ocimol	[27]
	Aromatic compounds 4-Allylphenol, anethole, anisaldehyde, benzyl alcohol, cuminaldehyde, estragole, ethyl cinnamate, methyl benzoate, methyl cinnamate, methyl eugenol, methyl salicylate, *p*-methoxycinnamaldehyde, phenethyl alcohol, phenyl acetaldehyde, safrole, benzaldehyde, *cis*-hex-3-enyl acetate	[21]
O. kilimandscharicum Guerke	**Monoterpene hydrocarbons** α-Phellandrene, α-pinene, α-terpinene, γ-terpinene, β-pinene, camphene, limonene, D-limonene, β-ocimene, (E)-β-ocimene, myrcene, ρ-cymene, terpinolene, thujene, α-terpinolene, isosylvestrene, γ-himachalene	[28, 29, 34]
	Oxygenated monoterpene α-Citral, α-terpineol, borneol, bornyl acetate, camphor, linalool, citronellol, geraniol, myrtenol, 1,8-cineole, eugenol, terpinen-4-ol, *trans*-sabinene hydrate, (Z)-sabinene hydrate, α-campholenal, isoborneol, endo-borneol, globulol	[29–31, 34]
	Sesquiterpene hydrocarbons α-Copaene, α-gurjunene, α-humulene, β-caryophyllene, β-copaene, β-cubebene, β-elemene, δ-cadinene, γ-muurolene, *trans*-caryophyllene, germacrene-D, germacrene-B, β-selinene	[31, 32, 34]
	Oxygenated sesquiterpenes Cubenol, caryophyllene oxide, spathulenol, α-cadinol, viridiflorol	[28, 32]
	Phenolic compounds Vanillin	[33]
O. gratissimum L.	**Monoterpene hydrocarbons** α-Pinene, *cis*-ocimene, *trans*-ocimene, β-pinene, α-terpinene, p-cymene, myrcene, α-phellandrene, A^3-carene, sabinene, limonene, γ-terpinene, terpinolene, cis-sabinene hydrate	[35–39]
	Oxygenated monoterpene Camphor, eugenol, methyl eugenol, 1,8-cineole, trans-sabinene hydrate, linalool, δ-terpineol, terpinen-4-ol, α-terpineol, thymol, carvacrol, eugenol, methyleugenol	[36, 39–41]
	Sesquiterpene hydrocarbons *trans*-Caryophyllene, germacrene-D, α-farnese, β-bisabolene, *cis*-β-ocimene, *trans*-β-ocimene, α-copaene, β-elemene, β-caryophyllene, α-humulene, germacrene-D, β-selinene, trans-β-Farnesene, β-bisabolene, γ-cadinene, δ-cadinene, γ-muurolene, β-cubebene, α-cubebene	[33, 36, 40, 41]
	Sesquiterpenes oxygenated β-Caryophyllene epoxide	[36]
	Aromatic compounds Methyl cinnamate	[36]
O. canum Sims.	**Terpenoids** Cyclosativen	[42]

Species	Class of compounds	Reference
	Monoterpenes hydrocarbon α-Thujene, α-pinene, camphene, β-pinene, limonene, γ-terpinene, terpinolene, sabinene, myrcene, α-terpinene, *p*-cymene, (Z)-β-ocimene, (E)-β-ocimene, perillene, (E)-β-epoxyocimene, α-phellandrene, α-terpinene, *cis*-β-ocimene, *cis*-sabinene hydrate, carene, tricyclene	[43, 45, 48]
	Oxygeneated monoterpenes Eugenol, linalool, 1,8-cineole, 5-isopropyl-2-methylbicyclo[3.1.0]hexan-2-ol, fenchone, *trans*-linalool oxide (furanoid), camphor, δ-terpineol, terpinen-4-ol, α-terpineol, fenchyl acetate (endo), isobornyl acetate, borneol, thymol, menthone, geranial, α-fenchyl acetate, bornyl acetate, exo-2-hydroxycineole acetate, verbenone, camphenol, myrtenyl acetate	[43, 46, 48]
	Sesquiterpenes hydrocarbon α-Copaene, β-elemene, β-caryophyllene, germacrene D, β-bisabolene, E, E-α-farnesene, δ-cadinene, α-cadinene, *trans*-α-bergamotene, aromadendrene, α-humulene, *epi*-bicyclosesquiphellandrene, bicyclogermacrene, δ-guaiene, calamenene, (E)-α-bisabolene, valencene, *trans*-α-bergamotene, γ-muurolene, α-muurolene, epi-α-muurolol, elemol, β-selinene, α-selinene, (Z,E)-α-farnesene, *trans* β-ocimene, (E)-β-farnesene	[45, 46, 49]
	Oxygenated sesquiterpenes α-Cadinol, spathulenol, (Z)-nerolidol, caryophyllene oxide, β-eudesmol, viridiflorol	[42, 46]
	Aromatic compounds Methyl eugenol, estragole, benzyl benzoate	[45, 47]
	Esters 1-Octen-3-yl acetate	[44]
	Others Naphthalene	[46]
O. tenuiflorum L. *syn. O. sanctum* L (purple type)	**Monoterpenes hydrocarbon** α-Pinene, β-pinene, β-terpinene, D-limonene, camphene, sabinene, myrcene, *p*-cymene, limonene, α-terpinene, *α*-thujene, *α*-myrcene, α-terpinolene, terpinolene, γ-terpinene, (E)-β-ocimene, β-myrcene, *α*-camphene, β-terpinolene, β-*cis*-ocimene	[50–52]
	Oxygeneated monoterpenes Linalool, menthol, methyl chavicol, α-citral, carvone, lavandulol, hotrienol, eugenol, 1,8-cineole, globulol, borneol, bornyl acetate, camphor, thymol, geranial, citronellol, E-linalool, β-citral, carvacrol, methyl chavicol, α-fenchyl acetate, myrtenol, terpinen-4-ol, *trans*-sabinene hydrate, *cis*-geraniol, *cis*-*α*-terpineol, *cis*-linalool oxide, eucalyptol, *cis*-linalool oxide (furanoid), *trans*-linaool oxide (furanoid), δ-terpineol	[52–54]
	Sesquiterpenes hydrocarbon Caryophyllene, β-farnesene, germacrene-D, isoledene, β-selinen, β-cubebene, β-elemene, β-caryophyllene, β-bourbonene, α-humulene, γ-muurolene, bicyclogermacrene, δ-cadinene, α-copaene, *trans*-caryophyllene, selinene, β-elemene, *β*-guaiene, β-bisabolene, α-guaiene, germacrene-B, valencene, (E)-β-farnesene, *trans*-α-bergamotene, α-cubebene, β-germacrene, α-farnesene, α-caryophyllene, α-selinene, (Z)-β-farnesene	[50, 52–54]

Species	Class of compounds	Reference
	Sesquiterpenes oxygenated α-Cadinol, alloaromadendrene, caryophyllene oxide, cubenol, t-cadinol, spathulenol, viridiflorol	[53, 54]
	Aromatic compounds *p*-Methoxycinnamaldehyde, estragole, benzaldehyde	[51, 54]
O. tenuiflorum (white type)	**Monoterpene hydrocarbons** α-pinene, camphene, β-pinene, limonene, (E)-β-ocimene, *p*-cymene, γ-terpinene, camphene hydrate, carene, terpinolene, sabinene hydrate, terpinene, ocimene, limonene, terpinene, phellandrene, myrcene, sabinene, camphene, thujene, tricyclene	[55–57]
	Oxygeneated monoterpenes Linalool, borneol, eugenol, methyl eugenol, 1,8-cineole, α-terpineol, geraniol, *trans*-linalool oxide (furanoid), δ-terpineol, terpineol, terpinen-4-ol, δ-terpineol, camphor, fenchone, *trans*-sabinene hydrate, eucalyptol	[56–59]
	Sesquiterpenes hydrocarbon β-Elemene, β-caryophyllene, α-humulene, germacrene D, β-selinene, α-selinene, α-cubebene, δ-cadinene, elemol, bicyclogermacrene, α-cadinene, copaene, β-elemen, α-guaiene, γ-muurolene, δ-cadinene, amorphene, cubebene, α-bisabolene, cadinene, β-bisabolene, muurolene, germacrene, humulene, farnesene, sesquiphellandrene, bergamotene, guaiene, elemene, bourbonene, zingiberene	[55–59]
	Sesquiterpenes oxygenated Spathulenol, caryophyllene oxide, viridiflorol, β-eudesmol, γ-eudesmol	[60, 61]
	Aromatic compounds Estragole	[55, 60]

Table 1.
Compositions of species-wise distribution of bioactive compounds in Ocimum species.

Chemical diversification is of special significance if at the genus or species level both terpenes and phenylpropenes can be formed in the essential oil. Most Lamiaceae preferentially accumulate mono-(and sesqui-)terpenes in their erratic oils but some genera show oils too rich in phenylpropenes [17].

The action of essential oils begins by entering the human body via three possible different ways including direct absorption through inhalation, ingestion or diffusion through the skin tissue.

Table 1 presents published compositions of species-wise distribution of bioactive compounds in *Ocimum* species.

6. Biological activities of *Ocimum* species

The genus *Ocimum* (family Lamiaceae), collectively known as basil, is composed of a diverse and rich source of essential oil containing plants. The main issues of concern with the use of herbal drugs remain safety, validation of claims and standardization of product. Different species and forms of *Ocimum* spp. vary in growth habit, color, and aromatic composition, making the true botanical identity of basil difficult (**Figure 2**). There exist the problems of significant variation in the content of *Ocimum* plants across and within species, with the implication of varied

Figure 2.
Different species of Ocimum (a) Ocimum basilicum L.; (b) Ocimum kilimandscharicum Guerke; (c) Ocimum gratissimum L.; (d) Ocimum canum Sims.; (e) Ocimum tenuiflorum L. syn. O. sanctum L. (purple type); (f) Ocimum tenuiflorum (white type).

biological activities. During the last two decades, it has been shown that *Ocimum* oil and its constituents possess different biological activities including antioxidant, antimicrobial, anticancer, and anti-inflammatory properties.

6.1 Oxidant activity

Anti-oxidants play an important function in protecting the body against free radicals. They achieve this by stopping the formation of new free radicals species, converting older ones to free radicals, less toxic molecules that can be easily mopped up and preventing radical chain reaction [62]. The principal function of anti-oxidants is in suspending the oxidation of other molecules, by inhibiting the initiation or propagation of oxidizing chain reactions by free radicals and thereby reducing oxidative damage to the human body [63]. Two great mechanisms of procedure have been proposed for antioxidants [64]. The initial is a chain breaking method by which the initial antioxidant provides an electron to the free radical stage in the systems. The second technique involves the destruction of ROS/ reactive nitrogen species initiators (secondary antioxidants) by cutting off chain starting catalysts (**Figure 2**). The potential role in the food industry and human health, antioxidants are getting acceptance all across the globe. Antioxidants are defined as a substance that easily in small amounts, can inhibit or prohibit the oxidation of readily oxidizable elements. The antioxidant is also defined as a substance qualified of inhibiting special oxidizing stimulants or a substance that serves with oxidizing agents prior to creating damage to other fragments or a substance that sequesters metal ions or even a substance efficient of the recovering system such as iron transporting protein [65].

Natural antioxidants have been studied intensively during the past years which are mainly phenolic compounds. Moreover, oxidation is a degenerative process in biological systems due to the endogenous reactive oxygen species (ROS). Reactive oxygen species (ROS) are chemical properties originate in the body during metabolism that is overmuch reactive and may have one or expanded unpaired electrons. Oxidative stress, i.e., an inequality between ROS and antioxidant defenses have deleterious aftereffect, such as the peroxidation of membrane lipids and the aggression on biomolecules (proteins, membrane enzymes, carbohydrates and DNA) [66].

Various *Ocimum* species and their extracts or essential oils have been determined to achieve antioxidant activity [67, 68]. Phenolic acids, hydroxycinnamates, and flavonoids are perhaps the major antioxidants [67]. Vitamin antioxidants (e.g., ascorbic acid and carotenoids) are secondary contributors to the overall antioxidant capacity [69]. In essential oils, unsaturated terpenes having a cyclohexadiene structure (e.g., terpinene) and secondary cyclic oxygenated terpenes (e.g., thymol) may lead to antioxidant capacity, while acyclic unsaturated oxygenated monoterpenes (e.g., linalool), aromatic oxygenated monoterpenes (e.g., eugenol), methylchavicol (estragole), sesquiterpene hydrocarbons (e.g., α-bergamotene, germacrene D, γ-cadinene, δ-cadinene, β-selinene, sesquiterpenes oxygenated (e.g., spathulenol) may act as pro-oxidants [68].

Different test methods have been applied, and it becomes to be understood between the complete essential oil and individual components [70–73]. A strong chelating effect on metals as iron or copper has been reported to diminish the presence of ROS obtained from reactions bar with these metals [74]. In addition, the antioxidant actions of *Ocimum* essential oils have also been found employing metal-independent oxidative processes [75] or controlling stable free radicals, such as 2,2-diphenyl-1-picrylhydrazyl (DPPH) radical [71, 76]. The DPPH test, a test widely used to measure the ability to donate hydrogen atoms [77], was applied to measure the antioxidant capacities of *Ocimum* species extracted by different solvent systems; these include the methanol extracts of *O. basilicum* L., *O. canum* Sims., *O. gratissimum* L., *O. kilimandscharicum* Guerke, *O. sanctum*, *O. tenuiflorum* [67, 71, 76, 78–80]; the ethanol extracts from *O. basilicum* L., *O. gratissimum* L., *O. kilimandscharicum* Guerke, *O. sanctum* [76, 79, 81–83].

Particular attention has been focused on the scavenger ability to inhibit the process of low-density lipoprotein (LDL) cholesterol oxidation since it represents a major prevention mechanism against atherosclerosis. Various experimental evidence, using in vitro and in vivo preclinical models, showed a strong action of *Ocimum* essential oils [84]. Moreover, Aqueous extracts *Ocimum sanctum* and 70% ethanol extracts *Ocimum basilicum* L. were able to reduce 5-lipoxygenase-driven cellular recruitment of leukocytes and the damaging consequences of their ability to release ROS while leaving unimpaired the generation of prostaglandins, which promote microvascular blood flow and act as immunomodulators [85–87].

In many cases, the antioxidative activity (**Figure 3**) of essential oils cannot be attributed to the main compounds; minor compounds and synergistic effect may significantly contribute to the activity.

6.2 Antimicrobial

For thousands of years, folk medicine has used *Ocimum* leaf for the treatment of infections. Such protective properties have been confirmed by several studies performed in the last decades using *Ocimum* essential oil [29] or isolated compounds. Gram-positive and Gram-negative bacteria, as well as antiprotozoal and also *anti-Trichomonas vaginalis*, resulted sensitive to the antiproliferative action of *Ocimum oil* and its derivatives [29, 88–90]. At present, the exact mechanism responsible for the antimicrobial activity of *Ocimum* oil and its derivatives is still not completely

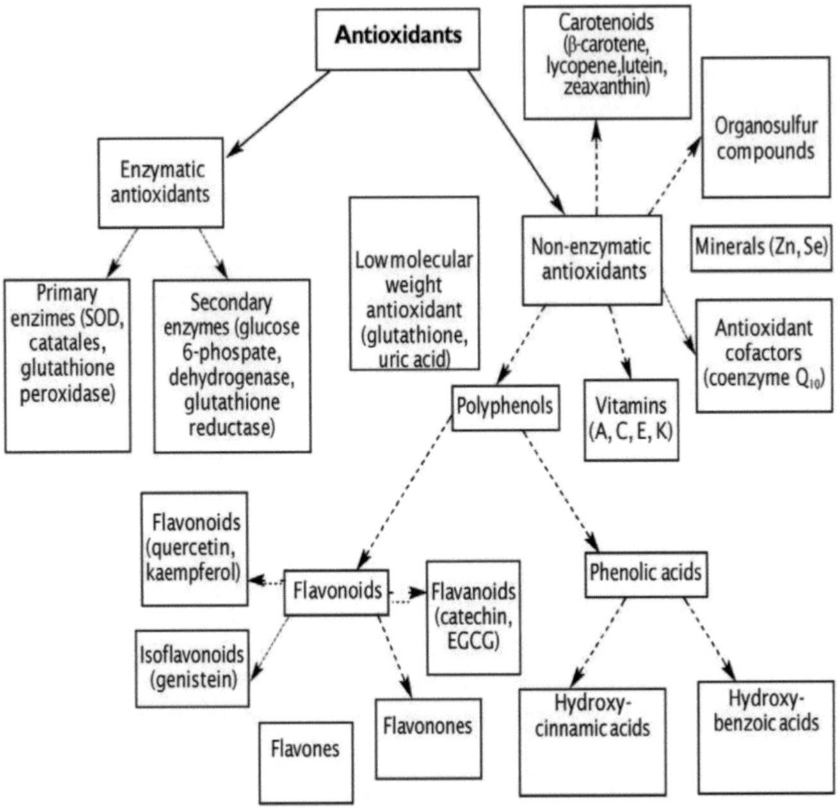

Figure 3.
Types of antioxidants.

clarified, although various modes of action in the bacterial cell have been discussed including degradation of the cell wall, damage to cytoplasmic membrane and membrane proteins, leakage of cell contents, coagulation of cytoplasm, and depletion of proton motive force [91–93].

The antibacterial and antifungal activities of *Ocimum* species have been studied on various bacteria and fungi [29, 94–96]. These studies indicate that essential oils are more efficient antifungals and antibacterials compared to the polar extracts [97–99]. *Ocimum sanctum* essential oils showed remarkable antimicrobial activity against bacteria and other microorganisms, such as periodontopathogens [100], mainly due to the presence of oxygenated monoterpenes in their chemical compositions [101].

The essential oil and methanol extracts of five *Ocimum* species have an appreciable activity against seven human pathogenic bacteria [29], essential oils of *Ocimum* species showed strong antimicrobial activity against all seven microorganisms tested. Oils of seven *Ocimum* taxa (*O. americanum* L., *O. basilicum* L., *O. campechianum* Mill., *O.* x *citriodorum* Vis., *O. kilimandscharicum* Baker ex Gürke and three botanical varieties and cultivars of *Ocimum basilicum* L.: 'Genovese', var. difforme and var. purpurascens)) showed strong antimicrobial activity against all 8 microorganisms tested by Carovic-Stanko et al. [102].

Among the antifungal activities, the *in vitro* antifungal activity of *O. basilicum* L. essential oil against Aspergillus flavus fungal growth and aflatoxin B1 production [103], essential oils of *O. basilicum* L. showed strong antifungal activity against *A. flavus*, and the main components were linalool, 1,8-cineol, eugenol, methyl cinnamate, α-cubebene, caryophyllene, β-ocimene and α-farnesene.

6.3 Anticancer activity

For a long time, the polyphenols of the *Ocimum* oil of the diet have been considered to play a role for the prevention of certain types of cancer in the Asian origin [104]. Even more than in *Ocimum* oil, constituents present in *Ocimum* leaf extract has shown strong antioxidant potency and inhibition of cancer cell proliferation, thus suggesting the protection against the genotoxic action of the ROS as one of the mechanisms explaining the anticancer effects of these compounds. Indeed, either methanol aqueous *Ocimum basilicum* L. leaf extract or the isolated constituents eugenol epoxide free radical scavenging activity and growth inhibition at low micromolar concentration on human breast cancer cell lines (MCF-7 and MDA-MB-231) [105, 106] and Human cancerous cell lines (HL60-promyelocytic blood leukemia cells) [107]. Such findings were further confirmed by other in vitro reports, testing the effects of *Ocimum basilicum* L. against four different humans cancer cell lines viz. human cervix adenocarcinoma HeLa cells, human melanoma FemX cells, human chronic myelogenous leukemia K562 cells, and human ovarian SKOV3 cells [108]. Furthermore, Karthikeyan et al. [109] demonstrated regression of tumors caused by orally administrated aqueous and ethanolic extracts of *Ocimum sanctum* in mice that developed spontaneous soft tissue sarcomas.

Monga et al. [110] studied the antimelanoma and radioprotective activity of essential oils obtained by 50% alcoholic aqueous leaf extract from five species of *Ocimum* viz. *Ocimum sanctum* (SE), *Ocimum gratissimum*, *Ocimum basilicum*, *Ocimum canum*, and *Ocimum kilimandscharicum*, were evaluated using C57BL and Swiss albino mice tumorigenesis; growth inhibition has in fact been associated with (a) reduction of tumor volume (b) blockage of messengers of pathways involved in cell proliferation was evident in all the oils but the greater was shown by that obtained from *Ocimum tenuiflorum* (syn. *O. sanctum*) compared to other *Ocimum* species. In various experiments and test systems, some mono-and sesquiterpenes showed activity, where camphor, 1,8-cineole and limonene were of greatest interest. Camphor, 1,8-cineole and limonene, the anti-inflammatory compound of *Ocimum kilimandscharicum* oil, showed a strong time-and dose-dependent cytotoxic effect on human ovarian cancer cell [111]. The potential antitumor effects of camphor have been shown previously [112, 113], and the mechanistic action of camphor against cancer included the improvement of immune function [114] and the radiosensitizing effect on transplantable mammary adenocarcinoma in mice [112]. Ursolic acid showed some potentiating effect on the anticancer activity of rosmarinic acid, cinnamic acid, caffeic acid, sinapic acid, and ferulic acid on various cell lines [106].

6.4 Cardiovascular protection

For decades, investigation on the health-promoting effects of Asian diet has been revealed that *Ocimum* oil consumption is a key factor in the cardiovascular protection found in Asian origin [115]. It is well established that the healthful properties of *Ocimum basilicum* L. oil depend largely on its Cardiac glycosides and catecholamines content [116]. But, many arguments prove that in *Ocimum* oil there are little bioactive components, much than Cardiac glycosides and catecholamines, effective for its cardiovascular protective properties: among them, the ethanolic fraction of *Ocimum* oil, and in specific omega-3 fatty acids have proved antioxidant, anti-platelet aggregation, vasodilatory, and anti-inflammatory effects, all engaged in this health beneficial action [117–119].

Oxidation of LDL cholesterol is one of the key steps in the induction of atherosclerotic lesions by increasing damage to the arterial side through several processes, including growth factor and chemotactic protein expression, inflammation, and build up local macrophages [120, 121] have indicated that ethanolic leaf extract of *Ocimum basilicum* L. oil strongly inhibits copper sulfate-induced oxidation of LDL, as a result of the step of different indicators of lipid oxidation [phospholipids (PL), cholesterol ester (CE), triacylglycerol (TG)].

Although the contraction of plasmatic cholesterol and LDL is the primary technique regulating the antiatherogenic activity of *Ocimum basilicum* L. extract, other implements are further identified [121]. It is well settled that local leukocyte and monocyte recruitment into the vessel wall is initial walk-in atherogenesis. This fact takes combined with the statement in the endothelial cells of adhesion fragments such as intercellular adhesion fragment-1 (ICAM-1) and vascular adhesion fragment-1 (VCAM-1). Aqueous extract of *Ocimum gratissimum* L. showed the capacity to reduce LPS-stimulated expression of BEAS-2B cell in human lung epithelial cells inhibiting its mRNA levels. Moreover, Li et al. [122], investigating the action of distillate and residue fractions of basil essential oil (viz. estragole, methyl eugenol, linoleic acid, α-cadinol, and α-bergamotene) in-a Raw 264.7 cells line, have demonstrated that residue fractions prevents the expression of gene and suppressed the production of cytokines (TNF-a, IL-b, IL-6) in LPS-induced Raw264.7 cells, which contribute to treating various disorders caused by extreme oxidative stress.

6.5 Anti-inflammatory

The inflammatory response involves long been compartmentalized into multiple attributes commonly termed redness, heat, pain, and edema. Inflammatory injuries lead to the discharge of a variety of fundamental mediators, cytokines, and chemokines that balance cellular infiltration that consequentially brings about resolving inflammatory response and restoration of tissue scrupulosity. However, immutable inflammatory stimuli or dysregulation of mechanisms of the resolution phase can lead to chronic inflammation [123, 124].

Ocimum extracts contain numerous constituents which could have anti-inflammatory effects. The anti-inflammatory effects of *Ocimum* oil phenolics, in RAW 264.7 macrophage cells have been described by Aye et al. [125]. When added to murine macrophages stimulated with bacterial lipopolysaccharide (LPS), *Ocimum* oil phenolics did not cause cytotoxicity in RAW 264.7 macrophage cells in vitro, as evaluated by a significant increase in the production of nitric oxide [125]. Additionally, NO is a significant inflammatory mediator generated by NOS (neuronal, inducible, and endothelial) under physiological and pathophysiological conditions [126]. It further serves as a crucial mediator during the inflammatory process. Enhanced NO production and iNOS expression contributes to the great cytotoxic function of LPS stimulated macrophages [127]. Thus, the reduction in NO production indicates the anti-inflammatory activities of the treatment in the cells. However, *Ocimum basilicum* L. ethyl acetate extract and butanol extract inhibited the growth of normal RAW 264.7 macrophage cells [128]. Also, *Ocimum basilicum* L. crude methanolic extract suppressed the induction of iNOS and the subsequent production of NO in LPS-stimulated RAW 264.7 macrophage cells [129, 130]. To test the anti-inflammatory activity of the *Ocimum basilicum* L. methanolic extracts has been determined by PBMC (peripheral blood mononuclear cells) in mitogenic lymphocyte proliferative assay, methanolic extracts enhanced the functional activity of these immune-competent cells, as evaluated by a significant inhibitory effects of methanolic leaf extracts, PHA activated PBMC proliferation could be suggestive of suppression of T cell proliferation [130]. This effect arose from its pivotal role in immune regulation [131, 132], T cell activation provides a target for pharmacological modulation aimed at achieving clinically useful immune-suppression [133].

Złote et al. [134] studied the capacity of phenolic-rich fraction obtained from the elicited basil leaves to inhibit the activity of two enzymes of inflammatory process (LOX and COX). This research found that a higher LOX and COX inhibition efficiency was positively correlated with the increased contents of rosmarinic, benzoic and *o*-coumaric acids determined after elicitation of basil. This result partially corresponds with the study of [135]. More recently, to gain insight into the mechanism of action and pharmacological value of the anti-inflammatory activity of aqueous and methanolic extracts of *Ocimum basilicum* L. in macrophage (RAW264.7) and human chondrosarcoma (SW1353) cell lines, and human primary chondrocytes to correlate their efficacy in terms of management of osteoarthritis (OA). Raina et al. [136] evaluated aqueous extract of *O. basilicum* L. significantly accustom the production of inflammatory mediators such as NO, PGE2, LTB4, and MMPs increased than the methanolic extract. The regulation of these inflammatory intermediaries is pivotal in OA, as it would have a direct effect on (1) chondrocyte survival, (2) production of proinflammatory cytokines, prostaglandins and leukotrienes, and (3) production of extracellular matrix-degrading enzymes such as MMPs. Due to the significant side-effects related to the use of NSAIDs, the check for natural products that would regulate the inflammatory cascade related to OA, without engaging chondrocyte survival, is extremely important. To investigate the anti-inflammatory effect of *Ocimum basilicum* L. oil, Rodrigues et al. [137] investigated an acute and chronic in vivo test as paw edema, peritonitis, and vascular permeability and granulomatous inflammation model. The anti-inflammatory mechanism of action was also analyzed by the participation of histamine and arachidonic acid pathways. These researchers found that the *Ocimum basilicum* L. essential oil and estragole significantly reduced paw edema induced by carrageenan and dextran. The smallest quantities of *Ocimum basilicum* L. essential oil (50 mg/kg) and estragole (30 mg/kg) revealed effectiveness in the decrease of paw edema created by histamine and arachidonic acid, vascular permeability inhibition and leukocyte emigration in the peritoneal fluid. These dosages were carried out of decrease the assured inflammatory process. The results followed between the *Ocimum basilicum* L. essential oil and estragole determine efficacy in anti-inflammatory activity, however, the essential oil is higher efficacious in the acute and chronic anti-inflammatory action. Dextran is a high molecular weight polysaccharide, which differently to carrageenan, induces anaphylactic reactions characterized by extravasation and formation of edema due to mastocyte degranulation with release of histamine and serotonin. Carrageenan induces an inflammatory response through opinion with sulfated polysaccharides, initially encouraging the free of chemical substances which encourage multifactorial facts, mostly concerning the free of substance P, bradykinin, histamine, serotonin, cytokines, and nitric oxide and, subsequently on commodity arise from the arachidonic acid pathway [138].

6.6 Antidiabetic

Diabetes mellitus is a chronic metabolic disorder caused by an absolute or relative lack of insulin and or reduced insulin activity which results in hyperglycemia and abnormalities in carbohydrate, fat and protein metabolism [139, 140]. The hypoglycemic effect of *O. tenuiflorum* L., *O. canum* Sims. and *O. gratissimum* L. in animals with alloxan-induced diabetes was applied to potentiation of glucose-induced insulin free and parallel increased peripheral uptake of glucose [141, 142]. Successive studies have reported a greater link of antidiabetic deal with the antioxidant effects of *Ocimum* oil. The character played by OS for diabetes complications such as retinopathy, nephropathy, and cardiovascular disease are well set up so that dietary antioxidant compounds was fixed to protect from the damages of oxidative stress and free radicals in diabetic cases [143].

In animal experimental designs of alloxan-induced diabetes, both antioxidant and hypoglycemic effects of *O. basilicum* L., *O. tenuiflorum* L., *O. canum* Sims. and *O. gratissimum* L. have been reported. By treating alloxan-diabetic rabbits [141, 142, 144], made a significant decrease in blood glucose levels as corresponded with diabetic control rabbits. Such a hypoglycemic work was related to its powerful antioxidant potentiality: in evidence, in interact, the rabbits studied with *Ocimum* oil showed further a renewal of the levels of malondialdehyde and most of the enzymatic and nonenzymatic endogenous antioxidants [141, 142, 144]. Similar results were achieved in alloxan-diabetic rats: the control, of *Ocimum tenuiflorum*-rich extracts showed significant hypoglycemic, hypolipidemic, and antioxidant effects in all the investigated diabetic rats [145, 146]. In an identical empirical design, *O. gratissimum* L. led to a decrease in the sugar level in plasma and a rise in superoxide dismutase, catalase, and glutathione peroxidase activities in liver and kidney. Furthermore, an opposed reaction against hepatic and renal toxicity in diabetic rats was also observed [147, 148]. Furthermore, the effects of *Ocimum sanctum* leaf polyphenols have been investigated also in insulin-secreting pancreatic β-cells, whose OS-induced alterations contribute to the pathogenesis of diabetes [149].

7. Future direction and conclusion

Despite the many appreciations of science and industry, present practice is filled with stress. Mobile devices and the web have vastly enhanced the pace of life so that many people feel that they are now going down in an endless-increasing ocean of data, while technical culture has overwhelmed us with growing vulnerability to unhealthy prepared and packed food and a profusion of pesticides, food container components, and many toxic modern chemicals. Urban citizens are nevertheless dealt with growing prosperity disparity, social segregation, excessive turbulence, air, water and soil pollution and disconnection from nature. Therefore, while industrialization experiences served to stronger lifespans and impressive expansions in human populations, it is now agreed that the extremest causes of death and disease on the globe are preventable lifestyle-related chronic diseases [150].

The biodiversity of essential oils containing the small molecular terpenoids remains an enchanted field of investigation, and the continuous usage of this reward in a broad field of studies suggested these demands in the consequent [151]. Screening, identifying, and dealing with this vast biodiversity will require a progressing progress of precise, rich-throughput experimental methods including new driving procedures.

The beneficial health effects of *Ocimum* oil compounds have also been proven by many randomized, crossover, controlled, human studies on biomarkers of health performed in the last decades. Several preclinical studies suggest that such beneficial effects may be mainly ascribed to the phenolic compounds. Further development of biotechnology with the genomic and metabolomic analyses and genetic engineering will advance a variety of fields involving bioactive compounds ranging from food and animal nutrition to plant protection. Although many biological activities as antimicrobial or antioxidative and other effects have been intensively studied and well documented. However, well studies are needed to further characterize the *in vivo* effects of individual *Ocimum* derivatives applied as specific agents or in a mixture, consisting of their safety analysis on mortals. Moreover, a better evidence of their molecular procedure of activity may appropriate the system to a better application in human pharmacology.

In inference, the tendency of the last moments in the treatment of herbal productions have indicated that nature prospects and trust in pure and healthful products including medicines, cosmetics, household products, since easily as foodstuffs of plant and animal origin have belonged to a vital issue. So, indeed a well-balanced risk-benefit assessment of bioactive essential oils is one of the major challenges and policymakers must be convinced that review on natural products as the volatile terpenoids in essential oils is a huge task to ensure the ultimate human and animal welfare [151].

Author details

Debjoy Bhattacharjya[1], Sinchan Adhikari[2], Arijit Biswas[2], Anil Bhuimali[1], Parthadeb Ghosh[2] and Soumen Saha[1,2*]

1 Cytogenetics and Plant Breeding Section, Department of Sericulture, Raiganj University, Raiganj, West Bengal, India

2 Cytogenetics and Plant Biotechnology Research Unit, Department of Botany, University of Kalyani, Kalyani, West Bengal, India

*Address all correspondence to: thustu@gmail.com

References

[1] Kunwar RM, Adhikari N. Ethnomedicine of Dolpa district, Nepal: The plants, their vernacular names and uses. Lyonia. 2005;**8**:43-49

[2] UNESCO. FIT/504-RAF-48 Terminal Report: Promotion of Ethnobotany and the Sustainable Use of Plant Resources in Africa; Paris. 1998. p. 60

[3] Idu M, Umweni AA, Odaro T, Ojelede L. Ethnobotanical plants used for oral healthcare among the Esan Tribe of Edo State, Nigeria. Ethnobotanical Leaflets. 2009;**13**:548-563

[4] Danesi F, Elementi S, Neri R, Maranesi M, D'antuono LF, Bordoni A. Effect of cultivar on the protection of cardiomyocytes from oxidative stress by essential oils and aqueous extracts of basil (*Ocimum basilicum* L.). Journal of Agriculture & Food Chemistry. 2008;**56**:9911-9917

[5] Joshi V, Bothara B, Surana J. Evaluation of aqueous extract of *Ocimum sanctum* in experimentally induced parkinsonism. Journal of Chemical and Pharmaceutical Research. 2011;**3**:478-487

[6] Velioglu YS, Mazza G, Gao L, Oomah BD. Antioxidant activity and total phenolics in selected fruits, vegetables, and grain products. Journal of Agricultural Food & Chemistry. 1998;**46**:4113-4117

[7] Cross CE, Halliwell B, Borish ET. Oxygen radicals and human disease. An International Medicine. 1987;**107**:526-545

[8] Liu RH. Potential synergy of phytochemicals in cancer prevention: Mechanism of action. Journal of Nutrition. 2004;**134**:3479S-3485S

[9] Xi P, Liu RH. Whole food approach for type 2 diabetes prevention. Molecular Nutrition & Food Research. 2016;**60**:1819-1836

[10] Balunasa MJ, Kinghorn D. Drug discovery from medicinal plants. Life Sciences. 2005;**78**:431-441

[11] Newman DJ, Cragg GM. Natural products as sources of new drugs from 1981 to 2014. Journal of Natural Products. 2016;**79**:629-661

[12] Dudareva N, Klempien A, Muhlemann JK, Kaplan I. Biosynthesis, function and metabolic engineering of plant organic volatile compounds. New Phytologist. 2013;**198**:16-32

[13] Herbert RB, editor. The Biosynthesis of Secondary Metabolites. 2nd ed. Chapman & Hall: London; 1989

[14] Nolaou KC, Chen JS, Corey EJ. Classics in Total Synthesis, Further Targets, Strategies, Methods III. Weinheim: Wiley; 2011. pp. 1-770

[15] Verpoorte R. Secondary metabolism. In: Verpoorte R, Alfermann AW, editors. Metabolic Engineering of Plant Secondary Metabolism. Springer: Dordrecht; 2000

[16] Wei A, Shibamoto T. Antioxidant/Lipoxygenase inhibitory activities and chemical compositions of selected essential oil. Journal of Agriculture and Food Chemistry. 2010;**58**:7218-7225

[17] Bradu BL, Sobti SN, Pushpangadan P, Khosla KM, Rao BL, Gupta SC. Development of superior alternate source of clove oil from 'Clocimum' (*Ocimum gratissimum* Linn.). In: Proc. 11th Int. Congr. of Essential Oils, Fragrances and Flavours. Vol. 3. 1989. pp. 97-103

[18] Chang X, Alderson PG, Wright CJ. Enhanced UV-B radiation alters basil (*Ocimum basilicum* L.) growth and stimulates the synthesis of volatile oils. Journal of Horticulture and Forestry. 2009;**1**:027-031

[19] Martino LD, Feo VD, Nazzaro F. Chemical composition and *in vitro* antimicrobial and mutagenic activities of seven Lamiaceae essential oils. Molecules. 2009;**14**:4213-4230

[20] Ismail M. Central properties and chemical composition of *Ocimum basilicum* essential oil. Pharmaceutical Biology. 2006;**44**:619-626

[21] Lee SJ, Umano K, Shibamoto T, Lee KG. Identification of volatile components in basil (*Ocimum basilicum* L.) and thyme leaves (*Thymus vulgaris* L.) and their antioxidant properties. Food Chemistry. 2005;**91**:131-137

[22] Hassanpouraghdam MB, Gohari GR, Tabatabaei SJ, Dadpour MR Inflorescence and leaves essential oil composition of hydroponically grown *Ocimum basilicum* L. Journal of the Serbian Chemical Society. 2010;**75**:1361-1368

[23] Lawrance BM, Mokheyee BD, Willis BJ. Developments in Food Sciences, Flavour and Fragrances; A World Perspective. The Netherlands: Elsevier; 1988

[24] Bassolé IH, Lamien-Meda A, Bayala B, Tirogo S, Franz C, Novak J, et al. Composition and antimicrobial activities of Lippia multiflora Moldenke, *Mentha x piperita* L. and *Ocimum basilicum* L. essential oils and their major monoterpene alcohols alone and in combination. Molecules. 2010;**15**:7825-7839

[25] Vani SR, Cheng SF, Chuah CH. Comparative study of volatile compounds from genus *Ocimum*. American Journal of Applied Sciences. 2009;**6**:523-528

[26] Özcan M, Chalchat JC. Essential oil composition of *Ocimum basilicum* L. and *Ocimum minimum* L. in Turkey. Czech Journal of Food Science. 2002;**20**:223-228

[27] Poonkodi K. Chemical composition of essential oil of *Ocimum Basilicum* L. (basil) and its biological activities—An overview. Journal of Critical Reviews. 2016;**3**:56-62

[28] Pandey AK, Singh P, Tripathi NN. Chemistry and bioactivities of essential oils of some *Ocimum* species: An overview. Asian Pacific Journal of Tropical Biomedicine. 2014;**4**:682-694

[29] Saha S, Dhar TN, Sengupta C, Ghosh PD. Biological activities of essential oils and methanol extracts of five *Ocimum* species against pathogenic bacteria. Czech Journal of Food Science. 2013;**31**:194-202

[30] Joshi RK. Chemical composition of the essential oil of camphor basil (*Ocimum kilimandscharicum* Guerke). Global Journal of Medicinal Plant Research. 2013;**1**:207-209

[31] Mahesh SP, Patil MB, Kumar R, Patil SR. Evaluation of aqueous extract of leaves if *Ocimum kilimandscharicum* Guerke wound healing activity in albino wistar rats. International Journal of Pharmtech Research Coden (USA). 2009;**1**:544-550

[32] Kashyap CP, Kaur R, Arya V, Vipin K. Therapeutic potency of *Ocimum kilimandscharicum* Guerke—A review. Global Journal of Pharmacology. 2011;**5**:191-200

[33] Dolly G, Nidhi S, Bps S, Shweta R, Shikha A. *Ocimum kilimandscharicum*: A systematic review. Journal of Drug Delivery & Therapeutics. 2012;**2**:45-52

[34] Charles DJ, Simon JE. Essential oil constituents of *Ocimum kilimandscharicum* Guerke. Journal of Essential Oil Research. 1992;**4**:125-128

[35] Matasyoh LG, Matasyoh JC, Wachira FN, Kinyua MG, Muigai AWT, Mukiama TK. Chemical composition and antimicrobial activity of the essential oil of *Ocimum gratissimum* L. growing in Eastern Kenya. African Journal of Biotechnology. 2007;**6**:760-765

[36] Ntezurubanza L, Scheffer JJC, Svendsen AB. Composition of the essential oil of *Ocimum gratissimum* grown in Rwanda. Planta Medica. 1987;**53**:421-423

[37] Chimnoi N, Reuk-ngam N, Chuysinuan P, Khlaychan P, Khunnawutmanotham N, Chokchaichamnankit D, et al. Characterization of essential oil from *Ocimum gratissimum* leaves: Antibacterial and mode of action against selected gastroenteritis pathogens. Microbial Pathogenesis. 2018;**118**:290-300

[38] Nguemtchouin MGM, Ngassoum MB, Chalier P, Kamga R, Ngamo LST, Cretin M. *Ocimum gratissimum* essential oil and modified montmorillonite clay, a means of controlling insect pests in stored products. Journal of Stored Products Research. 2013;**52**:57-62

[39] Nakamura CV, Ueda-Nakamura T, Bando E, Melo AFN, Cortez DAG, Filho BPD. Antibacterial activity of *Ocimum gratissimum* L. essential oil. Memórias do Instituto Oswaldo Cruz, Rio de Janeiro. 1999;**94**:675-678

[40] Vieiraa RF, Grayerb RJ, Patonb A, Simona JE. Genetic diversity of *Ocimum gratissimum* L. based on volatile oil constituents, flavonoids and RAPD markers. Biochemical Systematics and Ecology. 2001;**29**:287-304

[41] Pessoaa LM, Moraisb SM, BevilaquaaCML,LucianobJHS.Anthelmintic activity of essential oil of *Ocimum gratissimum* Linn. and eugenol against *Haemonchus contortus*. Veterinary Parasitology. 2002;**109**:59-63

[42] Belong P, Ntonga PA, Fils EMB, Dadji GAF, Tamesse JL. Chemical composition and residue activities of *Ocimum canum* Sims and *Ocimum basilicum* L essential oils on adult female *Anopheles funestus ss*. Journal of Animal & Plant Sciences. 2013;**19**:2854-2863

[43] Kalita J, Khan ML. Commercial potentialities of essential oil of *Ocimum* members growing in north East India. International Journal of Pharmacy & Life Sciences. 2013;**4**:2559-2567

[44] Ntonga PA, Baldovini N, Mouray E, Mambu L, Belong P, Grellier P. Activity of *Ocimum basilicum*, *Ocimum canum*, and *Cymbopogon citratus* essential oils *against Plasmodium falciparum* and mature-stage larvae of *Anopheles funestus s.s*. Parasite. 2014;**21**:33

[45] Selvi MT, Thirugnanasampandan R, Sundarammal S. Antioxidant and cytotoxic activities of essential oil of *Ocimum canum* Sims. from India. Journal of Saudi Chemical Society. 2012;**19**:97-100

[46] Chagonda LS, Makanda CD, Chalchat JC. The essential oils of *Ocimum canum* Sims (basilic camphor) and *Ocimum urticifolia* Roth from Zimbabwe. Flavour and Fragrance Journal. 2000;**15**:23-26

[47] da Silvaa VD, Almeida-Souzab F, Teles AM, Neto PA, Mondego-Oliveira R, Filho NEM, et al. Chemical composition of *Ocimum canum* Sims. essential oil and the antimicrobial, antiprotozoal and ultrastructural alterations it induces in *Leishmania amazonensis* promastigote. Industrial Crops and Products. 2018;**119**:201-208

[48] Olugbade TA, Kolipha-Kamara MI, Elusiyan CA, Onawunmi GO, Ogundaini AO. Essential oil chemotypes of three *Ocimum* species found in Sierra Leone and Nigeria. Medicinal & Aromatic Plants. 2017;**6**:284. DOI: 10.4172/2167-0412.1000284

[49] Tonzibo ZF, Chalchat JC, N'Guessan YT. Chemical composition of essential oils of *Ocimum canum* Sims from Côte d'Ivoire. Journal of Essential Oil Bearing Plants. 2008;**11**:530-535

[50] Khan A, Ahmad A, Akhtar F, Yousuf S, Xess I, Khan LA, et al. *Ocimum sanctum* essential oil and its active principles exert their antifungal activity by disrupting ergosterol biosynthesis and membrane integrity. Research in Microbiology. 2010;**161**:816-823

[51] Kicel A, Kurowska A, Kalemba D. Composition of the essential oil of *Ocimum sanctum* L. grown in poland during vegetation. Journal of Essential Oil Research. 2005;**17**:217-219

[52] Ijaz B, Hanif MA, Mushtaq Z, Khan MM, Bhatti IA, Jilani MJ. Isolation of bioactive fractions from *Ocimum Sanctum* essential oil. Oxidation Communications. 2017;**40**:158-167

[53] Saroj T, Krishna A. A comparison of chemical composition and yield of essential oils from shoot system parts of *Ocimum sanctum* found in semi-arid region of Uttar Pradesh. Agrotechnology. 2017;**6**:172. DOI: 10.4172/2168-9881.1000172

[54] Kumar A, Shukla R, Singh P, Dubey NK. Chemical composition, antifungal and antiaflatoxigenic activities of *Ocimum sanctum* L. essential oil and its safety assessment as plant based antimicrobial. Food and Chemical Toxicology. 2010;**48**:539-543

[55] Sims CA, Juliani HR, Mentreddy SR, Simon JE. Essential oils in holy basil (*Ocimum tenuiflorum* L.) as influenced by planting dates and harvest times in North Alabama. Journal of Medicinally Active Plants. 2014;**2**:33-41

[56] Pino JA, Rosado A, Rodriguez M, Garcia D. Composition of the essential oil of *Ocimum tenuiflorum* L. grown in Cuba. Journal of Essential Oil Research. 1998;**10**:437-438

[57] Brophy JJ, Goldsack RJ, Clarkson JR. The essential oil of *Ocimum tenuiflorum* L. (Lamiaceae) growing in northern Australia. Journal of Essential Oil Research. 1993;**5**:459-461

[58] Yamani HA, Pang EC, Mantri N, Deighton MA. Antimicrobial activity of Tulsi (*Ocimum tenuiflorum*) essential oil and their major constituents against three species of bacteria. Frontiers in Microbiology. 2016;**7**:681. DOI: 10.3389/fmicb.2016.00681

[59] Raina AP, Kumar A, Dutta M. Chemical characterization of aroma compounds in essential oil isolated from "holy basil" (*Ocimum tenuiflorum* L.) grown in India. Genetic Resources and Crop Evolution. 2013;**60**: 1727-1735

[60] Sharma V, Sharma A, Seth R. A study on Antidermatophytic potential of *Ocimum tenuiflorum* essential oil and chemical composition evaluation. International Journal of PharmTech Research. 2016;**9**:151-160

[61] Tangpao T, Chung HH, Sommano SR. Aromatic profiles of essential oils from five commonly used Thai basils. Food. 2018;**7**:175. DOI: 10.3390/foods7110175

[62] Ismail HI, Chan KW, Marion AA, Ismail M. Phenotic content and antioxidanat activity of cantoloupe (*Cucumis melol*) methanolic extracts. Food Chemistry. 2010;**119**:643-647

[63] Ismail A, Margan ZM, Foong CW. Total activity and phenolic content in

selected vegetables. Food Chemistry. 2004;**87**:581-586

[64] Rice-Evans CA, Diplock AT. Current status of antioxidant therapy. Free Radical Biology & Medicine. 1993;**15**:77-96

[65] Brewer MS. Natural antioxidants: Sources, compounds, mechanisms of action, and potential applications. Comprehensive Reviews in Food Science and Food Safety. 2011;**10**:221-247

[66] Barroso MF, Ramalhosa MJ, Alves RC, Dias A, Soares CMD, OlivaTeles MT, et al. Total antioxidant capacity of plant infusions: Assessment using electrochemical DNA-based biosensor and spectrophotometric methods. Food Control. 2016;**68**:153-161

[67] Hakkim FL, Arivazhagan G, Boopathy R. Antioxidant property of selected *Ocimum* species and their secondary metabolite content. Journal of Medicinal Plants Research. 2008;**2**:250-257

[68] Filip S, Vidović S, Vladić J, Pavlić B, Adamović D, Zeković Z. Chemical composition and antioxidant properties of *Ocimum basilicum* L. extracts obtained by supercritical carbon dioxide extraction: Drug exhausting method. The Journal of Supercritical Fluids. 2016;**109**:20-25

[69] Bhattacharya A, Aggarwal A, Sharma N, Cheema J. Evaluation of some anti-oxidative constituents of three species of *Ocimum*. International Journal of Life Sciences. 2014;**8**:14-17

[70] Pripdeevech P, Chumpolsri W, Suttiarporn P, Wongpornchai S. The chemical composition and antioxidant activities of basil from Thailand using retention indices and comprehensive two-dimensional gas chromatography. Journal of Serbian Chemical Society. 2010;**75**:1503-1513

[71] Kwee EM, Niemeyer ED. Variations in phenolic composition and antioxidant properties among 15 basil (*Ocimum basilicum* L.) cultivars. Food Chemistry. 2011;**128**:1044-1050

[72] Farouk A, Fikry R, Mohsen M. Chemical composition and antioxidant activity of *Ocimum basilicum* L. essential oil cultivated in Madinah Monawara, Saudi Arabia and its comparison to the Egyptian Chemotype. Journal of Essential Oil Bearing Plants. 2016;**19**:1119-1128

[73] Falowo AB, Mukumbo FE, Idamokoro EM, Afolayan AJ, Muchenje V. Phytochemical constituents and antioxidant activity of sweet basil (*Ocimum basilicum* L.) essential oil on ground beef from Boran and Nguni Cattle. International Journal of Food Science. 2019;**9**:1-8

[74] Suanarunsawat T, Ayutthaya WDN, Songsak T, Thirawarapan S, Poungshompoo S. Lipid-lowering and antioxidative activities of aqueous extracts of *Ocimum sanctum* L. leaves in rats fed with a high-cholesterol diet. Oxidative Medicine and Cellular Longevity. 2011;**1**:1-9

[75] Cohen MM. Tulsi-*Ocimum sanctum*: A herb for all reasons. Journal of Ayurveda & Integrative Medicine. 2014;**5**:251-259

[76] Jayasinghe C, Gotoh N, Aoki T, Wada S. Phenolics composition and antioxidant activity of sweet basil (*Ocimum basilicum* L.). Journal of Agriculture and Food Chemistry. 2003;**51**:4442-4449

[77] Subramanian M, Chintalwar GJ, Chattopadhyay S. Antioxidant and radioprotective properties of an *Ocimum sanctum* polysaccharide. Redox Report. 2005;**10**:257-264

[78] Mahapatra SK, Roy S. Phytopharmacological approach of

free radical scavenging and antioxidative potential of eugenol and *Ocimum gratissimum* Linn. Asian Pacific Journal of Tropical Medicine. 2014;7:S391-S397

[79] Nahak G, Mishra RC, Sahu RK. Phytochemical investigation and in vitro antioxidant evaluation of some *Ocimum* species. Journal of Pharmacy Research. 2011;**4**:2340-2343

[80] Tewari D, Pandey HK, Sah AN, Meena H, Chander V, Singh R, et al. Phytochemical, antioxidant and antidepressant evaluation of *Ocimum basilicum*, *O. tenuiflorum*, *O. kilimandscharicum* grown in India. Journal of Biologically Active Products from Nature. 2015;**5**:120-131

[81] Omodamiro OD, Jimoh MA. Antioxidant and antibacterial activities of *Ocimum gratissimum*. American Journal of Phytomedicine and Clinical Therapeutics. 2015;**3**:010-019

[82] Venuprasad MP, Kandikattu HK, Razack S, Khanum F. Phytochemical analysis of *Ocimum gratissimum* by LC-ESI–MS/MS and its antioxidant and anxiolytic effects. South African Journal of Botany. 2014;**92**:151-158

[83] Tanuj J, Vijay J. Antioxidant activity of ethanolic extract of *Ocimum Kilimandscharicum* using hydroxyl radical scavenging method. Journal of Drug Delivery and Therapeutics. 2017;7:66-68

[84] Jamshidi N, Cohen MM. The clinical efficacy and safety of Tulsi in humans: A systematic review of the literature. Evidence-Based Complementary and Alternative Medicine. 2017;**1**:1-13

[85] Samak G, Rao MS, Kedlaya R, Vasudevan DM. Hypolipidemic efficacy of *Ocimum Sanctum* in the prevention of atherogenesis in male albino rabbits. Pharmacology. 2007;**2**:115-127

[86] Mirje MM, Zaman SU, Ramabhimaiah S. Evaluation of the anti-inflammatory activity of *Ocimum sanctum* Linn (Tulsi) in albino rats. International Journal of Current Microbiology and Applied Sciences. 2014;**3**:198-205

[87] Güez CM, de Souza RO, Fischer P, de Moura Leão MF, Duarte JA, Boligon AA, et al. Evaluation of basil extract (*Ocimum basilicum* L.) on oxidative, antigenotoxic and anti-inflammatory effects in human leukocytes cell cultures exposed to challenging agents. Brazilian Journal of Pharmaceutical Sciences. 2017;**53**:e15098

[88] Eldin HME, Badawy AF. In vitro anti-*Trichomonas vaginalis* activity of *Pistacia lentiscus* mastic and *Ocimum basilicum* essential oil. Journal of Parasitic Diseases. 2015;**39**:465-473

[89] de Lima Silva L, Garlet QI, Koakoski G, de Abreu MS, Mallmann CA, Baldisserotto B, et al. Anesthetic activity of the essential oil of *Ocimum americanum* in *Rhamdia quelen* (Quoy & Gaimard, 1824) and its effects on stress parameters. Neotropical Ichthyology. 2015;**13**:715-722

[90] Caamal-Herrera SO, Carrillo-Cocom LM, Escalante-Réndiz DY, Aráiz-Hernández D, Azamar-Barrios JA. Antimicrobial and antiproliferative activity of essential oil, aqueous and ethanolic extracts of *Ocimum micranthum* Willd leaves. BMC Complementary and Alternative Medicine. 2018;**18**:55

[91] Dorman HJD, Deans SG. Antimicrobial agents from plants: Antibacterial activity of plant volatile oils. Journal of Applied Microbiology. 2000;**88**:308-316

[92] Burt S. Essential oils: Their antibacterial properties and potential applications in foods—A review.

International Journal of Food Microbiology. 2004;**94**:223-253

[93] Schmidt N, Mishra A, Lai GH, Wong GCL. Arginine-rich cell-penetrating peptides. FEBS Letter. 2010;**584**:1806-1813

[94] Londhe AM, Kulkarni AS, Lawand RV. *In-vitro* comparative study of antibacterial and antifungal activities: A case study of *Ocimum kilimandscharicum*, *Ocimum tenuiflorum* and *Ocimum gratissimum*. International Journal of Pharmacognosy and Phytochemical Research. 2015;**7**:104-110

[95] Miller AB, Cates RG, Lawrence M, Soria JAF, Espinoza LV, Martinez JV, et al. The antibacterial and antifungal activity of essential oils extracted from Guatemalan medicinal plants. Pharmaceutical Biology. 2014;**53**:548-554

[96] Pandey S. Antibacterial and antifungal activities of *Ocimum gratissimum* L. International Journal of Pharmacy and Pharmaceutical Sciences. 2017;**9**:26-31

[97] Wagura AG, Kimenju JW, Gichimu BM. Comparative antibacterial effects of raw and essential oils of *Ocimum gratissimum* L. against *Ralstonia solanacearum* (Smith). International Journal of Plant Pathology. 2011;**2**:144-152

[98] Chenni M, Abed DE, Rakotomanomana N, Fernandez X, Chemat F. Comparative study of essential oils extracted from Egyptian basil leaves (*Ocimum basilicum* L.) using hydro-distillation and solvent-free microwave extraction. Molecules. 2016;**21**:113

[99] Chouhan S, Sharma K, Guleria S. Antimicrobial activity of some essential oils-present status and future perspectives. Medicine. 2017;**4**:58

[100] Eswar P, Devaraj CG, Agarwal P. Anti-microbial activity of Tulsi {*Ocimum Sanctum* (Linn.)} extract on a periodontal pathogen in human dental plaque: An in vitro study. Journal of Clinical and Diagnostic Research. 2016;**10**:ZC53-ZC56

[101] Mondal S. Antimicrobial and Immunomodulatory effects of Tulsi (*Ocimum sanctum* Linn) [thesis]. Faculty of All India Institute of Medical Sciences, New Delhi, India for the award of the degree of Doctor of Philosophy. 2010. pp. 1-141

[102] Carovic-Stanko K, Orlić S, Politeo O, Strikić F, Kolak I, Milos M, et al. Composition and antibacterial activities of essential oils of seven *Ocimum* taxa. Food Chemistry. 2010;**119**:196-201

[103] El-Soud NHA, Deabes M, El-Kassem LA, Khalil M. Chemical composition and antifungal activity of *Ocimum basilicum* L. essential oil. Open Access Macedonian Journal of Medical Science. 2015;**3**:374-379

[104] Warrier PK, Nambair VPK, Ramankutty C. Indian Medicinal Plants: A Compendium of 500 Species. Arya Vaidya Sala, Kottakal, Kerala: Orient Longman, India; 1995

[105] Behbahani M. Evaluation of in vitro anticancer activity of *Ocimum basilicum*, *Alhagi maurorum*, *Calendula officinalis* and their parasite *Cuscuta campestris*. PLoS One. 2014;**9**:e116049. DOI: 10.1371/journal.pone.0116049

[106] Mohammadi M, Majd A, Nejadsattari T, Hashemi M. Antioxidant and anticancer activities of *Ocimum basilicum* L. *cv. Dark Opal* (Lamiaceae). Pharmacognosy Communications. 2014;**4**:48-58

[107] Naidu JR, Gunjan M, Chen Y, Sasidharan S. Evaluation of *in vitro* cytotoxic activity of *Ocimum basilicum* and *Mentha spicata* extracts. Asian Journal of Pharmaceutical and Clinical Research. 2016;**9**:131-134

[108] Zarlaha A, Kourkoumelis N, Stanojkovic TP, Kovala-Demertzi D. Cytotoxic activity of essential oil and extracts of *Ocimum basilicum* against human carcinoma cells. Molecular docking study of isoeugenol as a potent cox and lox inhibitor. Digest Journal of Nanomaterials and Biostructures. 2014;**9**:907-917

[109] Karthikeyan K, Gunasekaran P, Ramamurthy N, Govindasamy S. Anticancer activity of *Ocimum sanctum*. Pharmaceutical Biology. 1999;**37**:285-290

[110] Monga J, Sharma M, Tailor N, Ganesh N. Antimelanoma and radioprotective activity of alcoholic aqueous extract of different species of *Ocimum in* C_{57}BL mice. Pharmaceutical Biology. 2011;**49**:428-436

[111] de Lima VT, Vieirab MC, Kassuyac CAL, Cardosod CAL, Alvesa JM, Foglioe MA, et al. Chemical composition and free radical-scavenging, anticancer and anti-inflammatory activities of the essential oil from *Ocimum kilimandscharicum*. Phytomedicine. 2014;**21**:1298-1302

[112] Goel HC, Roa AR. Radiosensitizing effect of camphor on transplantable mammary adenocarcinoma in mice. Cancer Letters. 1988;**43**:21-27

[113] Banerjee S, Welsch CW, Rao AR. Modulatory influence of camphor on the activities of hepatic carcinogen metabolizing enzymes and the levels of hepatic and extrahepatic reduced glutathione in mice. Cancer Letters. 1995;**88**:163-169

[114] Ghant VK, Hiramoto NS, Solvason HB, Tyring SK, Spector NH, Hiramoto RN. Conditioned enhancement of natural killer cell activity, but not interferon, with camphor or saccharin-LiCl conditioned stimulus. Journal of Neuroscience Research. 1987;**18**:10-15

[115] Samant SS, Dhar U. Diversity, endemism and economic potential of wild edible plants of Indian Himalaya. International Journal of Sustainable Development & World Ecology. 2002;**4**:179-191

[116] Muralidharan A, Dhananjayan R. Cardiac stimulant activity of *Ocimum basilicum* Linn. extracts. Indian Journal of Pharmacology. 2004;**36**:163-166

[117] Dupasquier CM, Dibrov E, Kneesh AL, Cheung PKM, Lee KGY, Alexander HK, et al. Dietary flaxseed inhibits atherosclerosis in the LDL receptor-deficient mouse in part through antiproliferative and anti-inflammatory actions. American Journal of Physiology-Heart and Circulatory Physiology. 2007;**293**:2394-2402

[118] Amrania S, Harnafi H, Gadi D, Mekhfi H, Legssyer A, Aziz M, et al. Vasorelaxant and anti-platelet aggregation effects of aqueous *Ocimum basilicum* extract. Journal of Ethnopharmacology. 2009;**125**:157-162

[119] Janbaz KH, Hamid I, Gilani AH, Qadir MI. Spasmolytic, bronchodilator and vasodilator activities of Aqueousmethanolic extract of *Ocimum basilicum*. International Journal of Agriculture & Biology. 2014;**16**:321-327

[120] Packard RR, Maganto-Garcia E, Gotsman I, Tabas I, Libby P, Lichtman AH. CD11c(+) dendritic cells maintain antigen processing, presentation capabilities, and CD4(+) T-cell priming efficacy under hypercholesterolemic conditions

associated with atherosclerosis. Circulation Research. 2008;**103**:965-973

[121] Bravo E, Amrani S, Aziz M, Harnafi H, Napolitano M. *Ocimum basilicum* ethanolic extract decreases cholesterol synthesis and lipid accumulation in human macrophages. Fitoterapia. 2008;**79**:515-523

[122] Li H, Ge Y, Luo Z, Zhou Y, Zhang X, Zhang J, et al. Evaluation of the chemical composition, antioxidant and anti-inflammatory activities of distillate and residue fractions of sweet basil essential oil. Journal of Food Science and Technology. 2017;**54**:1882-1890

[123] Perretti M, D'Acquisto F. Annexin A1 and glucocorticoids as effectors of the resolution of inflammation. Nature Reviews. 2009;**9**:63-70

[124] D'Acquisto F, Maione F, Pederzoli-Ribeil M. From IL-15 to IL-33: The never-ending list of new players in inflammation. Is it time to forget the humble aspirin and move ahead? Biochemical Pharmacology. 2010;**79**:525-534

[125] Aye A, Jeon YD, Lee JH, Bang KS, Jin JS. Anti-inflammatory activity of ethanol extract of leaf and leaf callus of basil (*Ocimum basilicum* L.) on RAW 264.7 macrophage cells. Oriental Pharmacy and Experimental Medicine. 2019;**19**:217-226

[126] Kim HW, Roh DH, Yoon SY, Kang SY, Kwon YB, Han HJ, et al. The anti-inflammatory effects of low- and high-frequency electroacupuncture are mediated by peripheral opioids in a mouse air pouch inflammation model. Journal of Alternative and Complementary Medicine. 2006;**12**:39-44

[127] Son JK, Son MJ, Lee EK, Moon TC, Son KH, Kim CH, et al. Ginkgetin, a biflavone from Ginko biloba leaves, inhibits cyclooxygenases-2 and 5-lipoxygenase in mouse bone marrow-derived mast cells. Biological and Pharmaceutical Bulletin. 2005;**28**:2181-2184

[128] Happar M, Tursun A, Wen-ting Z, Lu-feng C, Tohti I, Reyim N, et al. Efect of extracts of *Ocimum basilicum* L. on mouse macrophage raw 264.7 and 5-lipoxygenase. Chinese Traditional Patent Medicine. 2013;**35**:1599-1604

[129] Benedec D, Pârvu AE, Oniga I, Toiu A, Tiperciuc B. Effects of *Ocimum Basilicum* L. extract on experimental acute inflammation. Revista Medico-chirurgicala a Societatii de Medici si Naturalisti din Iasi. 2007;**111**:1065-1069

[130] Selvakkumar C, Gayathri B, Vinaykumar KS, Lakshmi BS, Balakrishnan A. Potential anti-inflammatory properties of crude alcoholic extract of *Ocimum basilicum* L. in human peripheral blood mononuclear cells. Journal of Health Science. 2007;**53**:500-505

[131] Murphy KM, Reiner SL. The lineage decisions of helper T cells. Nature Reviews Immunology. 2002;**2**:933-944

[132] Ranjan D, Chen C, Johnston TD, Jeon H, Nagabhushan M. Curcumin inhibits mitogen stimulated lymphocyte proliferation, NFkB activation, and IL-2 signaling. Journal of Surgical Research. 2004;**121**:171-177

[133] Van Den Brande JM, Peppelenbosch MP, Van Deventer SJ. Treating Crohn's disease by inducing T lymphocyte apoptosis. Annals of the New York Academy of Sciences. 2002;**973**:166-180

[134] Złotek U, Szymanowska U, Karaś M, Świeca M. Antioxidative and anti-inflammatory potential of phenolics from purple basil (*Ocimum basilicum* L.) leaves induced by jasmonic, arachidonic

and β-aminobutyric acid elicitation. International Journal of Food Science and Technology. 2016;**51**:163-170

[135] Gawlik-Dziki U, Świeca M, Sugier D. Enhancement of antioxidant abilities and the lipoxygenase and xanthine oxidase inhibitory activity of broccoli sprouts of biotic elicitors. Acta Scientiarum Polonorum, Hortorum Cultus. 2012;**11**:13-25

[136] Raina P, Deepak M, Chandrasekaran CV, Agarwal A, Wagh N, Kaul-Ghanekar R. Comparative analysis of anti-inflammatory activity of aqueous and methanolic extracts of *Ocimum basilicum* in RAW264.7, SW1353 and human primary chondrocytes. Journal of Herbal Medicine. 2016;**6**:28-36

[137] Rodrigues LB, Martins AOBPB, Ceśario FRAS, Castro FF, Albuquerque TR, Fernandes MNM, et al. Anti-inflammatory and antiedematogenic activity of the *Ocimum basilicum* essential oil and its main compound estragole: In vivo mouse models. Chemico-Biological Interactions. 2016;**257**:14-25

[138] Coelho CF, Vieira RP, Lopes-Martins PSL, Aparecida S, Teixeira AUB, Gouvea IM, et al. The effect of inhaled nitric oxide on the carrageenan-induced paw edema. Histology and Histopathology. 2015;**30**:117-124

[139] Boyle JP, Honeycutt AA, Venkat Narayan KM, Hoerger TJ, Geiss LS, Chen H, et al. Projection of Diabetes Burden Through 2050: Impact of changing demography and disease prevalence in the U.S. Diabetes Care. 2001;**24**:1936-1940

[140] Wild S, Roglic G, Green A, Sicree R, King H. Global prevalence of diabetes: Estimates for the year 2000 and projections for 2030. Diabetes Care. 2004;**27**:1047-1053

[141] Mousavi L, Salleh RM, Murugaiyah V, Asmawi MZ. Hypoglycemic and anti-hyperglycemic study of *Ocimum tenuiflorum* L. leaves extract in normal and streptozotocin-induced diabetic rats. Asian Pacific Journal of Tropical Biomedicine. 2016;**6**:1029-1036

[142] El-Beshbishy HA, Bahashwan SA. Hypoglycemic effect of basil (*Ocimum basilicum*) aqueous extract is mediated through inhibition of α-glucosidase and α-amylase activities: An in vitro study. Toxicology and Industrial Health. 2012;**28**:42-50

[143] Lean ME, Noroozi M, Kell L, Burns J, Talwar D, Sattar N, et al. Dietary flavonols protect diabetic human lymphocytes against oxidative damage to DNA. Diabetes. 1999;**48**:176-181

[144] Bihari CG, Manaswini B, Keshari PS, Kumar TS. Phytochemical investigation & evaluation for antidiabetic activity of leafy extracts of various *Ocimum* (Tulsi) species by alloxan induced diabetic model. Journal of Pharmacy Research. 2011;**4**:28-29

[145] Parasuraman S, Balamurugan S, Christapher PV, Petchi RR, Yeng WY, Sujithra J, et al. Evaluation of antidiabetic and antihyperlipidemic effects of hydroalcoholic extract of leaves of *Ocimum tenuiflorum* (Lamiaceae) and prediction of biological activity of its Phytoconstituents. Pharmacognosy Research. 2015;**7**:156-165

[146] Egesie UG, Ibu ABAJO, Egesie OJ. Safety and hypoglycaemic properties of aqueous leaf extract of *Ocimum Gratissimum* in Streptozotocin induced diabetic rats. Nigerian Journal of Physiological Sciences. 2006;**21**:31-35

[147] Ezeani C, Ezenyi I, Okoye T, Okoli C. *Ocimum basilicum* extract exhibits antidiabetic effects via inhibition of hepatic glucose mobilization and carbohydrate

metabolizing enzymes. Journal of Intercultural Ethnopharmacology. 2017;**6**:22-28

[148] Okoduwa SIR, Umar IA, James DB, Inuwa HM. Anti-diabetic potential of *Ocimum gratissimum* leaf fractions in fortified diet-fed Streptozotocin treated rat model of Type-2 diabetes. Medicine. 2017;**4**:1-21

[149] Hannan JML, Marenah L, Ali L, Rokeya B, Flatt PR, Abdel-Wahab YHA. *Ocimum sanctum* leaf extracts stimulate insulin secretion from perfused pancreas, isolated islets and clonal pancreatic β-cells. Journal of Endocrinology. 2006;**189**:127-136

[150] WHO. Overview—Preventing chronic diseases: A vital investment. 2005

[151] Franz C, Baser KHC, Windisch W. Essential oils and aromatic plants in animal feeding—A European perspective. A review. Flavour and Fragrance Journal. 2010;**25**:327-340

Chapter 10

Medicinal Properties of Selected Asparagus Species: A Review

Polo-Ma-Abiele Hildah Mfengwana
and Samson Sitheni Mashele

Abstract

Asparagus species are naturally distributed along Asia, Africa, and Europe and are known to have numerous biological properties. This review article was aimed to provide an organized summary of current studies on the traditional uses, phytochemistry, and pharmacological and toxicological studies of *Asparagus laricinus Burch.*, *Asparagus africanus Lam.*, *Asparagus officinalis L.*, *Asparagus racemosus Willd.*, and *Asparagus densiflorus (Kunth)* Jessop to attain and establish new insights for further researches. Information used in this review was obtained from electronic database including PubMed central, Google scholars, Science direct, Scopus, and Sabinet. Based on the present findings, the existing literature still presents some breaches about the mechanism of action of various constituents of these plants, and their relation to other plant compounds in poly-herbal formulations, as well as their long-term use and safety. More in-depth studies are still needed for active compounds and biological activities of *Asparagus laricinus*, *Asparagus africanus*, and *Asparagus densiflorus*. Therefore, innumerable opportunities and possibilities for investigation are still available in novel areas of these plants for future research stud¬ies. It can be concluded that all selected Asparagus species have tremendous potential to improve human health and the pharmacological activities of these plants can be attributed to bioactive phytochemicals they possess.

Keywords: Asparagaceae, *Asparagus africanus lam.*, *Asparagus densiflorus (kunth) Jessop*, *Asparagus laricinus Burch.*, *Asparagus officinalis L.*, *Asparagus racemosus Willd.*, pharmacological actions, phytochemistry

1. Introduction

Historically, plants were used for numerous purposes for mankind in general, inter alia, feeding and catering, culinary spices, medicine, various forms of cosmetics, symbols in worship and for a variety of ornamental goods. They are still being used for these purposes. The traditional medicines are sold in market places and prescribed by traditional healers at their home [1] particularly in the rural areas where herbal medicine is the main source of the healthcare system. South Africa is blessed with a vast variety of plants since it has such a large diversity of more than 20,000 types of species. The research and scientific community find this to be a great source of interest [2]. Since the 1990s, great interest is being shown in plants that can be used as important sources of new medicines and herbs, which have become mainstream throughout Africa [3].

It is estimated that three quarters of the world of mankind relies on herbal and traditional medicine as a basis for primary healthcare [4]. It was discovered that between 12 and 15 million South Africans still rely on more than 700 indigenous types of plants for the supply of their traditional herbal medicines [5]. Up to 60% of the South African population consults one of an estimated 200,000 traditional healers in rural areas [6]. These herbal medicines which are extracted from plants and used for medicinal purposes often result in acute toxicity. For example, it is estimated that between 8000 and 20,000 people die every year in South Africa due to the fact that these medicinal plants are used incorrectly [7, 8]. The Food and Drug Administration [FDA] indicates that both serious and moderate adverse events from many botanical and others traditional medicinal products are significantly underreported, and that the annual number of such cases is at least 50,000 each year [9, 10].

Different research studies to elucidate and validate the ethnobotanical value of medicinal plants have been conducted and reported by investigators world-wide, with findings that were established from the use of various methods, and also under diverse conditions.

In this review study, five native Asparagus species (namely *Asparagus laricinus Burch.*, *Asparagus africanus Lam.*, *Asparagus officinalis L.*, *Asparagus racemosus Willd.* and *Asparagus densiflorus Kunth) Jessop*) were evaluated for their historical, etymological, morphological, phytochemical and pharmacological aspects. The findings of this review study are summarized, and the medicinal properties of the chosen Asparagus species are documented in this review study.

2. Methodology

2.1 Search criteria

Original articles, research papers published in journals and in PubMed central, Google scholars on plants of interest (*Asparagus laricinus Burch.*, *Asparagus africanus Lam.*, *Asparagus officinalis L.*, *Asparagus racemosus* and *Asparagus densiflorus (Kunth) Jessop)*, and medicinal uses were studied, and related articles and papers were also taken into consideration. The two species, *Asparagus laricinus Burch.* and *Asparagus africanus Lam.* were the first choice according to the research studies at the laboratory of authors for their pharmacological activities and toxicology. The other three species of interest were randomly selected from an enormous number of Asparagus species retrieved when the key words "Asparagus species with medicinal properties" were used in the search.

2.2 Data analysis

The available literature was especially studied for historical, etymological, morphological, phytochemical and pharmacological aspects of *Asparagus laricinus Burch.*, *Asparagus africanus Lam.*, *Asparagus officinalis L.*, *Asparagus racemosus* and *Asparagus densiflorus (Kunth) Jessop*. Priority was given to ethnobotanical reports, laboratory work and clinical trials carried out on all five species. Finally, results were obtained from all collected data and literature studied.

3. Asparagus species with medicinal properties

The genus Asparagus is an herbaceous plant comprising approximately 150 species around the world which are comprised of herbs, shrubs and vines [11]. Asparagus forms part of *Asparagaceae*, which is a monogeneric family and was formerly included in the *Liliaceae* family. Asparagus species are naturally distributed along Asia, Africa and Europe [12]. Most of these species have economic value as ornamental plants, such as *Asparagus plumosus*, and for their medicinal properties from plants such as *Asparagus laricinus Burch*. Asparagus species have numerous biological properties, such as being antioxidant, anti-inflammatory, antibacterial, antihepatotoxic, immunostimulant, and reproductive agents. Among large number of asparagus species that are used as medicine, five of them have been chosen as they have been investigated for their anticancer activity, namely: *Asparagus laricinus Burch.; Asparagus africanus Lam.; Asparagus officinalis L.; Asparagus racemosus*; and *Asparagus densiflorus (Kunth) Jessop*.

4. *Asparagus laricinus Burch.*

Vernacular names
English: Wild asparagus;
Afrikaans: Bergkatbos, Bergkatdoring, Fynkatbos, Katdoring, Langbeenkatdoring; Zulu: Ibutha, Setswana: Lesitwana [13].
Synonym: *Protasparagus laricinus (Burch.*) Oberm.

4.1 Historical aspects

Asparagus laricinus Burch. is a very hardy, evergreen, shrubby Asparagus with fine, feathery foliage and silvery, zigzag branchlets. It has myriads of tiny white, nectar-rich flowers that develop during spring and summer. These flowers are fragrant and attract insects and birds. Birds in the garden are attracted by its bright red and black berries. It may be grown in the sun or semi-shade and is a very useful plant for a security hedge as it is extremely spiny. It is fire-resistant and, if the stems burn, it shoots quickly from the base. The new shoots can be eaten as a vegetable. It grows in sun or shade and in all types of soil except water-logged soils. It can grow up to 1.5 m. *Asparagus laricinus Burch.* is native to Botswana and South Africa, Lesotho and Swaziland. They are used to treat tuberculosis, sores, red water, uterine infection, general alignments, umbilical cord inflammation, and serve as a diuretic.

4.2 Phytochemical active principals

Roots and leaves of *Asparagus laricinus Burch.* had tannins, saponins, terpenes, steroids. However, only roots showed the presence of alkaloids [14], while leaves are devoid of alkaloids [15]. The leaves further had flavonoids, glycosides, steroids and carbohydrates. The stems are rich in saponins, tannins, and flavonoids, with a lack of steroids, glycosides and carbohydrates [15]. The *Asparagus laricinus Burch.* aqueous roots extract contained 4.2 g/l GAE (Gallic acid equivalent) total phenolic content, while leaves and stem aqueous extract showed the phenolic concentration of 0.572 mg/GAE and 0.277 mg/GAE, respectively. It was apparent that leaves had more phenolic content than the stem, and this was supported by the number of active phytochemicals identified from both parts of the plant. Fuku et al. [14] isolated and identified three compounds from the *Asparagus laricinus Burch.* roots: indole-3-carbinol, α-sitosterol and ferulic acid.

4.3 Pharmacological actions

Secondary metabolites produced by plants for plant protection do not only benefit plants, but they also have health benefits for human beings. These compounds result in antimicrobial medicines [16], anti-inflammatory drugs, anticancer drugs, and plant-based anti-oxidants. Phytochemical screening was performed on the leaves and roots of *Asparagus laricinus Burch.*, and parts had tannins, saponins, terpenes, and steroids. However, only roots showed the presence of alkaloids. It was also shown that flavonoids which are known to have an ability to inhibit microbial growth also scavenge antioxidants. The leaf extract contained steroids, these being important compounds as sex hormones. Both leaves and stem extracts showed that they contain saponins, which ultimately has a suppressive effect on inflammation [17]. This is a main reason why *Asparagus laricinus Burch.* is used in traditional medicine. *In vivo* anti-inflammatory activity studies of this plant are being conducted in the Unit of Drug discovery, CUT [18].

Tannins are generally found in most plant parts: bark, wood, leaves, fruits and roots, and can have a toxic effect on filamentous fungi, yeasts and bacteria [19]. No alkaloids were found in this study. Leaf extracts further showed positive antibacterial activity on *S. aureus*, *S. saprophyticus*, *E. cloacae* and *B. subtilis*. Inhibition of *Staphylococcus aureus* by the *Asparagus laricinus Burch.* plant extract demonstrates huge potential for using this plant to extraction in the treatment of microbial infections, especially in the light of the growing antibiotic resistance in micro-organisms. The presence of phenols correlates with the antibacterial and antioxidant activities of the leaf extract of *Asparagus laricinus Burch.*, as demonstrated by Ntsoelinyane and Mashele [15].

In recent times, there has been a growing interest in finding antioxidants which occur naturally to replace synthetic antioxidants, many of which are being restricted due to their carcinogenicity [20, 21]. Free radical scavenging molecules such as flavonoids, tannins, alkaloids, quinones, amines, vitamins and other metabolites have anti-inflammatory, anti-carcinogenic, antibacterial and antiviral activities [22]. *Asparagus laricinus Burch.* aqueous extracts of roots and leaves showed positive antioxidant activity with DPPH assay [15]. Flavonoids contain anion radicals and produce membrane bound enzymes [23]. This could be the reason for the mechanisms of antioxidative action of *Asparagus laricinus Burch.* leaf extract. The antioxidant that is found in the plant extract may also be due to polyphenols as phenolics being present [24]. The aqueous leaves extract of *Asparagus laricinus Burch.* produced significant activity as an antioxidant, and this could be due to the presence of ferulic acid; and as it is a known to protect cells from oxidative stress.

Using the Ames test on *Salmonella typhimurium* strains: TA97, TA98, TA100 and TA102 without any metabolic activation, Mashele and Fuku [25] evaluated the mutagenic and antimutagenic properties of the aqueous root's extracts of this plant. The extract was non-mutagenic towards all strains, had moderate inhibitory effect on TA100, and had low inhibitory effects on TA102 and TA97 [7]. Root aqueous extract showed an indirect mutagenic effect toward TA102 after metabolic activation, but not in TA97, TA98 and TA100. However, it was found that the Ames test, without S9 (liver extract of a rat, hamster or human) metabolic activation, could only detect direct mutagens, while S9 metabolic activation allowed the detection of indirect mutagens which were mostly caused by conjugation reactions of metabolic oxidation systems. Cytotoxicity activity on Vero cells was also elucidated. The cytotoxicity tests indicated no cytotoxic effect below 500 μg/ml concentration of the *A. laricinus Burch.* aqueous extract [7].

The phytoconstituents detected from *Asparagus laricinus Burch.* may have caused the cytotoxic activity, although their precise mode of action is poorly understood. Only a few compounds were isolated from the roots of *Asparagus laricinus Burch.*: indole-3-carbinol, α-sitosterol and ferulic acid. β-Sitosterol have numerous therapeutic and chemo-preventive uses in the medical field [26, 27]. Prostate cancer is being treated by Indole-3-carbinol [28]. Anticancer activity on breast (MCF7), renal (TK10) and melanoma (UACC62) using roots aqueous and ethanol extracts was shown by Mashele and Kolesnikova [29], who revealed that ethanol extracts were very active while aqueous extracts were weakly active. However, ethanol roots extract only showed the presence of tannins while the aqueous roots extract showed a number of active phytochemicals. These results should be investigated further to elucidate the aforementioned difference. It may be that the presence of other active compounds somehow affected the ability of tannins by neutralizing their activity in the aqueous root's extracts. Another possibility is that active compounds from the roots were present and these were missed during the phytochemical screening of this plant.

Mokgawa evaluated possible toxic effects of dried roots, stem and leaves of *Asparagus laricinus Burch.* extracts using Sprague Dawley rats as animal models [18]. Histological evaluation could not reveal any pathological changes in both aqueous and ethanoic extracts across all levels of dosages. Full blood count results could not point in the direction of toxicity, adverse effects or hazards as indicated by statistically similar results between the exposed and unexposed groups, using both aqueous and ethanol extracts at different concentrations [18]. According to results obtained by Mokgawa, histological assessment has proven that both aqueous and ethanolic extracts of *Asparagus laricinus Burch.* had no detrimental or adverse effects on vital organs of Sprague Dawley rats [18]. Tissue damage, lesions or inflammation were not observed on the kidney, liver or spleen of treatment groups in comparison to the control group. The pattern was observed across increasing doses of aqueous and ethanolic extracts. It was, therefore, concluded that toxicological evaluation of *Asparagus laricinus Burch.* extracts may be considered relatively free of toxicity when given orally, because it did not cause death, damage or inflammation to tissues, nor did it produce any remarkable biochemical and hematological adverse effects in both male and female Sprague Dawley rats [18]. Further studies may also be conducted to demonstrate *in vivo* efficacy against cancer as studies to date were done using cell lines *(in vitro* studies).

4.4 Reflections and future recommendations

Only preliminary screening of phytochemicals was done on crude extracts. Isolation of active pure compounds was only done on roots (3 compound identified) and not on leaves, even though leaves showed so much active compounds. This compound identification still needs to be done and testing of them has not been done either. Both leaves and roots extracts showed the presence of saponins. However, the identification and isolation of those specific saponins has not been done. Intensive work still needs to be performed regarding the mutagenicity or genotoxicity of the plant extracts for the confirmation of the safety of *Asparagus laricinus Burch.*, as the root cytotoxicity results were promising, while the safety of the leaves also needs to be investigated. The toxicological study of the roots of *Asparagus laricinus Burch.* confirmed that the plant extract did not cause any harm *in vivo* and can thus be considered as non-toxic. However, the *in vivo* anticancer activity of the root extract has not been done in order to confirm or corroborate the results obtained in the screening study that was conducted. Both *in vitro* and *in vivo* anticancer, cytotoxicity and mutagenicity studies still need to be done on the leaf extract. The ability of the crude extract of this plant as an antibacterial agent was confirmed, and findings supported the use of this plant against infections. However, not all ethnobotanical claims of this plant have been confirmed as the anti-TB activity, anti-inflammatory activity and its ability as a diuretic still needs to be elucidated.

5. *Asparagus africanus Lam.*

Vernacular names

English: Wild Asparagus climbing asparagus fern, bush asparagus;

Afrikaans: Haakdoring, Katdoring Wag-'n-bietjie, Wag-n'-bietjie, Wag-'n-bietjie-doring; Xhosa: Ubulawu Ubumhlope, Umthunzi; Zulu: Isigoba, Isigobo; Sesotho: Lelala-tau-le-leholo, Leunyeli; Banda: ngorozo; Kirundi: imburabano, umunsabe; Maasai: embere e papa; Afeen Oromo: Seriti.

Synonym: *Protasparagus africanus* (Lam.) Oberm.

Scientific classification

Kingdom: plantae; **Subfamily:** asparagoideae; **Genus:** Asparagus—asparagus; **Order:** asparagales; **Species:** *Asparagus africanus* Lam.—African Asparagus; **Family:** Liliaceae.

5.1 Historical aspects

Asparagus africanus Lam., commonly known as African Asparagus, is a monocot. *Asparagus africanus Lam* (Liliaceae) is an erect armed herb that grows to a height of up to 6 m. The plant is found in many parts of tropical Africa and can grow between 700 and 3800 m above sea level. The fruit consists of a rounded berry which has a width of about 5–6 mm and it contains only one seed. It starts off as green colored but eventually becomes orange and somewhat shriveled as it matures. They grow during most parts of the year.

Their roots are traditionally used for: the relief of pain, rheumatism and chronic gout, hematuria, hemorrhoids, headache, backache, stomach pain, sore throat and otitis. It is also used to treat malaria, central nervous system related conditions, tuberculosis, venereal diseases and as an aid during childbirth [30].

5.2 Phytochemical active principals

The main contents of the plant are carbohydrates and saponins which have small quantities of flavonoids and tannins. Three steroidal saponins were isolated from the roots of *Asparagus africanus Lam.* [31]. Two compounds namely 2 beta-, 12 alpha-dihydroxy-(25R)-spirosta-4, 7-dien-3-one, lignan (+)-nyasol, and (Z)-(+)-4,4-(3-ethenyl-1-propene-1,3-diyl) bisphenol) were isolated by Oketch-Rabah et al. [32] from *Asparagus africanus Lam* roots.

5.3 Pharmacological actions

In traditional medicine, *Asparagus africanus Lam.* is used for treating headaches, backaches, and stomach pains and also is used to assist in childbirth and for hematuria, hemorrhoids, malaria, leishmaniasis, bilharziasis, syphilis, and gonorrhea [32, 33]. External application of the root is used for the relief of pain, rheumatism and chronic gout [34]. It is further used a diuretic, for sore throats and otitis [35]. The focus of many anti-infective drugs as well as alternative sources of antimalarial agents in various parts of the world has been on the use of medicinal plants [36–38]. During *in vitro* studies on extracts from the root of *Asparagus africanus Lam.*, it has been found that they can be used as a counter activity against four different malaria schizont strains [32].

Oketch-Rabah et al. [32], were able to isolate two antiprotozoal compounds, a sapogenin (muzanzagenin) and lignan ((+) nyasol), which they reported to be responsible for the antimalarial activity. Even though this plant has displayed promising antiplasmodial activities, no remarkable in vivo studies have been found up to now which can strengthen the preclinical study profile. There has been only one report by Dikasso et al. [39], on the in vivo anti-malarial activity of extracts of hydroalcoholic from *Asparagus africanus Lam.* in mice which have been infected with *Plasmodium berghei* [39]. The extract displayed parasite suppressive effects on

P. berghei infected rats in a dose dependent manner. However, the effects on Packed Cell Volume and body temperature observed were inconclusive [39].

Asparagus species are known to have steroidal saponins as their major bioactive constituents. The presence of saponins and carbohydrates from *Asparagus africanus Lam.* showed significant analgesic and anti-inflammatory activities as reported by Hassan et al. [30]. Saponins are also known to have broad spectrum of pharmacological and antimicrobial activities [40]. According to Madikizela et al. [41] *Asparagus africanus Lam.* leaves showed very active antimycobacterial activity on *Mycobacterium aurum A+*, with moderate antibacterial activity against *Klebsiella pneumonia* [41], due to the saponins present in this plant. The methanolic extract obtained from the roots of *Asparagus africanus Lam.* were considered to be none toxic as there was no mortality caused on rats after a dose of 5000 mg/kg was administered by mouth [42], and these findings were corroborated by Kedebe et al. [43], using hydro-alcoholic extracts of *Asparagus africanus Lam.*

Three steroidal saponins, which are the most probable components of estrogen, have been separated from the roots of *Asparagus africanus Lam.* [31], and *Asparagus officinalis L.* that was reported to have uterine contractile properties. Steroidal saponins have been found to be one of the active principles of the majority of anti-fertility agents [44, 45]. *In vitro* and *in vivo* studies of the extracts of ethanol of leaves and roots *of Asparagus africanus Lam.* displayed the ability of both extracts to have a potential acetylcholine effect on uterine contraction [46]. These results suggested the possibility of interaction of the extracts with endogenous acetylcholine to induce an abortifacient effect. Thus, this plant should not be used during pregnancy because of the possibility of unintentionally abortion.

Saponins isolated from this plant have displayed anti-inflammatory activities against several experimental types of inflammation in mice and rats [47]. During the initial inflammation process, histamine and serotonin are released resulting in inflammation signs such as edema, pain, redness and heat. In a study by Kebede et al. [43], rats were injected with edemagenic agents to trigger edema (sign of inflammation), root extracts of *Asparagus africanus Lam.* were administrated to the rats, and the expected edema was not observed as the plant inhibited an antihistaminic agent. The extract activity was then found to be more pronounced in the first phase of the rat edema (within 90 min), thus making it possible for the extract to contain antihistaminic activity [43].

5.4 Reflections and future recommendations

The roots of this plant are traditionally used for the relief of pain, rheumatism and chronic gout, hematuria, hemorrhoids, headache, backache, stomach pain, sore throat and otitis. In addition, they are used to treat malaria, central nervous system related conditions, tuberculosis, venereal diseases, and aid in childbirth. *In vitro* and *in vivo* studies performed on this plant confirmed the ethnobotanical claims of this plant and active compounds were isolated and identified. However, the mechanism of action of these compounds as anti-inflammatory, antimycobacterial, antiplasmodial and anti-infertility agent has not been performed to date. Toxicological studies showed that the plant was not toxic. However, its ability to induce mutagenicity was not determined.

6. *Asparagus officinalis L.*

Vernacular name

English: Asparagus, Garden asparagus, White asparagus, Sparrow grass and Common asparagus; **Arabic:** Ehlilaj aswad, Helion, Dhagboth, Akla, al theeb;

Chinese: Shi diao baiGerman: Spargel; **French:** Asperge; **Italian:** Asparagio; **Japanese:** Oranda-kiji-kakushi; **Portuguese:** Espargo; **Spanish:** Espárrago, Esparraguera, and **Swedish:** Sparris.

Synonyms: Asparagus caspius Hohen.; Asparagus longifolius Fisch. ex Steud.; *Asparagus officinalis* var. caspius (Hohen.) Asch. & Graebn.; *Asparagus officinalis subsp. officinalis*; Asparagus polyphyllus Steven ex Ledeb.

Scientific classification

Kingdom: plantae; **Subfamily:** asparagoideae; **Clade:** Angiosperms; **Genus:** Asparagus—asparagus; **Order:** asparagales; **Species:** *Asparagus officinalis L.*—garden asparagus; **Family:** Asparagaceae.

6.1 Historical aspects

Asparagus officinalis L. is a perennial herb which grows to a height of between 60 and 150 cm. It has thick swollen root stock. Its stem is multi-branched, with 2–6 needle-like shoots in whorls. Flowers are wide (4–6 mm), 6-lobed, perianth regular with a whitish–greenish yellow. Leaves are rudimentary, scale-like and its axillary shoots are needle-like and whorled. Fruits are round, initially green and, when ripe, forms an orange, 6–10 mm wide berry [17, 49]. The plant was distributed in Central and Southern Europe, the Middle East, Western Siberia and Northern Africa. It was then cultivated in many places. It is now distributed in Eastern Africa, Asia, Europe, Northern and Southern America [48]. Asparagus stalks are commonly eaten as a vegetable. Roots and seeds have been used as a treatment for various illnesses and as a diuretic, despite the lack of clinical evidence.

6.2 Phytochemical active principals

Chemical constituents of *Asparagus officinalis L.* contain steroid saponins including asparagosides A, B, D, F, Ge3w2q H, I, the bitter steroid saponins, amino acids, fructans (asparagose and asparagosine), ferulic acid and flavonoids (quercetin, rutin, hyperoside, and isoquercitrin) [36, 49, 50]. Shao et al. [51], further isolated two oligofurostanosides *Asparagus officinalis L.* seeds, and their structures were identified as 3-O-[alpha-Lrhamnopyranosyl-(1→2)-(alpha-L-rhamnopyranosyl-(1→4))-beta-D-glucopyranosyl]-26-O-[beta-D-gluco pyranosyl]-(25R)-22 alpha-methoxyfurost-5-ene-3 beta,26-diol(methyl protodioscin), and with the corresponding 22 alpha-hydroxy analogs (protodioscin). New asparagusic acid anti-S-oxide methyl ester (a new acetylenic compound) and asparagusic acid syn-S-oxide methyl ester, 2-hydroxyasparenyn {3,4″-trans-2-hydroxy-1-methoxy-4-[5-(4-methoxyphenoxy)-3-penten-1-ynyl]-benzene}, and eleven known compounds, [asparenyn, asparenyol, (±)-1-monopalmitin, ferulic acid, 1,3-O-di-p-coumaroylglycerol, 1-O-feruloyl-3-O-p-coumaroylglycerol, blumenol C, (±)-epipinoresinol, linoleic acid, 1,3-O-diferuloylglycerol, and 1,2-O-diferuloylglycerol, were separated from an ethyl acetate-soluble fraction of the methanol extract of the aerial parts of *Asparagus officinalis L.* [48]. Two major anthocyanins (A1 and A2) were also separated from peels of the spears of *Asparagus officinalis L.*. However, A1 was identified as cyanidin 3-[3″-(O-beta-d-gluco pyranosyl)-6″-(O-alpha-l-rhamnopyranosyl)-O-beta-dglucopyranoside], while A2 was recognized to be cyanidin 3-rutinoside, which was found to be in higher plants [48].

Sun et al. [52], recognized a new steroidal saponin, yamogenin II, with a unique aglycone moiety, and a structure of (25S)-spirostan-5-ene-3β-ol-3-O-α-Lrhamnopyranosyl-(1,2)-[α-L-rhamno pyranosyl-(1,4)]-β-D-glucopyranoside from the dried stems of *Asparagus officinalis L.*. Furthermore, more saponins were isolated from the plant included (25R)-furost-5-en-3β,22,26-triol-3-O-[α-L-rhamnopyranosyl-(1→4)-β-D-glucopyranoside]-26-O-β-D-glucopyranoside, (25R)-furostane-3β,22,26-triol-3-O-[α-L-rhamnopyranosyl-(1→4)-β-Dglucopyranoside]-26-O-β-D-glucopy ranoside, and (25S)-furostane -3β,22,26-triol-3-O-[α-L-rhamnopyranosyl-(1→4)-β-D-glucopyranoside]-26-O-β-D-glucopyranoside, and 3-O-[{α-L-rhamnopyranosyl-(1→2)} {α-L-rhamnopyranosyl-(1→4)}-β-Dglucopyranosyl]-(25S)-spirost-5-ene-3β-ol [53]. Nutritional analysis showed that the plant contained water 93.5%, total protein 1.91%, fat 0.16%, carbohydrates 2.04%, total dietary fiber 1.31%, and total nitrogen 0.31% [54]. The amino acid and mineral contents were found to be much higher in the leaves than the shoots [54].

6.3 Pharmacological actions

Asparagus officinalis L. is believed to have laxative, diuretic and contraceptive effects, and as a remedy for neuritis, rheumatism, cancer, toothache relieve, face acne lesion, as well as to stimulate hair growth [55]. According to research findings, the aqueous extract of *Asparagus officinalis L.* showed some antidiabetic effect after diabetic rats were treated with the extract, and their elevated blood glucose was suppressed [56]. The extract further displayed dangerous antioxidant activity in *in vitro* and *in vivo* assays. *Asparagus officinalis L.* crude saponins from the shoots (edible part) of asparagus, were found to have antitumor activity as they promoted the growth of HepG2 cells, and of human leukemia HL-60 cells in a way which caused it to become dose-dependent.

Shao et al., separated two oligofurostanosides from the seeds of *Asparagus officinalis L.* with cytotoxic activity [51]. They repressed the growth of human leukemia HL-60 cells in culture and macromolecular synthesis in a manner which promoted dose-dependence. Saponins from old stems of asparagus (SSA) exerted potential repressive activity on tumor growth and metastasis of breast, colon and pancreatic cancer cells. Sakaguchi et al. [57], found that anthocyanins A1 and A2 separated from the spears of *Asparagus officinalis L.* were found to act as antioxidants. The saponin fraction of the *Asparagus officinalis L.* exerted antifungal activity [58, 59]. The intake of asparagus also improved antioxidant status (superoxide dismutase and catalase enzymes) and prevented lipid peroxidation [60]. This corroborated with the findings by Hafizur et al. [56]. The hypolipidemic effect of n-butanol extract from asparagus by-products was evaluated in mice fed a high-fat diet, and the results were positive [56].

The antibacterial potential of the ethanolic extracts was determined against *Escherichia coli*, *Pseudomonas aeruginosa*, *Staphylococcus aureus* and *Bacillus cereus*, and the activity was seen only on *Escherichia coli*, while there was no antimicrobial activity in the same concentration against other tested pathogenic bacteria [61]. The taking in of asparagus alleviated some clinical symptoms (stool consistency, stool blood, and spleen hypertrophy) during active colitis. Other pharmacological effects of *Asparagus officinalis L.* were anti-fatigue effects, enhanced anoxia tolerance, induced analgesia and improved memory, and decreased the prevalence of lipid peroxide in plasma, liver and brains of the rats [62].

Jang et al. examined *Asparagus officinalis L.* for its inhibitory effects against both cyclo-oxygenase-1 and -2, thus having anti-inflammatory potential [48], due to linoleic acid identified as the most active compound in the plant [48, 54]. Aqueous extract of *Asparagus officinalis L.* resulted in relaxation of spontaneous contractions of separated smooth muscle of rabbit jejunum [57]. *Asparagus officinalis L.* also induced diuretic effects [63].

6.4 Reflections and future recommendations

Roots and seeds have been used as a treatment for various illnesses and as a diuretic, despite the lack of clinical evidence. So many active chemical constituents were isolated and identified. However, not all pharmacological activity of those isolated compounds was done. *In vivo* anticancer activity is desirable to confirm the in vitro findings. Studies have claimed that *Asparagus officinalis L.* has other pharmacological effects which were not reported as ethnobotanical uses of this plant, such as anti-fatigue effects, enhanced anoxia tolerance, induced analgesia and improved memory, as well as the decreased contents of lipid peroxide in plasma, liver and brains of the rats. However, this was not linked to the active compound present from the plant that could be responsible for those pharmacological activities.

7. *Asparagus racemosus Willd.*

Vernacular names

English name: Wild Asparagus, Indian Asparagus, Hundred Roots, *Asparagus racemosus Willd*; **Hindi:** Satavar.

Synonyms: Indeevari, Sukshamapatra, Bahusuta, Shatmooli, Narayani, Bhiru, Virya, Madabhanjani, Shatpadi, Shatvirya.

Scientific classification

Kingdom: Plantae; **Sub-family:** Asparagoideae; **Clade:** Angiosperms; **Genus:** Asparagus; **Order:** Asparagales; **Species:** *Asparagus racemosus Willd.*; **Family:** Asparagaceae.

7.1 Historical aspects

Asparagus racemosus Willd. has been used traditionally for ages as a female reproductive tonic as it prevents abortion and promotes the health of the mother and growing fetus when used in antenatal care. This plant also increases lactation and is useful for the treatment of gynecological diseases when used in postnatal care [64]. Research has supported other reported ethnobotanical uses of this plant for female reproductive system-related health issues. This plant can be described as a climbing shrub which is thorny and has woody stems. The leaves become minute scales and spines. Fruits are round and are a purple black color. Roots are succulent and tuberous and taper at both ends. It is distributed throughout India, and almost commonly found in areas up to an altitude of 4000 feet in the Himalayas and in Ceylon [64]. Tuberous roots of the plant are the parts used [65]. The tubers are eaten as a sweetmeat. The root contains juice which, when fresh used with honey as a demulcent in bilious dyspepsia or diarrhea. It is used in the preparation of medicated oils for external application to sufferers of nervous and rheumatic infections, and urinary tract infections [64].

7.2 Phytochemical active principals

The main active sections of *Asparagus racemosus Willd.* are steroidal saponins (Shatavarins I–IV), which are the phytoestrogen compounds which are present in the roots of this plant [66–69]. Shatavarin IV is a glycoside of sarsasapogenin having two molecules of rhamnose and one molecule of glucose. It also contains mucilage and starch. The 8-methoxy-5,6,4′-trihydroxyisoflavone, a new isoflavone, was separated from the roots of *Asparagus racemosus Willd.* by Saxena and Chaurasia [70]. A novel oligospirostanosid 1,3-O-[α-L-3-rhamnopyronosyl-(1 → 2)-α-L-rhamnopyronosyl(1 → 4)-O-β-D-glycopyranosyl]25(S)-5β-Spirostan-3β-ol also known as immunoside was isolated, and it was biologically evaluated as an immunomodulatory agent [71]. Wiboonpun et al. [72], isolated a new antioxidant compound named Racemofuran, together with known compounds asparagamine A and racemosol. Three steroidal saponins (Racemosides A, B and C) were also isolated from the methanolic extract of fruit of *Asparagus racemosus Willd.*. Polycyclic alkaloid like asparagamine A, and disaccharide in roots are also reported in other research studies [73–75]. *Asparagus racemosus Willd.* is also reported to have alkaloids, proteins, starch, tannin, flavonoids, glycosides of quercetin, rutin and hyperoside in roots and flowers [76]. Quercetin 3-glucuronide is present in leaves [77]. There were few trace minerals like zinc (53.15), manganese (19.98), copper (5.29), and cobalt (22.00 microgram per gram) together with calcium, magnesium, potassium, zinc, and selenium [68, 78]. The callus culture of *Asparagus racemosus Willd.* has shown synthesis of sarsasapogenin [79]. However, no report has been received on the chemistry of the contents of its fruit.

7.3 Pharmacological actions

The healing qualities of *Asparagus racemosus Willd.* are useful to a wide array of ailments. Ayurvedic (Indian traditional medicines database) literature considers it a strong drug which can improve memory intelligence and physical strength and maintain youthfulness [48, 80]. *Asparagus racemosus Willd.* can also be used as a uterine sedative. In addition, a glycoside, Shatavarin 1, separated from the roots of *Asparagus racemosus Willd.* has been found to be responsible for the competitive blocking of oxytocin-induced contractions, *in vitro* as well as *in vivo* [19, 64]. In India, it is recognized as a female tonic. In spite of being a rejuvenating herb, it is recognized as being used in female infertility, as it increases libido, is able to cure inflammation of sexual organs, and can be used to moisten dry tissues of the sexual organs. It further enhances folliculogenesis and ovulation, prepares the womb for conception, prevents miscarriages, acts as post-partum tonic by increasing lactation, normalizing uterus and changing hormones. It is also used in leucorrhoea and menorrhagia [65, 81].

The roots of *Asparagus racemosus Willd.* have been described as bitter-sweet, emollient, cooling, nervine tonics, preventing constipation, and may be used as an aphrodisiac, diuretic, carminative and antiseptic [82]. The powdered dried root exhibits galactagogic properties as there was an increase in milk secretion during lactation [83]. While active it has resulted in the action of released corticosteroids or an increase in prolactin. The other study also agreed with the galactagogic effect of this plant, because an alcoholic extract of *Asparagus racemosus Willd.*, increased the prolactin levels in female rats (Kumar et al., 2008). It served as a potential stimulator for early restoration of milk production without any adverse effects [84]. The juice of fresh roots of *Asparagus racemosus Willd.* is recommended for duodenal ulcers [85]. The plant can also be used to treat skin diseases, wounds and as a demulcent in dyspepsia [64]. The aqueous root extract possesses immunoadjuvant potential [37]. In the roots of *Asparagus racemosus Willd.* antioxidant and anti-ADH (Antidiuretic hormone) activity were found to be present [72, 86], and there was antitumor and anticancer activity [87, 88].

Asparagus racemosus Willd. displayed a preventative action on DMBA (-7,12-Dimethylbenz[a]anthracene) induced mammary carcinogenesis in rats. Rats which were fed on *Asparagus racemosus Willd.* diet displayed a decline in both tumor incidence and mean number of tumors per tumor bearing animal [76]. Studies also showed that the plant has anti-ulcerogenic activity [89], anti-inflammatory activity and antimicrobial activity [90]. Antimicrobial activity was used against *Escherichia coli*, *Shigella dysenteriae*, *Shigella sonnei*, *Shigella flexneri*, *Vibrio cholerae*, *Salmonella typhi*, *Salmonella typhimurium*, *Pseudomonas pectida*, *Bacillus subtilis* and *Staphylococcus aureus*, and sensitivity was observed in all strains under study [90]. As *Asparagus racemosus Willd.* is believed to have no antibacterial action, protection offered by *Asparagus racemosus Willd.* against sepsis by altering function of macrophages, shows its potential immunomodulatory property [91, 92].

Methanolic extract of roots displayed important antitussive activity on sulfur dioxide-induced coughs in mice [93]. An aqueous solution of the crude alcoholic extract of the roots displayed significant antiprotozoal activity against *Entamoeba histolytica in vitro* [94]. An examination was made on rat liver mitochondria for the possible antioxidant effects of crude extract and purified aqueous fraction of *Asparagus racemosus Willd.* against member damage induced by the free radicals generated during gamma radiation [86, 95]. It also raised the urinary concentration of magnesium, which is considered as one of the suppressors of crystallization [96]. Aqueous and butanol fractions displayed less prominent effects on the release of, especially at lower glucose concentration [97]. *Asparagus racemosus Willd.* further showed the potential of anti-HIV, (Human immunodeficiency virus) and its active principles are being investigated [98].

It can be concluded that *Asparagus racemosus Willd.* has immense importance in the folk medicine. In Ayurveda, *Asparagus racemosus Willd.* has been described as perfectly safe for long term use, even during pregnancy and lactation. To support this theory, an *in vivo* study was conducted and the systemic administration of higher doses of all extracts did not display any abnormality behavioral patterns in mice and rats [99], neither did it produce mortality even up to higher oral dosages of 64 g/kg [100].

7.4 Reflections and future recommendations

Intensive research has been performed on the active compounds from this plant. The plant was reported to have anticancer activity. However, cell lines used, and solvents used were not mentioned. While antitumor activity was reported on the mammary carcinogenesis only, more research using other cell lines is required to explore the antitumor and anticancer activity of this plant root's extract. The potential of this plant as cancer inducing agent has not been thoroughly elucidated as *in vivo* studies showed the safety of this plant. However, the exposure duration was not long enough. It is well known that the process of carcinogenesis is very slow, thus the adverse effects after 10 years of using *Asparagus racemosus Willd.* has not been determined scientifically. It will, therefore, be important to understand the mutagenicity of the plant before we can conclude that it is hundred percent safe. Mechanism of action of the active compounds from this plant and human trials are required, as different metabolic reactions in humans may influence the activity of the compound.

8. *Asparagus densiflorus* (kunth) jessop

Vernacular names

Sprenger's asparagus fern, bushy asparagus, asparagus fern and smilax. Inwele in Zulu.

Synonyms: Asparagopsis densiflora Kunth, Prot*asparagus densiflorus*.

Scientific classification

Kingdom: Plantae; **Subfamily:** Asparagoideae; **Clade:** Angiosperms; **Genus:** Asparagus; **Order:** Asparagales; **Species:** A. densiflorus; **Family:** Asparagaceae.

8.1 Historical aspects

South African asparagus was observed for the first time in an illustration from 1686, while plants from the 'Sprengeri' group began to be cultivated as early as 1888. Today they are grown in all parts of the word, and are hardy, drought-tolerant and quite salt-tolerant plants which are used for plants foliage and as garden plants. The South

African Asparagus species and the European Asparagus species are related, and they display a very interesting structure botanically. They do not have true leaves at all, but these are actually cladodes which may actually be modified branches, while the spines are formed from modified branches or from modified leaves. The most popular forms form part of the emerald ferns of the *Asparagus densiflorus (Kunth) Jessop* 'Sprengeri' group [101]. They form large cushions which have long, arching stems more or less densely covered with dark green, leaves which have the appearance like needles. The plant appears fernlike, but its flowers and fruit clearly place it among the angiosperms.

Asparagus densiflorus (Kunth) Jessop is a delicate, fern-like perennial plant which has arching stems which grow up to 1 m long and have a scrambling habit. A large cushion of dark green needle-like leaves is formed by the plant. It is often proved to be of use for medicinal purposes and can also be used as ground cover in partial or light shade, but it flourishes in full sun if watered regularly. It has very small, hardly observable spines which is unlike most Asparagus fern species. It is also attractive as an indoor or patio plant in large containers or hanging baskets. Unlike most Asparagus fern species, it only has very small, hardly noticeable spines. The roots of the plants are extensive which contain numerous grape sized tubers. These provide food in nature for extensive periods of drought in summer. The root system is used extensively for binding soil on slopes [101].

8.2 Phytochemical active principals

The phytochemical analysis of the aqueous and ethanolic extracts was carried out for the presence of flavonoids, tannins, phenolics, saponins, cardiac glycosides, terpenoids, quinones, amino acids, carbohydrates and alkaloids [46]. Both extracts showed a lack of amino acids, and were found to contain flavonoids, tannins, phenolics, saponins, cardiac glycosides and carbohydrates. In addition, the ethanol extract was found to contain terpenoids and alkaloids, whereas the aqueous was found to contain quinones. Estimated flavonoid and phenolic content of *Asparagus densiflorus (Kunth) Jessop* aqueous plant extracts were 900 and 380 μg/ml [102].

8.3 Pharmacological actions

An infusion of the plants' leaves may be used to treat pain in the abdomen, as a general tonic and to boost immunity. Further it may be used as a cleansing agent to rid the body of "poison" and "dirty blood". Thrust and ulcers in the mouth associated with HIV may also be treated by this plant. According to Davids et al. [103], traditional health practitioners (THPs) reported that *Asparagus densiflorus (Kunth) Jessop* is one of the "strongest" plants used for HIV. Moreover, according to Singh et al. [102], *Asparagus densiflorus (Kunth) Jessop* is considered as one of the ethnomedicinal plants. However, its ethnomedicinal actions had never been discussed. No literature was found on the correlation or link of phytochemical active compounds and pharmacological activity of this particular plant. *Asparagus densiflorus (Kunth) Jessop* aqueous and ethanoic leaf structures were screened for their antibacterial activity against *Enterobacter aerogenes*, *Clostridium perfringens* and *Salmonella typhimurium*. However, it was found that the aqueous extract showed a potential to inhibit growth of all three selected micro-organisms, while the ethanol extract inhibited only the growth of *Enterobacter aerogenes* [46].

8.4 Reflections and future recommendations

Only the preliminary phytochemical screening was performed on the crude extract of this plant, but no deeper research has been done on the activity of this plant in order to confirm its ethnobotanical claims as a plant used to treat thrush, ulcers in the mouth as well as for HIV. Thrush results from an overgrowth of normal flora in the mouth. The anti-fungal activity of this plant has not been studied yet, together with its anti-HIV activity, as the plant has been reported to be used ethnobotanically for HIV. Further studies to isolate the active compounds, elucidate the safety of this plant and to fully confirm its pharmacological activity are needed.

9. Conclusions

The genus Asparagus is an herbaceous plant comprising approximately 150 species around the world, and consisting of herbs, shrubs and vines. Asparagus species possess bioactive properties, such as: antioxidant, anti-inflammatory, antibacterial, antihepatotoxic, immunostimulant, and reproductive agents. In the present review study, five native Asparagus species (namely: *Asparagus laricinus*

Plant species	**Active phytochemicals**	**Mode of Action**	**Beneficial role in human health**
Asparagus laricinus Burch	*Asparagus laricinus Burch* roots and leaves have tannins, saponins, terpenes and steroids. The stems are rich in saponins, tannins, and flavonoids, with a lack of steroids, glycosides and carbohydrates. Only roots have alkaloids and indole-3-carbinol, α-sitosterol and ferulic acid were isolated from the roots	*Asparagus laricinus Burch.* has flavonoids that inhibited microbial growth and scavenged antioxidants. Both leaves and stem extracts showed that they contain saponins, which ultimately has a suppressive effect on inflammation. This plant was not toxic when administered orally to rats, it was non-mutagenic, however, root aqueous extract showed an indirect mutagenic effect toward *Salmonella typhimurium* TA102 strain after metabolic activation, but not in TA97, TA98 and TA100 strains. β-Sitosterol isolated from this plant is known for its therapeutic and chemo-preventive uses in the medical field, while prostate cancer is being treated by Indole-3-carbinol. This plant had anticancer activity on breast (MCF7), renal (TK10) and melanoma (UACC62)	*Asparagus laricinus Burch.* is used to treat tuberculosis, sores, red water, uterine infection, general alignments, umbilical cord inflammation, and serve as a diuretic
Asparagus africanus Lam.	The main contents of the plant are carbohydrates, saponins, flavonoids and tannins. Two compounds namely; 2 beta-, 12 alpha-dihydroxy-(25R)-spirosta-4,7-dien-3-one, lignan (+)-nyasol, and (Z)-(+)-4,4-(3-ethenyl-1-propene-1,3-diyl) bisphenol) were isolated from *Asparagus africanus Lam* roots, but their mode of actions has not been documented	This plant has acetylcholine effect on uterine contraction and antiplasmodial activity. Two antiprotozoal compounds; a sapogenin (muzanzagenin) and lignan ((+) nyasol), which were reported to be responsible for the antimalarial activity, have been isolated from *Asparagus africanus Lam.* The presence of saponins and carbohydrates from *Asparagus africanus Lam.* showed significant analgesic, anti-inflammatory activities and antimicrobial activities	*Asparagus Africanus Lam.* is used for treating headaches, backaches, stomach pains, malaria, to treat sexual transmitted infections and also used to assist in childbirth. It is also used to treat central nervous system related conditions, tuberculosis, venereal diseases and as an aid during childbirth. External application of the root is used for rheumatism and chronic gout. It is further used a diuretic, for sore throats and otitis

Plant species	Active phytochemicals	Mode of Action	Beneficial role in human health
Asparagus officinalis L.	*Asparagus officinalis L.* contain steroid saponins, amino acids, fructans, ferulic acid and flavonoids (quercetin, rutin hyperoside, and isoquercitrin), oligofurostanosides, new isolated asparagusic acid, two major anthocyanins. Carbohydrates, proteins, dietary fiber and nitrogen where found in this plant.	*Asparagus officinalis L.* showed antidiabetic activity through the suppression of blood glucose levels. The saponins from this plant inhibits the growth of leukemic cells, had antifungal activity, and scavenged antioxidants. The extracts of this plant further displayed antibacterial, anti-inflammatory potential *in vitro* also induced diuretic effects *in vivo*	*Asparagus officinalis L.* is believed to have laxative, diuretic and contraceptive effects, and as a remedy for neuritis, rheumatism, cancer, toothache relieve, face acne lesion, as well as to stimulate hair growth
Asparagus racemosus Willd.	*Asparagus racemosus Willd.* is reported to have alkaloids, proteins, starch, tannin, flavonoids, glycosides of quercetin, rutin and hyperoside in roots and flowers. There were few trace minerals identified from this plant species such as zinc, manganese, copper, cobalt, calcium, magnesium, potassium, zinc and selenium. The main active sections of *Asparagus racemosus Willd.* are steroidal saponins, racemofuran, polycyclic alkaloid, together with known compounds asparagamine A and racemosol	The glycoside from *Asparagus racemosus Willd.* blocks the oxytocin-induced contractions and have galactagogic properties that leads to increase in milk secretion during lactation. The roots of *Asparagus racemosus Willd.* showed significant antioxidant, anticancer, anti-inflammatory, anti-ADH and antimicrobial activities Methanolic extract of roots displayed important antitussive activity and antiprotozoal activity against *Entamoeba histolytica in vitro*	*Asparagus racemosus Willd.* is used to improve memory intelligence, physical strength, as a uterine sedative, and to maintain youthfulness. It is used as a female tonic to prevent abortion, to promote the health of the mother and growing fetus when used in antenatal care and acts as post-partum tonic by normalizing uterus and changing hormones. Despite being a rejuvenating herb, it is used in female infertility, as it increases libido, ovulation, can be used to moisten dry tissues of the sexual organs and it's able to cure inflammation of sexual organs. This plant also increases lactation and is useful for the treatment of gynecological diseases when used in postnatal care. The juice of fresh roots is recommended for duodenal ulcers. The plant can also be used to treat urinary tract infections, skin diseases, and wounds, in preventing constipation, and as an aphrodisiac, diuretic, carminative, and antiseptic. It is also used for nervous and rheumatic infections

Plant species	Active phytochemicals	Mode of Action	Beneficial role in human health
Asparagus densiflorus (Kunth) Jessop	This plant has flavonoids, tannins, phenolics, saponins, cardiac glycosides, terpenoids, quinones, amino acids, carbohydrates and alkaloids. In addition, the aqueous extracts have quinones	*Asparagus densiflorus (Kunth) Jessop* aqueous and ethanolic leaves extracts can inhibit antibacterial activity against *Enterobacter aerogenes*, *Clostridium perfringens* and *Salmonella typhimurium*. However, other pharmacological activities of this plant have not been scientifically investigated and documented	An infusion of the plant leaves is used to treat thrush and ulcers in the mouth, for abdominal pains, as a tonic to boost immunity, as a cleansing agent to rid the body of "poison" and "dirty blood"

Table 1.
Summary of phytochemicals, mode of action and the role of selected asparagus species in human health.

Burch., *Asparagus africanus Lam.*, *Asparagus officinalis L.*, *Asparagus racemosus Willd.* and *Asparagus densiflorus (Kunth)* Jessop) were evaluated for their historical, etymological, morphological, phytochemical and pharmacological aspects. The phytochemicals, mode of action and the role of selected Asparagus species in human health have been summarized in **Table 1**.

Conflict of interest

"The authors declare no conflict of interest."

Author details

Polo-Ma-Abiele Hildah Mfengwana[1] and Samson Sitheni Mashele[2*]

1 Department of Health Sciences, Central University of Technology, Free State, Bloemfontein, South Africa

2 Unit for Drug Discovery Research, Central University of Technology, Free State, Bloemfontein, South Africa

*Address all correspondence to: smashele@cut.ac.za

References

[1] Von Maydell HJ. Trees and Shrubs of the Sahel. 2nd ed. Germany: Verlag Josef Margraf. Weikersheim; 1996. pp. 562-563

[2] Cherry M. South Africa—Serious about biodiversity science. PLoS Biology. 2005;**3**(5):145-150. Available from: http://journals.plos.org/plosbiology/article?id=10.1371/journal.pbio.0030145 [Accessed: October 15 2016]

[3] Pavithra PS, Janani VS, Charumathi KH, Indumathy R, Potala S, Verma RS. Antibacterial activity of plants used in Indian herbal medicine. International Journal of Green Pharmacy. 2010;**4**:22-28

[4] Efferth T, Kaina B. Toxicities by herbal medicines with emphasis to traditional Chinese medicine. Current Drug Metabolism. 2011;**12**(10):989-996

[5] Meyer JJM, Afolayan AJ, Taylor MB, Engelbrecht L. Inhibition of herpes simplex virus type 1 by aqueous extracts from shoots of *Helichrysum qureonites* (Asteraceae). Journal of Ethnopharmacology. 1996;**52**:41-43

[6] Liu WJH. Traditional Herbal Medicine Research Methods. 1st ed. New York: John Wiley & Sons; 2011. pp. 111-123

[7] Debbie S, Ladds G, Duez P, Williamson E, Chan K. Pharmacovigilance of herbal medicine. Journal of Ethnopharmacology. 2012;**140**(3):513-518

[8] Statssa. Mortality and causes of death in South Africa: Findings from death notification. 2008. Available from: https://www.statssa.gov.za/publications/P03093/P030932006.pdf [Accessed: 13 November 2016]

[9] Hazell L, Shakir SAW. Under reporting of adverse drug reactions: Systematic review. Drug Safety. 2006;**29**(5):385-396

[10] Lard-Whiteford S, Nguyen D, Stevens T, Amchin W. Guidance for Industry on Complementary and Alternative Medicine Products and their Regulation by the Food and Drug Administration. U.S. Department of Health and Human Services—Food and Drug Administration. Available from: http://www.fda.gov/downloads/RegulatoryInformation/Guidances/UCM145405.pdf; 2006 [Accessed: 13 November 2016]

[11] Brummitt RK, Harder DK, Lewis GP, Lock JM, Polhill RM, Verdcourt B. Leguminosae subfamily Papilionoideae. Flora of Zimbabwe. 2007;**3**(3):96-98

[12] Thakur S, Sharma DR. Review on medicinal plant: Asparagus adscendens roxb. International Journal of Pharmaceutical Sciences and Health Care. 2015;**5**(3):82-97

[13] Foden W, Potter L. Asparagus laricinus Burch: National Assessment—Red List of South African Plants. 2005. Available from: http://redlist.sanbi.org/species.php?species=728-59 [Accessed: 19 October 2016]

[14] Fuku S, Al-Azzawi AM, Madamombe-Manduna IT, Mashele S. Phytochemistry and free radical scavenging activity on Asparagus laricinus. International Journal of Pharmacology. 2013;**9**(5):312-317

[15] Ntsoelinyane PH, Mashele S. Phytochemical screening, antibacterial and antioxidant activities of *Asparagus laricinus* leaf and stem extracts. Bangladesh Journal of Pharmacology. 2014;**9**:10-14

[16] Briskin DP. Medicinal plants and phytomedicines. Linking plant biochemistry and physiology to human health. Plant Physiology. 2000;**124**:507-514

[17] Just MJ, Recio MC, Giner RM, Cullar MJ, Manez S, Bilia AR. Antiinflammatory activity of unusual Lupane saponins from Bupleurum fruticescens. Planta Medica. 1998;**64**:404-407

[18] Mokgawa SD. The toxicological study of Asparagus laricinus in rats. In: Paper Presentation at Medical Laboratory Professionals Congress; Organized by Unit of Drug Discovery at Central University of Technology; Port Elizabeth – SA. 2015. p. 23. <cut.ac.za> [Accessed: 15 May 2015]

[19] Scalbert A. Antimicrobial properties of tannins. Phytochemistry. 1991;**30**:3875-3883

[20] Mosquera OM, Correa YM, Buitrago DC, Nio J. Antioxidant activity of twenty-five plants from colombian biodiversity. Memorias do Instituto Oswaldo Cruz. 2007;**102**:631-634

[21] Sasaki YS, Kawaguchi A, Kamaya M, Ohshita K, Kabasawa K, Iwama KT, et al. The comet assay with 8 mouse organs: Results with 39 currently used food additives. Mutation Research. 2002;**519**:103-119

[22] Sala A, Del Carmen Recio M, Giner RM. Anti-inflammatory and antioxidant properties of Helichrysum italicum. The Journal of Pharmacy and Pharmacology. 2002;**54**(3):365-371

[23] Li MJ, You JY, Liu ZY, Zhang HQ. Microwave-assisted extraction of flavonoids from flos sophorae. Guangxi Normal University. 2003;**21**:103-104

[24] Thabrew MI, Hughes RD, McFarlane IG. Antioxidant activity of Osbeckia aspera. Phytotherapy Research. 1998;**12**:288-290

[25] Mashele S, Fuku SL. Evaluation of the antimutagenic and mutagenic properties of *Asparagus laricinus*. Medical Technology SA. 2011;**25**(2):33-36

[26] Baskar AA, Ignacimuthu S, Paulraj MG, Al Numair KS. Chemopreventive potential of β-sitosterol in experimental colon cancer model—An in vitro and in vivo study. BMC Complementary and Alternative Medicine. 2010;**10**:10-24

[27] Zak A, Zeman M, Vitkova D, Hrabak P, Tvrzicka E. β-Sitosterol in the treatment of hypercholesterolemia. Casopís lékařů českých. 1990;**129**:1320-1323

[28] Garikapaty VP, Ashok BT, Chen YG, Mittelman A, Latropoulos M, Tiwari RK. Anti-carcinogenic and anti-metastatic properties of indole-3 carbinol in prostate cancer. Oncology Reports. 2005;**13**:89-93

[29] Mashele SS, Kolesnikova N. In vitro anticancer screening of asparagus laricinus extract. Pharmacology. 2010;**2**:246-252

[30] Hassan HS, Ahmadu AA, Hassan AS. Analgesic and anti-inflammatory activities of *Asparagus africanus* root extract. African Journal of Traditional, Complementary, and Alternative Medicines. 2008;**5**(1):27-31

[31] Debella AJ, Hasliger EJ, Kunert OJ, Michl G, Abebe D. Steroidal saponins from *Asparagus africanus*. Phytochemistry. 1999;**5**(8):1069-1075

[32] Oketch-Rabah HA, Dossaji SF, Christensen SB. Antiprotozoal compounds from *Asparagus africanus*. Journal of Natural Products. 1997;**60**(10):1017-1022

[33] Desta B. Ethiopian traditional herbal drugs. Part II: Antimicrobial activity of 63 mechanical plants. Journal of Ethnopharmacology. 1993;**39**(2):129-139

[34] Lewu FB, Afolayan AJ. Ethnomedicine in South Africa: The role of weedy species. African Journal of Biotechnology. 2009;**8**(6): 929-934

[35] Oliver B. Medicinal Plants in Nigeria. Ibadan, Nigeria: Bulletin by Nigerian College of Arts, Science and Technology; 1960. pp. 17-41

[36] Anthony MP, Burrows JN, Duparc S, JMoehrle J, Wells TNC. The global pipeline of new medicines for the control and elimination of malaria. Malaria Journal. 2012;**11**:316-310

[37] Ginsburg H, Deharo E. Call for using natural compounds in the development of new antimalarial treatments—An introduction. Malaria Journal. 2011;**10**(Suppl. 1):S1-S10

[38] Wells TNC. Natural products as starting points for future anti-malarial therapies: Going back to our roots? Malar Journal. 2011;**10**(Suppl 1):S3-S10

[39] Dikaso D, Makonnem E, Debella A, Abebe D, Urga K, Makonnem N, et al. In vivo antimalarial activity of hydro alcoholic extracts from *Asparagus africanus* Lam in mice infected with *P. berghei*. Ethiopian Journal of Health Development. 2006;**20**(2):117-121

[40] Sparg SG, Light ME, van Staden J. Biological activities and distribution of plant saponins. Journal of Ethnopharmacology. 2004;**94**:219-243

[41] Madikizela B, Ndhlala AR, Finnie JF, van Staden J. Antimycobacterial, anti-inflammatory and genotoxicity evaluation of plants used for the treatment of tuberculosis and related symptoms. Journal of Ethnopharmacology. 2014;**153**(2):386-391

[42] Dubois HP, Geiling EMH. Textbook of Toxicity. London, UK: Oxford University Press; 1959. pp. 302-303

[43] Kebede S, MekbeAfework M, Debella A, Ergete W, Makonnen E. Toxicological study of the butanol fractionated root extract of *Asparagus africanus* Lam., on some blood parameter and histopathology of liver and kidney in mice. BMC Research Notes. 2016;**9**(1):49. DOI: 10.1186/s13104-016-1861-5

[44] Fransworth MR, Bingel AS, Cordell GA, Crane FA, Fong HS. Potential value of plants as sources of new anti-fertility agent I. Journal of Pharmaceutical Sciences. 1975;**64**:535-598

[45] Goldfien A. The gonadal hormones and inhibitors. In: Katzung BG, editor. Basic and Clinical Pharmacology. 6th ed. East Norwalk, VA: Appleton and Lange; 1995. pp. 608-636

[46] Tafessel G, Mekonnen Y, Makonnen E. Antifertility effect of aqueous and ethanol extracts of the leaves and roots of *Asparagus africanus* in rats. African Health Sciences. 2006;**6**:81-85

[47] Lacaille-Dubois MA, Wagner H. Review of the biological and pharmacological activities of saponins. Phytomedicine. 1996;**2**(4):363-386

[48] Jang DS, Cuendet M, Fong HHS, Pezzuto JM, Kinghorn AD. Constituents of *Asparagus officinalis* evaluated for inhibitory activity against cyclooxygenase-2. Journal of Agricultural and Food Chemistry. 2004;**52**(8):2218-2222

[49] Grubben GJH, Denton OA. In Plant Resources of Tropical Africa. Vegetables PROTA Foundation, Earth Limited., Netherlands. 2004;**2**, pp. 96-97. Available from: http://edepot.wur.nl/417517

[50] Nature Gate. *Asparagus officinalis*. 2016. Available from: www.luontoportti.com/suomi/en/kukkakasvit /asparagus [Accessed: 27 October 2016]

[51] Shao Y, Poobrasert O, Kennelly EJ, Chin CK, Ho CT, Huang MT, et al. Steroidal saponins from *Asparagus officinalis* and their cytotoxic activity. Planta Medica. 1997;**63**(3):258-262

[52] Sun Z, Huang X, Kong L. A new steroidal saponin from the dried stems of *Asparagus officinalis* L. Fitoterapia. 2010;**81**(3):210-213

[53] Dawid C, Hofmann T. Structural and sensory characterization of bitter tasting steroidal saponins from Asparagus spears (*Asparagus officinalis* L.). Journal of Agricultural and Food Chemistry. 2012;**60**(48):11889-11900

[54] Al-Snafi AE. The pharmacological importance of *Asparagus officinalis*—A review. Journal of Pharmaceutical Biology. 2015;**5**(2):93-98

[55] Leung AY, Foster S. Encyclopedia of Common Natural Ingredients Used in Food, Drugs and Cosmetics. 2nd ed. New York: John Wiley Sons, Inc.; 1996. pp. 52-54

[56] Hafizur RM, Kabir N, Chishti S. *Asparagus officinalis* extract controls blood glucose by improving insulin secretion and β-cell function in streptozotocin-induced type 2 diabetic rats. The British Journal of Nutrition. 2012;**108**(9):1586-1595

[57] Sakaguchi Y, Ozaki Y, Miyajima I, Yamaguchi M, Fukui Y, Iwasa K, et al. Major anthocyanins from purple asparagus (*Asparagus officinalis*). Phytochemistry. 2008;**69**(8):1763-1766

[58] Makoto S, Masayuki S, Makiko M, Watanabe K. An antifungal saponin from white asparagus (*Asparagus officinalis* L) bottoms. Journal of the Science of Food and Agriculture. 1996;**72**:430-434

[59] Shimoyamada M, Suzuki M, Sonta H, Maruyama M, Okubo K. Antifungal activity of the saponin fraction obtained from *Asparagus officinalis* L and its active principle. Agricultural and Biological Chemistry. 1990;**54**:2553-2557

[60] Garc'ıa MD, Puerta RD, S'aenz MT, Marquez-Mart'ın A, Fern'andez-Arche MA. Hypocholesterolemic and hepatoprotective effects of "triguero" Asparagus from Andalusia in rats fed on a high cholesterol diet. Evidence-Based Complementary and Alternative Medicine. 2012;(9):1-6. DOI: 10.1155/2012/814752 [Accessed: 26 October 2016]

[61] Khorasani A, Sani W, Philip K, Taha RM, Rafat A. Antioxidant and antibacterial activities of ethanolic extracts of *Asparagus officinalis* cv. Mary Washington: Comparison of in vivo and in vitro grown plant bioactivities. African Journal of Biotechnology. 2010;**9**(49):8460-8466

[62] Ye MR, Li R, Liao HF, Liao XZ, Huang GY. Pharmacological study on *Asparagus officinalis* Linn. Zhongguo Zhong Yao Za Zhi. 1994;**19**(4):240-242

[63] Schilcher H, Rau H. Nachweis der aquaretischen Wirkung von Birkenblatterund Goldrutenauszugen im Tierversuch. Urologe B. 1988;**28**:274-280

[64] Kinage P, Chaudhari D. Shatavari: One solution for various female health issues—A review. World Journal of Pharmaceutical Sciences. 2016;**5**(5):1105-1114

[65] Sharma K, Bhatnagar M. *Asparagus racemosus* (Shatavari): A versatile female tonic. International Journal of Pharmaceutical and Biological Archive. 2011;**2**(3):855-863

[66] Anonymous. La-Medicca (India) Private Limited. *Asparagus racemosus*. 2007. Available from: http://www.la-medicca.com/raw-herbs-asparagus-racemosus.html [Accessed: 26 October 2016]

[67] Hayes PY, Jahidin AH, Lehmann R, Penman K, Kitching W, Voss JJD. Steroidal saponins from the roots of *Asparagus racemosus*. Phytochemistry. 2008;**69**:796-804

[68] Mohanta B, Chakraborty A, Sudarshan M, Dutta RK, Baruah M. Elemental profile in some common medicinal plants of India: Its correlation with traditional therapeutic usage. Journal of Radioanalytical and Nuclear Chemistry. 2003;**258**(1):175-179

[69] Sekine T, Fukasawa N, Kashiwagi Y, Ruangrungsi N, Murakoshi I. Structure of asparagamine A: Novel polycyclic alkaloid from *Asparagus racemosus*. Chemical & Pharmaceutical Bulletin. 1994;**42**(6):1360-1362

[70] Saxena VK, Chaurasia SA. New isoflavone from the roots of *Asparagus racemosus*. Fitoterapia. 2001;**72**(3):307-309

[71] Sachan AK, Das DR, Dohare SL, Shuaib M. *Asparagus racemosus* (Shatavari): An overview. International Journal of Chemical and Pharmaceutical Sciences. 2012;**1**(3):937-941

[72] Wiboonpun N, Phuwapraisirisan P, Tip-pyang S. Identification of antioxidant compound from *Asparagus racemosus*. Phytotherapy Research. 2004;**18**:771-773

[73] Dutta DC. Textbook of Gynecology. 4th ed. New Delhi, India: New Central Book Agency; 2007. pp. 168-169

[74] Negi JS, Singh P, Joshi GP, Rawat MS, Bisht VK. Chemical constituents of Asparagus. Pharmacognosy Reviews. 2010;**4**(8):215-220

[75] Sekine TN. TIFFNal structure and relative stereochemistry of a new polycyclic alkaloid, asparagamine A, showing anti-oxytocin activity, isolated from *Asparagus racemosus*. Journal of the Chemical Society, Perkin Transactions 1. 1995;**1**(4):391-393

[76] Agrawal A, Sharma M, Rai SK, Singh B, Tiwari M, Chandra R. The effect of the aqueous extract of the roots of *Asparagus racemosus* on hepatocarcinogenesis initiated by diethylnitrosamine. Phytotherapy Research. 2008;**2**(22):1175-1182

[77] Sharma SC. Constituents of the fruits of *Asparagus racemosus* wild. Die Pharmazie. 1981;**36**(10):709-711

[78] Singh A, Sinha B. Asparagus racemosus and its phytoconstituents: Updated review. Asian Journal of Pharmaceutical Research. 2014;**4**(4):230-240

[79] Kar DK, Sen S. Sarsasapogenin in callus culture of *Asparagus racemosus*. Current Science. 1985;**54**(12):585-590

[80] Sharma PC, Yelne M, Dennis TJ. Database on Medicinal Plants Used in Ayurveda. New Delhi, India: CCRAS Publications; 2002. pp. 418-421

[81] Gautam M, Diwanay S, Gairola S, Shinde Y, Patki P, Patwardhan B. Immunoadjuvant potential of *Asparagus racemosus* aqueous extract in experimental system. Journal of Ethnopharmacology. 2004;**91**:251-255

[82] Choudhary BK, Kar A. Mineral contents of *Asparagus racemosus*. Indian Drugs. 1992;**29**:623-627

[83] Kumar S, Mehla RK, Dang AK. Use of shatavari (*Asparagus racemosus*) as a galactopoietic and therapeutic herb—A review. Agricultural Reviews. 2008;**29**(2):132-138

[84] Dalvi SS, Nadkarni PM, Gupta KC. Effect of *Asparagus racemosus* (*Shatavari*) on gastric emptying time in normal healthy volunteers. Journal of Postgraduate Medicine. 1990;**36**:91-94

[85] Kishore P, Pandey PN, Pandey SN, Dash S. Treatment of duodenal ulcer with *Asparagus racemosus* Linn. Indian

Journal of Research in Homeopathy. 1980;**15**:409-415

[86] Kamat JP, Boloor KK, Devasagayam TPA, Venkatachalam SR. Antioxidant properties of *Asparagus racemosus* against damage induced by γ-radiation in rat liver mitochondria. Journal of Ethnopharmacology. 2000;**71**:425-435

[87] Rao AR. Inhibitory action of Asparagus racemosus on DMBA-induced mammary carcinogoenesis in rats. International Journal of Cancer. 1981;**28**:607-610

[88] Shao Y, Chin CK, Ho CT, Ma W, Garrison SA, Huang MT. Antitumor activity of the crude saponins obtained from asparagus. Cancer Letters. 1996;**104**:31-36

[89] Datta GK, Sairam K, Priyambada S, Debnath PK, Goel RK. Lymphatic transport of orally administered drugs. Indian Journal of Experimental Biology. 2002;**40**:1173-1177

[90] Mandal SC, Nandy A, Pal M, Saha BP. Evaluation of antibacterial activity of *Asparagus racemosus* wild root. Phytotherapy Research. 2000;**14**:118-119

[91] Gautam M, Saha S, Bani S, Kaul A, Mishra S, Patil D. Immunomodulatory activity of *Asparagus racemosus* on systemic Th1/Th2 immunity: Implications for immune adjuvant potential. Journal of Ethnopharmacology. 2009;**121**(2):241-247

[92] Thatte U, Chhabria S, Karandikar SM, Dahanukar S. Immunotherapeutic modification of *E. coli* induced abdominal sepsis and mortality in mice by Indian medicinal plants. Indian Drugs. 1987;**25**:95-97

[93] Mandal SC, Kumar CKA, Mohana LS, Sinha S, Murugesan T, Saha BP. Antitussive effect of *Asparagus racemosus* root against sulfur dioxide-induced cough in mice. Fitoterapia. 2000;**71**(6):686-691

[94] Roy RN, Bhagwager S, Chavan SR, Dutta NK. Preliminary pharmacological studies on extracts of root of *Asparagus racemosus* (Satavari) wild. The Indian Journal of Medical Research. 1971;**6**:132-138

[95] Visavadiya NP, Soni B, Madamwar D. Suppression of reactive oxygen species and nitric oxide by *Asparagus racemosus* root extract using in vitro studies. Cellular and Molecular Biology. 2009;**55**:1083-1095

[96] Christina AJ, Ashok K, Packialashmi M. Antilithiatic effect of *Asparagus racemosus* wild on ethylene glycol-induced lithiasis in male albino wistar rats. Methods and Findings in Experimental and Clinical Pharmacology. 2005;**27**(9):633-638

[97] Hannan JM, Marenah L, Ali L, Rokeya B, Flatt PR, Abdel-Wahab YH. Insulin secretory actions of extracts of *Asparagus racemosus* root in perfused pancreas, isolated islets and clonal pancreatic beta-cells. The Journal of Endocrinology. 2007;**192**(1):159-168

[98] Sabde S, Bodiwala HS, Karmase A, Deshpande PJ, Kaur A. Anti-HIV activity of Indian medicinal plants. Journal of Natural Medicines. 2011;**65**:662-669

[99] Jetmalani MH, Sabins PB, Gaitonde BB. A study on the pharmacology of various extracts of Shatavari—*Asparagus racemosus* (wild). The Indian Journal of Medical Research. 1967;**2**:1-10

[100] Narendranath KA, Mahalingam S, Anuradha V, Rao IS. Effect of herbal galactagogue (Lactare) a pharmacological and clinical observation. Medical-Surgical Nursing. 1986;**26**:19-22

[101] Jamieson HG. Asparagus densiflorus—The Flowering Plants of Africa. 2002. Available from: http://www.plantzafrica.com/plantab/asparagdens.htm [Accessed: 15 October 2016]

[102] Singh P, Wazbir J, Barlaskar H, Kaur A, Bhardwaj A, Kaur L. Antibacterial and phytochemical analysis of ethnomedicinal plants. Research Journal of Pharmaceutical, Biological and Chemical Sciences. 2014;**5**(5):1203-1208

[103] Davids D, Blouws T, Aboyade O, Gibson D, De Jong JT, Van't Klooster C, et al. Traditional health practitioners' perceptions, herbal treatment and management of HIV and related opportunistic infections. Journal of Ethnobiology and Ethnomedicine. 2014;**10**(77):1-14. DOI: 10.1186/1746-4269-10-77

Chapter 11

Phytochemicals and Their Antifungal Potential against Pathogenic Yeasts

Cristina de Andrade Monteiro
and Julliana Ribeiro Alves dos Santos

Abstract

The rate of fungal infections is increasing rapidly, and pathogenesis of their species is poorly understood. Among fungi, *Candida* species are a major cause of morbidity and mortality worldwide and thus represent a serious threat to public health. In addition, *Cryptococcus* spp. are yeasts responsible for serious lung infections and meningitis. Polyenes, fluoropyrimidines, echinocandins, and azoles are used as commercial antifungal agents to treat fungal infections. However, the presence of intrinsic and developed resistance against azole antifungals has been extensively documented. The re-emergence of classical fungal diseases has occurred because of the increment of the antifungal resistance phenomenon. In this way, the development of new satisfactory therapy for fungal diseases persists as a major challenge of present-day medicine. The urgent need includes the development of alternative drugs that are more efficient and tolerant than those traditional already in use. The identification of new substances with potential antifungal effect at low concentrations or in combination is also a possibility. This chapter briefly examines the infections caused by *Candida* and *Cryptococcus* species and focuses on describing some of the promising alternative molecules and/or substances that could be used as antifungal agents, their mechanisms of action, and their use in combination with traditional drugs.

Keywords: medicinal plants, yeast infections, antifungal agents, antifungal activity, phytochemicals

1. Introduction

Fungal infections are considered a serious health problem, especially in people with weakened immune systems, and are a main cause of morbidity and mortality worldwide [1].

However, the impact of these "opportunistic" diseases on human health is not widely highlighted [2]. Due to this, research related to fungi occurs slowly compared to those caused by other pathogens.

Among the different mycotic infections, those caused by *Candida* and *Cryptococcus* are the most threatening due to severity of the disease and higher worldwide occurrence [3]. The pathogenicity of fungal infections proceeds in well-organized steps. For example, *Candida* cell surface adhesion factors first promote its adherence to host surface, followed by releasing of various hydrolytic enzymes and other virulence factors for invasion and damage of the host tissues [4].

Candida species can cause a variety of infections from the mildest to the most severe being candidemia the most frequent hospital infection accounting for up to 15% of bloodstream infections. *Candida* species are the main causative agents in 50–70% of systemic fungal infections [5].

Cryptococcus species are other yeasts of medical importance, with more than 39 species, among which *Cryptococcus gattii* and *Cryptococcus neoformans* are the most clinically relevant [6–8]. However, other species such as *Cryptococcus albidus* and *Cryptococcus laurentii* are emerging pathogens involved in several types of infections [6, 9–11].

These yeasts are present in several environmental niches, such as woody sites (decomposing tree trunks, mainly eucalyptus, and soil), vegetable remains, domestic dust, and bird excrement, more precisely in *Columba livia* [12–14]. The source of the infection is exogenous and occurs primarily by inhalation or by direct inoculation into the tissue after trauma of desiccated spores or yeasts. It is believed that the only source of infection is environmental, since there are no reports of transmission between animals and humans or between humans [15].

The main virulence factors of *Cryptococcus* species are growth capacity at 37°C, polysaccharide capsule, melanin synthesis, and production of urease and antioxidant enzymes, causing primary or opportunistic cryptococcosis, such as pulmonary, cutaneous, and meningitis diseases [6, 8, 13, 16–19]. Cryptococcosis is the third opportunistic infection associated with AIDS [20].

In addition to delays in yeast diagnosis, there is currently a limited antifungal armamentarium in use against yeast diseases including only four chemical classes: polyenes, triazoles, echinocandins, and flucytosine. Antifungals act by binding specific components of fungal plasma membrane or its biosynthetic pathways or even cell wall components [21]. However, most of the antifungal agents used in the clinic is fungistatic and often led to the development of resistance by fungal species. Modern early antifungal treatment strategies, such as prophylaxis and empirical and preemptive therapy, result in long-term exposure to antifungal agents, which is a major driving force for the development of resistance.

Among the available antifungal agents, azoles are the preferred and most frequently used drugs for treatment of *Candida* and *Cryptococcus* infections. Fluconazole (FLZ), a type of azole, is often preferred in treatments of *Candida* infections because of its low cost and toxicity, in addition to availability in varied formulations [22]. However, there are many reports that described resistance development among *Candida* species, especially in relation to azoles.

Infectious Diseases Society of America recommends the treatment of cryptococcosis through FLZ and amphotericin B (AMB) with or without combination with 5-flucytosine (5-FC), followed by prolonged maintenance with fluconazole. Other azole compounds such as itraconazole (ITC), voriconazole, and posaconazole may be used as an alternative to FLZ in cases of contraindication or inefficacy of the latter [23, 24]. However, there has been a progressive increase in isolates of *Cryptococcus* spp. resistant to FLZ, which complicates the management of cryptococcal meningitis [25]. On the other hand, AMB and 5-FC are not available in all countries and are, respectively, nephrotoxic and hepatotoxic, limiting the anti-cryptococcal therapeutic [24].

Considering the limited availability of antifungals in use and the emergence of resistance, the control of *Candida* and *Cryptococcus* infections is a challenge in the modern clinic. In this way there is a continuous need for the search for new substances with new mechanisms of action with the aim of developing novel broad spectrum antifungal drugs with better efficacy.

In this way, plants stand out as the major producers of promising substances, the phytochemicals. Identification of new molecules with antifungal potential for the manufacture of new drugs, more effective and less toxic, is essential to facing the challenge. The use of phytochemicals alone or in combination with traditional drugs represents an important alternative to conventional therapy. The combination of drugs usually requires lower doses of antimicrobials. This reduction might lead to a toxicity decrease, which results in a higher tolerance to the antimicrobial by the patient.

2. Pathogenic yeast infections: a serious health problem

In the last two decades, fungal infections have shown a significant increment. In addition to the increase in the number of patients with compromised immune system, factors such as increasing number of patients using catheters, the use of broad-spectrum antibiotics, the rising number of patients requiring organ transplantations, as well as those with hematological malignancies and diabetes also contribute to this phenomenon [26, 27].

Even though fungal infections cause significant amount of human morbidity and mortality, the impact of these "opportunistic" diseases on human health is not widely highlighted [2]. Due to this, the research into the pathophysiology of human fungal infections is slow in comparison to other disease-causing pathogens. Recently, an editorial published in the journal *Nature Microbiology* [28] ratified the importance of not neglecting fungi. The call proposed a reflection on fungi and how these microorganisms have been neglected, even with studies already consolidated showing their medical relevance.

The most frequent fungal diseases affecting populations in the world are candidiasis [29–34] and cryptococcosis [8, 20, 25]. There are several types of candidiasis as mucosal candidiasis, cutaneous candidiasis, onychomycosis, systemic candidiasis [35, 36], and pulmonary candidiasis. An important fact is that candidiasis is an infection that can affect both immunocompromised and healthy people [37, 38]. Candidemia is the most relevant and prevalent nosocomial fungal infection associated with a high mortality rate (up to 49%) in patients with a compromised immune system [39, 40]. The association of *Candida* with bloodstream infections depends on patient's condition, age, and geographic region. Candidemia is such an important infection that in 10–40% of cases, it is associated with sepsis or septic shock [41].

Candida albicans continues to be the most prevalent species isolated from fungal infections [27, 42–44]. However, the prevalence of other *Candida* species has increase substantially. These species are *C. parapsilosis*, *C. tropicalis*, *C. krusei*, *C. glabrata*, *C. guilliermondii*, *C. orthopsilosis*, *C. metapsilosis*, *C. famata*, and *C. lusitaniae* [44–46].

Candida species presents high degree of flexibility, being able to grow in extremely different environments regarding to the availability of nutrients, temperature variation, pH, osmolarity, and amount of available oxygen [47]. This fact associated with the high resistance capacity of species to antifungals, their virulent features, and capability of forming biofilms with other species [48, 49] makes the genus *Candida* a serious risk to human health [50]. Thus, *Candida* species are highly adaptable and possess numerous strategies to survive in conditions that can affect their overgrowth and alter their susceptibility profiles.

Cryptococcus spp. may remain latent in the lungs, leading to asymptomatic infection, or may cause multifocal lung disease. The latency period of *Cryptococcus* can range from 6 weeks to more than 1 year after inhalation [51]. The fungus presents neurotrophism and can migrate to the central nervous system (CNS) through hematogenous dissemination and, when crossing the blood-brain barrier, can cause meningoencephalitis [13, 18]. Episodes of mental confusion in patients with cryptococcosis have been described [52, 53]. Neurocryptococcosis is the most severe form of the disease with high mortality rates in the absence of adequate treatment [18, 23]. The mortality due to cryptococcosis is higher than the mortality caused by tuberculosis and similar to that caused by malaria [54].

Another clinical manifestation is cutaneous cryptococcosis, which is rare and usually secondary to hematogenous dissemination. Cutaneous lesions are characterized by an infiltrative plaque of a solid tumor mass that can present ulcerative and necrotic lesion [17]. Pulmonary and cutaneous lesions due to nodular features may be misdiagnosed as tumor lesions [55]. In addition to the respiratory tract, CNS and skin, other sites may be affected: prostate, eyes, adrenal glands, lymph nodes, bone marrow, and liver [51].

Until now, there are three proposals to explain fungal neurotropism. The first is that neuronal substrates present in the basal ganglia promote cryptococcal growth and survival, and, thus, perivascular spaces may serve as a niche for *Cryptococcus*, as described by [56] in a healthy female patient who had evidence of *Cryptococcus* infection within the perivascular spaces of the parenchyma. The second proposal describes that it is possible that there are specific neuronal receptors that can attract *Cryptococcus* to the CNS [57]. The third hypothesis, one of the most widespread, is that the fungus uses neurotransmitters such as dopamine that aids in the synthesis of melanin [19, 57, 58].

Besides the clinical importance of fungal infections caused by theses pathogenic yeasts, interestingly, climatic abnormalities due to phenomena such as La Niña and El Niño have recently been described as important in the distribution and occurrence of mycoses in countries influenced by them [59].

3. Traditional antifungal agents against yeasts

In the last two decades, there has been an increasing, but limited, discovery of antifungal agents [47]. These include azoles, such as fluconazole, itraconazole, ketoconazole (KTC), miconazole, and clotrimazole, polyenes (amphotericin B [AMB] and nystatin), allylamines, thiocarbamates, morpholines, 5-fluorocytosine, and echinocandins (for instance, caspofungins) [21]. However, fungal cells and human cells are eukaryotic, so antifungal compounds target both cell types, resulting in considerable side effects in patients and fewer available targets for drug action. Antifungals target three cellular components of fungi (**Figure 1**). Azoles inhibit ergosterol biosynthesis by interfering with the enzyme lanosterol 14-α-demethylase in endoplasmic reticulum of the fungal cell. This enzyme is involved in the transformation of lanosterol into ergosterol, a component that is part of the plasma membrane structure of the fungus (**Figures 1** and **2**). Thus, as the concentration of ergosterol is reduced, the cell membrane structure is altered, thereby inhibiting fungal growth [60].

Azoles comprise a five-member azole ring containing two (imidazole) or three nitrogen atoms (triazole) attached to a complex side chain [61, 62]. Imidazoles include KTC, miconazole, econazole, and clotrimazole, and triazoles include FLZ, ITC, voriconazole (synthetic triazole derivative of FLZ of second generation), and posaconazole (hydroxylated analog of itraconazole) [63].

AMB and nystatin bind to ergosterol causing the disruption of the membrane structure and promoting extravasation of intracellular constituents such as ions and sugars and, consequently, cell death [21] (**Figure 1**).

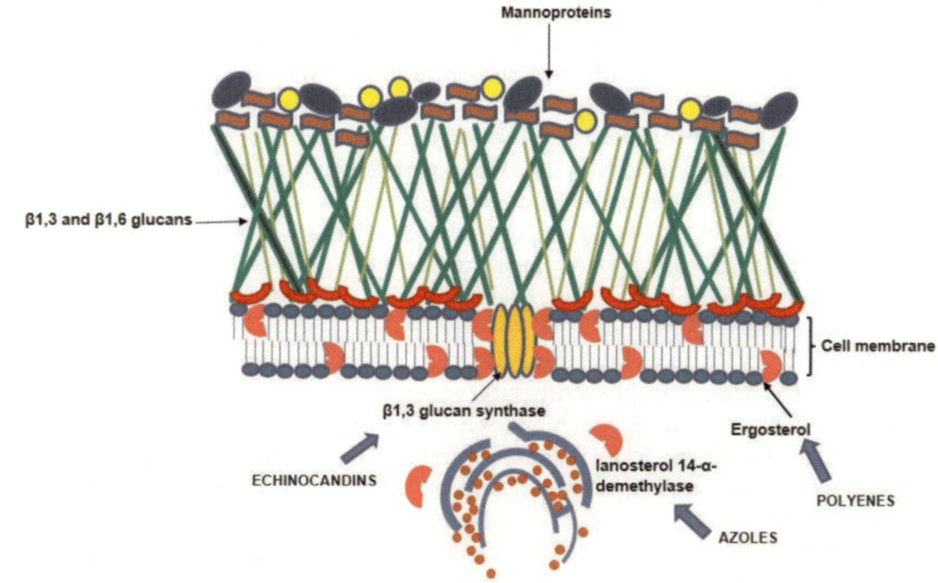

Figure 1.
Mechanisms of action of some traditional antifungal agents on cellular targets. Azoles inhibit the ergosterol synthesis in the endoplasmic reticulum of the fungal cell by interfering with the enzyme lanosterol 14-α-demethylase. Polyenes act by binding to ergosterol present at the cell membrane. Echinocandins inhibit (1,3) β-D-glucan synthase, thereby preventing glucan synthesis.

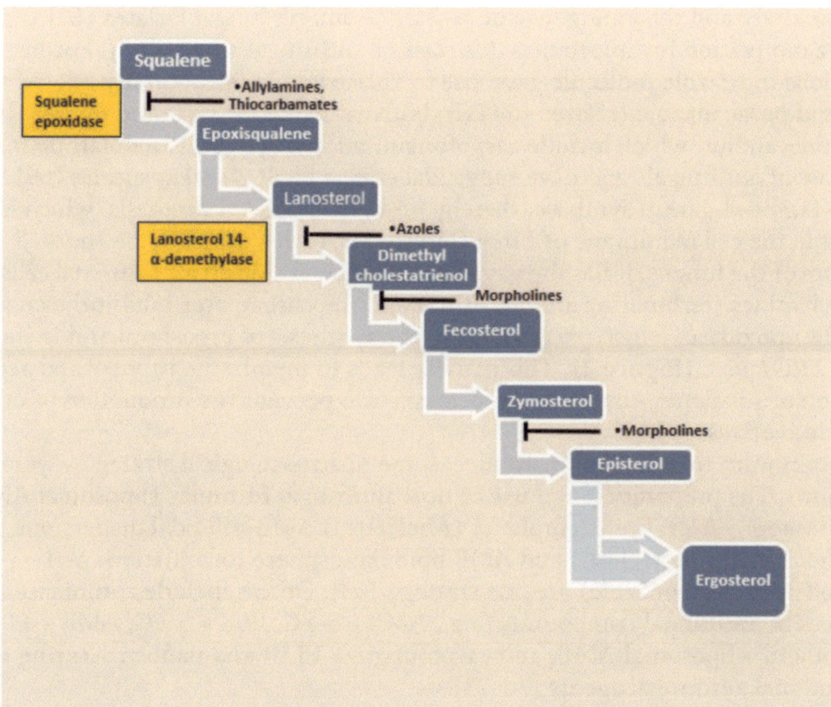

Figure 2.
Specific point of action of antifungal drugs in the ergosterol biosynthesis pathway.

Pyrimidine analogs include 5-fluorocytosine and 5-fluorouracil (5FU). The first has fungistatic properties and enters the fungal cell through cytosine permease, inhibiting the thymidylate-synthetase enzyme and interfering with DNA. 5-fluorouracil, which in turn can be phosphorylated to 5-fluorodeoxyuridine monophosphate, can be incorporated into RNA molecules [63]. Due to toxicity [64]; stronger side effects, such as hepatic impairment; interference with bone marrow function; and rapid occurrence of resistance especially among *Candida* species, the clinical use of 5-FC is preferred in association with AMB [65, 66]. In addition, the nephrotoxicity and hepatotoxicity of AMB and 5-FC, respectively, and the unavailability of these antifungals in many countries have limited their use in cryptococcal therapeutic [24].

Host's immunity, type of infection, site of origin of the samples, toxicity, bioavailability of the drug, and the sensitivity/resistance profile of the isolates interfere in the choice of the type of agent to be used [22]. AMB is considered the gold standard drug for most mycoses that affect patients at risk [67], although it has high toxicity. Azoles have fungistatic properties that affect cell growth and proliferation [65]. Among azoles, KTC was one of the firsts to emerge and was the first alternative to AMB [68]. Currently, FLZ is the drug of choice for most *Candida* and *Cryptococcus* infections [64] and is the most recommended antifungal agent for use in invasive candidiasis [47, 49].

For cryptococcosis, the choice of treatment depends on the patient's immunological status and mainly on the clinical of the disease, if it is just a pulmonary manifestation or if the infection is systemic. Fluconazole is recommended in cases of lung disease with mild to moderate symptoms. Amphotericin B with or without combination with 5-flucytosine is the recommended therapy for more serious infections such as meningoencephalitis, followed by prolonged maintenance with fluconazole [23, 24].

Although azoles are generally well-tolerated, they have limitations such as hepatotoxicity and the emergence of resistance among fungal isolates [69] which provide motivation for improving this class of antifungal agents [68]. For instance, alterations in triazole molecule gave rise to voriconazole (structurally related to FLZ) and posaconazole (related to ITC), both available for systemic therapy [66].

Echinocandins, which include caspofungin, micafungin, and anidulafungin, are a new class of antifungals and have fungicidal effects in all *Candida* species [66]. They inhibit (1,3) β-d-glucan synthase, thereby preventing glucan synthesis, which is present in the cell membrane of fungi (**Figure 1**). As this drug acts on the wall structure of the fungus, it has the advantage of a lower side effect in animal cells [47].

Allylamines (terbinafine and naftifine) and thiocarbamates inhibit the enzyme squalene epoxidase, which participates in the synthesis of ergosterol and is encoded by the *ERG1* gene (**Figure 2**). This activity leads to membrane rupture and accumulation of squalene. Allylamine effects can also prevent the production of other sterol derivatives.

To minimize toxicity and resistance, some pharmacological strategies were developed. The preparation and use of new antifungal formulas (liposomal AMB (Ambisome®), AMB lipid complexes (Abelcet®), AMB colloidal dispersions (Amphocil®/Amphotech®), and AMB lipid nanosphere formulations and β-cyclodextrin itraconazole) are one strategy [68]. Others include combination therapies of antifungal compounds (e.g., AMB + 5-FC, FLZ + 5-FC, AMB + FLZ, caspofungin + liposomal AMB, and caspofungin + FLZ) and nanostructuring of conventional antifungal agents [70–73].

However, all traditional antimycotic drugs have at least one restriction related to their use. Some do not have a broad spectrum of action or are fungistatic. Others have high toxicity and low bioavailability with significant side effects [74].

Therefore, limitations of treatment and drug resistance associated with pathogenicity of the clinical isolates support the urgent need to identify substances that are more effective, with new mechanisms of action in the fight against *Candida* and *Cryptococcus* infections.

4. Resistance in pathogenic yeasts: a significant problem

Most antifungals target sterols or the enzymes that synthesize them. However, the fungistatic nature of many of these antifungals and emergence of clinical drug resistance limits their success. Increased drug resistance in fungi is a problem that cannot be avoided, particularly for FLZ, which is the preferred antifungal for treating yeast infections [75].

The number of people at risk for fungal infections has been increasing, resulting in an increased use of antifungal agents, even as prophylaxis. Thus, besides the existence of some non-*albicans Candida* (NAC) species presenting inherent resistance to azoles, higher minimum inhibitory concentrations (MICs) for antifungals against *C. albicans* strains have been observed [76]. The World Health Organization (2014) categorizes antimicrobial resistance as that developed by the microorganism to an antimicrobial drug, which was initially effective in treatment of such infections. Low-dose prophylactic administration of azole derivatives, such as FLZ, for prolonged periods to prevent the occurrence of opportunistic infections in immunosuppressed patients also results in resistant phenotypes [27, 75]. Therapeutic failures and empiric treatment are facts which are likely to collaborate to the increased incidence of fungal infections.

In the last decade, a number of new clinical problems have arisen, requiring new guidelines regarding the treatment of cryptococcosis, mainly because clinical data have suggested that cryptococcal strains have become more resistant to drugs [23, 25]. Some relates say that clinical *Cryptococcus* isolates are frequently less susceptible to fluconazole than environmental isolates. However, Chowdhary et al. [77] evaluated the susceptibility profile of environmental and clinical strains of *C. gattii* and observed that environmental samples were less susceptible to fluconazole, itraconazole, and voriconazole in comparison to clinical isolates.

Heteroresistance is also a worrying phenomenon. It consists of the ability of a subpopulation of microorganism to adapt to high concentrations of the drug, resulting in resistant homogenous populations. However, heteroresistant strains return to the initial phenotype when the stimulus with the drug is withdrawn [78].

Some mechanisms for cellular and molecular resistance to FLZ in yeasts are described. In *Candida* and *Cryptococcus*, the first is related to the induction of multidrug pumps, which decrease the concentration of drug available in the intracellular compartment of yeast cells. Various genes belonging to the ATP-binding cassette superfamily or to the major facilitator superfamily encode efflux pumps were identified in *C. albicans*. Overexpression of some transporter genes or of their regulated genes can confer cross-resistance to various azoles [21]. In *C. gattii* and *C. neoformans*, *AFR*1, *MDR*1, and *AFR*2 genes encode ABC transporters that expel the azole out of the fungal cell, thereby causing resistance to these drugs [79].

A second mechanism of resistance involves modification of the target enzyme encoded by the *ERG11* gene, also known as cytochrome P_{450} lanosterol 14-α-demethylase (Cyp51). Mutations in this gene prevent azoles from binding to enzyme sites. Another mechanism of resistance is related to mutations in the *ERG3* gene which does not convert 14-α-methylfecosterol into 14-α-methyl-3,6-diol in the ergosterol synthesis pathway. This substitution causes azoles to have no fungistatic effects on the fungal cell membrane [21].

Transcriptional regulation is also important for the development of resistance mechanisms. YAP1, a protein, is important for the mechanism of *C. neoformans* heteroresistance to fluconazole and oxidative stress. Mutant strains of *C. neoformans* that lost protein YAP1 became hypersensitive to a variety of oxidizing agents and mainly to fluconazole [80].

Resistance to polyenes (AMB) in fungus is less common and in *C. albicans* is associated with the substitution of ergosterol with a precursor molecule or a general reduction of sterols in the plasma membrane [81]. Reduction of membrane ergosterol renders *Cryptococcus neoformans* and *Aspergillus* spp. less susceptible to amphotericin B [82]. Enzymes encoded by *ERG3* and *ERG2* genes participate in ergosterol biosynthesis and have the main alterations related to AMB resistance because mutations in their genes modify ergosterol content required for the action of polyenes [83].

The main resistance mechanism to echinocandins is related with point mutations in gene that encodes the major subunit of the glucan synthase enzyme (Fks subunit) (**Figure 1**) and can provide resistance to all echinocandin [84]. Other *Candida* species also present this resistance mechanism such as *C. tropicalis*, *C. parapsilosis*, *C. glabrata*, *C. krusei*, *C. guilliermondii*, and *C. dubliniensis* [85, 86].

Resistance to 5-FC can be of two types: primary, occurring via cytosine permease (encoded by the *FCY2* gene) whose mutation decreases drug uptake [87], and secondary, related to alterations in cytosine deaminase (encoded by *FCY1*) or uracil phosphoribosyltransferase (encoded by *FUR1*) activities. Cytosine permease is responsible by conversion of 5-FC to 5-fluorouridine or to 5-fluorouridine monophosphate (5-FUMP) [88]. Resistance is easily developed in fungal isolates from patients who are receiving the drug. However, other molecular mechanisms related to resistance to 5-FC must exist because most of them have not been observed in *C. albicans* [89].

The increase in the drug-resistant *Candida* and *Cryptococcus* strains to commercial antifungals has caught the attention of clinicians and researchers to medicinal plant products (commonly referred as phytochemicals). The use of phytochemicals with greater antifungal potential and different mechanisms of action may be useful in reducing the phenomenon of resistance. Lately, they have become a significant alternative for discovery of commercially viable, economically cheaper, and safe phytomedicines.

5. Medicinal plants as a source of antifungal agents

Global Action Fund for Fungal Infections (GAFFI), an international organization working to reduce infections and deaths associated with fungi, has reported that approximately 300 million people in the world suffer from a serious fungal infection every year and that among them over 1.35 million deaths are registered [90].

Despite the introduction of new and novel antifungal drugs, their production and impact are slow, and the development of antifungal resistance has forced the attention of researchers toward herbal products, mainly phytochemicals, in search of development of safe and economically viable antifungals.

Populations around the world have used folk medicine as an alternative therapy for various disorders. Currently, many species have been extensively studied in an attempt to discover new biologically active compounds with novel structures and mechanism of action for the development of new drugs.

Medicinal plants are commonly preferred because of their wide level of functional chemical groups with comparatively poor toxic substances, low-cost extracts, fewer side effects, and easy accessibility to people. Various bioactive compounds have been abundantly found such as phytochemicals.

Leaves, as well as the seeds and fruits of plants, have higher levels of phenolic compounds. The concentration of these compounds also depends on the nature of the chemical used as solvent in the extraction process as well as on the growth and storage conditions [91].

The biological activity of plant products has been evaluated against fungi. The ethanol extract, *Lonicera japonica* aerial parts, a medicinal plant of folk medicine of China that used to treat some diseases, showed a very strong antimicrobial activity against *Candida* species and potent wound healing capacity [92]. Methanolic extract of *Lannea welwitschii* leaves was antimicrobial against clinical yeasts. A preliminary phytochemical screening of extracts revealed tannins, flavonoids, alkaloids, and glycosides as compounds [93]. *Pyrostegia venusta* crude flower extracts, fractions, and pure compounds showed an effective broad spectrum antifungal activity [94].

An extract of *Piper betle* leaves inhibited the growth of *Candida* species [95], and four different extracts of *Strychnos spinosa* showed anti-*Candida* activity [96]. Hydro-methanolic extracts of leaves from *Juglans regia* and *Eucalyptus globulus* and methanol extract of *Cynomorium coccineum* demonstrated excellent antimycotic property against *Candida* strains [91, 97]. Akroum [98] showed antifungal activity in an acetylic extract of *Vicia faba* against *C. albicans* in vitro and reduced mortality rates in *Candida*-infected mice that were treated with the extract.

Berberine, a protoberberine-type isoquinoline alkaloid isolated from the roots, rhizomes, and stem bark of natural herbs, such as *Berberis aquifolium*, *Berberis vulgaris, Berberis aristata*, *Hydrastis canadensis*, *Phellodendron amurense, Coptis chinensis,* and *Tinospora cordifolia*, was described as powerful reducer of the viability of in vitro biofilms formed by fluconazole-resistant *Candida tropicalis* cells [99].

Ethanolic and aqueous extracts from different plants from Brazilian Cerrado commonly used in folk medicine such as *Eugenia dysenterica* and *Pouteria ramiflora* were promising against *C. tropicalis*, *C. famata*, *C. krusei*, *C. guilliermondii*, and *C. parapsilosis*. A phytochemical screening of active extracts from these plants disclosed as main components flavonoids and catechins [100]. Crude extract and fractions (n-butanolic and ethyl acetate ones) from *Terminalia catappa* leaves showed antifungal properties against *Candida* spp.; hydrolysable tannins (punicalin, punicalagin), gallic acid (GA), and flavonoid C-glycosides were the active components found in butanolic fraction [101].

Bottari et al. [102] determined the antimicrobial activity of the aqueous and ethanolic leaf extracts of *Carya illinoensis*. Both extracts had MIC values against seven *Candida* reference strains between 25 and 6.25 mg/mL. Phenolic acids (gallic acid and ellagic acid), flavonoids (rutin), and tannins (catechins and epicatechins) were likely responsible, in part, for the activity against *Candida* strains. Further, the extracts inhibited the production of *C. albicans* germ tubes.

5.1 Phytochemicals: polyphenols as substances most found in plants

Several woody plant produce medicinal phytochemicals such as polyphenols that are low molecular weight naturally occurring organic compounds containing one or more phenolic groups [103]. Further, polyphenols perform various substantial functions in plant physiology and, therefore, can be found, in lesser or greater quantity, in all of them.

Phenolic acids, flavonoids, tannins, and coumarins are some examples of phenolic compounds found in and extracted from medicinal plants [104] (**Table 1**). Research has shown that polyphenols have potentially healthy effects in

Phytochemicals	Bioactive compounds	Properties	Plant sources
Flavonoids	Flavan-3-ol	Against *Candida*	*Syzygium cordatum*
	Baicalein, gallotannin	Against *Candida*	*Scutellaria baicalensis*
Coumarins	Ulopterol	Against *M. canis*	*Skimmia laureola*
	Prenyletin; prenyletin-methyl-ether	Against *T. rubrum*; *T mentagrophytes*	—
	Osthenol	*C. albicans*, *Fusarium solani*, *A. fumigatus*	—
	5,8-Dihydroxyumbelliprenin	*T. interdigitale*, *M. gypseum*	*Ferula foetida*
Saponins	Colchiside	Phytopathogenic fungi	*Dipsacus asper* roots
Terpenes or terpenoids	Triterpenes	Against dermatophytes	Ethyl acetate leaf extract of *Satureja khuzestanica*
Lectins	Lectins	*Fusarium oxysporum*	Seed from native Amazon species
Tannins	Punicalagin Punicalin	Against *Candida* spp.	*Terminalia catappa*
	Punicalagin	*T. mentagrophytes*; *T. rubrum*; *M. canis*; *M. gypseum*	*Punica granatum*
	Ellagic acid, gallagic acid, punicalins, punicalagin	*C. albicans*, *Cryptococcus neoformans*, *Aspergillus fumigatus*	*Punica granatum*
	Lambertianin C, sanguiin H-6	*Geotrichum candidum*	*Rubus idaeus*

Table 1.
Phytochemicals with antifungal compounds derived from plants.

humans, working primarily as anticancer, antihypertensive, anti-allergen, anti-inflammatory, antioxidant, and antimicrobial agents. The antimicrobial activity of polyphenols has been extensively investigated mainly against bacteria [104]. Nevertheless, the antifungal activity of most of the phenolic compounds remains unknown. There are few studies on the mechanism of action of the substance, cytotoxicity, the synergism with traditional antifungals drugs, and their anti-virulence activities.

Those with the most promising antifungal activity isolated from natural sources include flavonoids, tannins, coumarins, quinones, lignans, and neolignans [105] (**Table 1**).

Flavan-3-ols, flavonols, and tannins have received the most attention among the known polyphenols, attributable to their large spectrum of efficacy and high antimicrobial property. Structurally, flavonoids are aromatic compounds with 15 carbon atoms (C15) on their basic skeleton; they consist in tricyclic phenolic compounds with two aromatic rings on their structure (C6–C3–C6) [105]. Flavonoids are a class of natural compounds with several known protective activities, including antifungal activity. The flavonoids include subclasses such as chalcones, flavones, isoflavones, flavonols, flavanols (flavan-3-ol), and anthocyanidins [106].

The activity of flavonols such as quercetin, myricetin, and kaempferol has been described in *C. albicans*. For instance, quercetin, myricetin, and kaempferol from propolis have showed activity against *Candida* species [107]. The flavanol subclass (flavan-3-ol) and gallotannin, extracted from *Syzygium cordatum*, also showed inhibitory properties on the growth of *C. albicans* [108]. Serpa et al. [109] isolated baicalein, belonging to a subclass of flavones, from *Scutellaria baicalensis*, and induced apoptosis in *C. albicans* (**Table 1**), and apigenin, a flavone isolated from propolis, showed antifungal potential. Flavonoids as much as coumarins and lignans have shown an antifungal potential against several species of dermatophytes [105].

Other important groups of polyphenolic compounds present in various plant parts, such as the roots, flowers, leaves, fruits, and seeds, are tannins. They are divided into hydrolyzable (ellagitannins) and condensed tannins (proanthocyanidins) and gallotannins [110]. They have the ability to precipitate macromolecules such as proteins [111] as well as have antimicrobial properties. However, the mechanisms underlying the antimicrobial action of tannins in different microorganisms are still under investigation [111].

Ellagitannins constitute a complex class of polyphenols characterized by one or more hexahydroxydiphenoyl (HHDP) which can be linked in various ways to the glucose molecule [112]. Ellagic acid, gallagic acid, punicalins, and punicalagins isolated from ethyl acetate and butanolic fractions of *Punica granatum* revealed antifungal activity against *C. albicans*, *Cryptococcus neoformans*, and *Aspergillus fumigatus* [113] (**Table 1**).

Ellagitannins isolated from *Ocotea odorifera*, a plant commonly used in Brazil in folk medicine, have a potential against *C. parapsilosis* [114]. Two ellagitannins isolated from raspberry (*Rubus idaeus* L.) fruit, lambertianin C and sanguiin H-6, showed fungistatic activity both in vitro and in situ against *Geotrichum candidum* [115]. Dos Santos et al. [111] verified that encapsulated tannins from *Acacia mearnsii* have moderate activity against *Aspergillus niger* (ATCC 9642) and *C. albicans* (ATCC 34147).

Coumarins have a C6-C3 skeleton, possessing an oxygen heterocycle as part of the C3 unit [105]. These compounds are known to play a role in disease and pest resistance, as well as UV tolerance. The antifungal activity of 40 coumarins was tested against reference strains of *Candida albicans*, *Aspergillus fumigatus*, and *Fusarium solani*, but among them only osthenol showed the most effective antifungal activity (Table 1). The authors argue that the action of osthenol can be related to the presence of an alkyl group at C-8 position [116].

Another coumarin derivative, 4-acetetatecoumarin, was effective in inhibiting *Aspergillus* spp., acting on the factors of virulence and affecting the structure of the fungal wall. Diversinin, a coumarin isolated from the petroleum ether extract of *Baccharis darwinii*, demonstrated antifungal activity against *T. rubrum*, *T. mentagrophytes*, and *M. gypseum*, being fungicidal. Another coumarin derivative, 5,8-dihydroxyumbelliprenin, isolated from *Ferula foetida*, was active against *M. gypseum* and *Trichophyton interdigitale* [105] (**Table 1**).

Phenylpropanoids are other naturally occurring compounds categorized as coumarins, phenylpropanoic acid, and lignans frequently studied for their anti-*Candida* properties [117]. Navarro-Garcia et al. [118] and Raut et al. [119] found that a coumarin (scopoletin) and two phenylpropanoic acids (salicylaldehyde and anisyl alcohol) have antifungal property against *C. albicans*, with MICs of 25, 31, and 31 μg/mL, respectively.

Shahzad et al. [103] observed the effectiveness of pyrogallol and curcumin (CUR) against various *C. albicans* clinical isolates. In addition, curcumin inhibited the adhesion capability of cells and demonstrated anti-biofilm activity. Curcumin is a flavonoid found in turmeric (Curcuma longa L.). Pure curcumin had potential activity against *Cryptococcus gattii* both in vitro and in vivo [120]. According to Ferreira et al. [121], the essential oil from *Curcuma longa* L. can reduce the colony diameter, germination, and sporulation of *Aspergillus flavus*.

Alalwan et al. [122] undertook a series of adsorption experiments with varying concentrations of curcumin and showed that 50 μg/mL could prevent adhesion of *C. albicans* SC5314 to denture materials. Curcumin-silver nanoparticles also showed potential anticandidal activity against fluconazole-resistant *Candida* species isolated from HIV patients with MIC range of 31.2–250 μg/mL [123].

Gallic acid is a polyphenol natural compound found in many medicinal plant species that has been shown to have anti-inflammatory and antibacterial properties. GA was found to have a broad spectrum of antifungal activity against dermatophyte and *Candida* strains. Authors verified that GA reduced the activity of sterol 14-α-demethylase P450 (CYP51) and squalene epoxidase in the *T. rubrum* membrane.

Teodoro et al. [124] demonstrated that acetone fraction from *Buchenavia tomentosa* aqueous extract and its major compound gallic acid had the ability to inhibit reference strains *C. albicans* ATCC 18804 and *Candida albicans* SC 5314 adherence and to disrupt 48 h-biofilm.

5.2 Essential oils as potential antifungal

In the eagerness to research and develop new substances to suppress the development of pathogenic fungi from natural plant substances, knowledge about the biological activities of essential oils has been growing. Essential oils' pharmacological activities, mainly related to their complex chemical composition and high concentrations of phenols, make these compounds particularly interesting for both the treatment and the prevention of fungal infections. Natural phenolic substances are among the most antifungal active substances present in essential oils, generally showing low toxic effects in animals [125]. They consist in a complex mixture of monoterpene and sesquiterpene hydrocarbons and oxygenated derivatives such as alcohols, aldehydes, ketones, and phenylpropanoids.

Essential oils are also called volatile oils or ethereal oils, as they have a high degree of evaporation when exposed to air. The presence of terpenes contributes to the complex constitution with the action against microorganisms being directly related to this characteristic [126]. Since ancient times, Mondello et al. [127] proposed that tea tree oil could be used in antifungal therapy, because it showed efficacy against multidrug-resistant *Candida* species in vitro and against mucosal candidiasis in vivo; they have also showed that terpinen-4-ol was the main substance presented in the oil which contribute to the anticandidal activity.

Several oils have demonstrated activity against *Candida* species. Sharifzadeh et al. [128] observed that essential oils from *Trachyspermum ammi* have anticandidal effects against isolates resistant to FLZ. Herbal essences from *Foeniculum vulgare*, *Satureja hortensis*, *C. cyminum*, and *Zataria multiflora* were tested against *C. albicans.* Essential oils from *Z. multiflora* showed the best anticandidal activity [129].

Carica papaya essential oils have inhibitory effects against *Candida* species, detected by agar diffusion and microdilution assays [130]. Minooeianhaghighi et al. [131] verified that a combination of essential oils from *Cuminum cyminum* and *Lavandula binaludensis* showed growth inhibition of *C. albicans* isolates, at very low concentrations (between 3.90 and 11.71 μg/mL). Essential oils from *Cymbopogon nardus* have also shown antimicrobial potential against *Candida* species, with inhibition of hyphal growth in *C. albicans* at concentrations between 15.8 and 1000 μg/mL. This oil also inhibited growth of filamentous fungus from the environment. Main compounds of *C. nardus* essential oil were the oxygen-containing monoterpenes: citronellal, geranial, geraniol, citronellol, and neral [126]. In addition to inhibiting biofilm formation [132], essential oils from *Artemisia judaica* have been shown to inhibit the formation of germination tubes in *C. albicans* and have shown that at a very low concentration (0.16 μL/mL), it inhibited 80% of *Candida* filamentation. Kose et al. [133] demonstrated the fungicidal potential of essential oils from *Centaurea baseri* against *Candida* species, with an MIC of 60 μg/mL.

Among the monoterpenes there is thymol (2-isopropyl-5-methylphenol) [134]. It is the most abundant constituent in essential oils from *Thymus vulgaris* (thyme) [135] and the major component of essential oils from *Origanum vulgare* (oregano) [136]. Thymol showed antifungal activity, fungistatic and fungicidal one, against *Candida* strains. Authors verified an MIC of 39 μg/mL against *C. albicans* and *C. krusei* and MIC of 78 μg/mL against *C. tropicalis*. Probably thymol acts by binding to ergosterol in the plasma membrane, thereby increasing ion permeability and resulting in cell death because an eightfold increase (from 39.0 to 312.5 μg/mL) in thymol MIC values against *C. albicans* was seen in the presence of exogenous ergosterol. A combination of thymol and nystatin resulted in synergy [137].

Terpenoids have shown synergistic effects with FLZ, so it may be useful as a candidate antifungal chemotherapeutic agent. In addition, terpenoids exhibit a very good antimycotic activity of filamentous-form growth of *C. albicans* at nontoxic concentrations [138]. Further, in experiments realized by [139], rubiarbonol G, a triterpenoid from *Rubia yunnanensis*, showed potent antimicrobial activity against *C. albicans*, with an MIC of 10.5 μg/mL.

The antifungal potential of terpenes, geraniol, and citronellol has been investigated previously, with effective inhibitory activity against *C. albicans* [138] and filamentous fungi of the *Aspergillus* species [140]. In addition, Mesa-Arango et al. [67] showed that oxygenated monoterpenes in the citral chemotype, such as geraniol, citral, and citronellal, have antifungal activity against *C. parapsilosis*, *C. krusei*, *Aspergillus flavus*, and *Aspergillus fumigatus*.

Terpenes' anti-biofilm activity and the efficacy of thymol, geraniol, and carvacrol in the treatment of *Candida* infections associated with the use of hospital devices have been related [141]. Effects of carvacrol on *Candida* cells can be associated with alterations in the cytoplasmic membrane and induction of apoptosis [108].

Although the process of discovering bioactive molecules is complex and time-consuming, involving isolation, identification, and optimization of pharmacokinetic and pharmacodynamic properties, as well as the selection of lead compounds for further drug development, data related here showed that plants are a promising source of active molecules with antifungal properties. Biological assays have shown that plant extracts or essential oils and their bioactivity molecules inhibit ATCC and clinical strains of fungi species, including those with resistance to drugs employed in medical practice. In addition, some are able to inhibit and control the main virulence factors of fungi species, such as the formation and proliferation of hyphae and filamentation and, more importantly, the eradication of mature biofilms.

Eugenol (4-allyl-2-methoxyphenol) is a phenolic compound and the main constituent of the essential oil isolated from the *Eugenia caryophyllata*. There are reports of some pharmacological effects of eugenol, such as antifungal and antibacterial agent, and its anti-*Candida* action seems to be related to the generation of oxidative stress concomitantly with lipid peroxidation of the cell membrane of *Candida albicans* yeast and the generation of reactive oxygen species [142]. Eugenol also showed antifungal effects against both *Cryptococcus gattii* and *C. neoformans* cells by causing morphological alterations, changes of cellular superficial charges, and oxidative stress. Thymol and carvacrol can represent alternative, efficient, and cost effective drugs for anti-biofilm therapy for *Cryptococcus* species.

Eugenol showed activity against *Alternaria* spp. and *P. chrysogenum*, by agar diffusion method [143] and, along with other monoterpenes such as carvacrol and isoeugenol, exhibited strong antifungal activity against *Rhizopus stolonifer* and *Absidia coerulea* [144].

5.3 Synergistic action between phytochemicals and antifungals

Resistance mechanisms are developed by fungi to the treatment with conventional drugs in addition to toxic side effects to human cells showed by these drugs; researchers' efforts in developing new strategies to improve treatment effectiveness of fungal infection are growing, with an interest in plants and folklore medicine.

The knowledge about synergistic effects of plant extracts or their compounds with traditional agents is nowadays a type of study that has aroused interest. Some in vitro screening assays have evidenced that plant extracts are less toxic than existing antifungal agents and, in combination with them, could reduce toxicity and increase antifungal potential [21, 145].

Accordingly, combination antifungal therapy offers the possibility of broadening the spectrum of drug activity, reducing toxicity, and decreasing fungal resistance [146].

Although combination of medications requires a careful evaluation of the synergistic, antagonistic, and agonist properties of the drugs involved [147], the use of drug combinations in treatment of infections by fungi is a common preferred strategy clinically. In many cases of fungal infection, combination therapy has been used successfully [21]. For some examples, see **Table 2**.

There are two main hypotheses about the type of interaction resulting from the combination of fluconazole and amphotericin B, based on the mechanisms of action of these drugs. In the theory of depletion, the interaction between fluconazole and amphotericin B would result in antagonism due to pre-exposure to fluconazole, which would lead to depletion of the membrane ergosterol, and thus there would be a decrease in the available sites for amphotericin B. In the second theory, the synergism, amphotericin B would lead to the formation of pores, which would facilitate the greater access of azole to the intracellular space, which by inhibiting the enzymes involved in ergosterol biosynthesis would increase the antimicrobial

Combination of antifungals	Target	References
AMP B + posaconazole AMP B + caspofungin AMP B + fluconazole	*Candida* biofilms *Candida* biofilms Cryptococcosis in murine model	[148] [54] [149]
Micafungin + fluconazole Micafungin + voriconazole Micafungin + AMP B Micafungin + isavuconazole	*Candida* infections	[150] [151] [152]
Flucytosine + voriconazole	*Candida* infections	[148]
Minocycline + fluconazole	*Candida albicans* biofilms	[148]
Posaconazole + caspofungin	*Candida* infections	[153] [154]
Terbinafine + azole	*Candida* growth	[155] [156]
Echinocandin + azole	Invasive candidiasis	[157]
AMP B + flucytosine	Invasive candidiasis	[158]
Natamycin + 5-fluorouracil	*Fusarium* species ocular isolates	[159]

AMP B: amphotericin B.

Table 2.
Various regimes of combinatorial antifungal therapy showing better efficacy in combination than that of independent drugs (adapted from [21]).

efficacy. According to these theories, the combination of fluconazole and amphotericin B could involve different interactions [160–162].

Considering the difficulties regarding to the treatment of candidiasis and cryptococcosis, the combination of antifungals represents an important alternative to conventional therapy. The synergistic effects of drugs are primarily attributable to cell wall damage by one antifungal. Thus, this component potentiates the activity of other drugs exactly against some constituent of plasma membrane. Alternatively, compromised cell wall with an increased permeability could facilitate movement of drugs across the cell membrane to their targets. Or, the synergistic action of different drugs occurs because they act on different targets of the same pathway, which can happen, for example, with the combination of azoles and allylamines.

The objective of this strategy is to maximize the antifungal effects. Tangarife-Castaño et al. [163] reported synergy between essential oils or plant extracts associated with antifungal drugs when used as anti-*C. albicans* agents. The best synergistic effects were obtained from the combination between itraconazole and *P. bredemeyeri* extract against *C. albicans*.

Synergistic potential was observed when methanolic extract of *T. catappa* leaves was combined with nystatin or AMB against reference strains of *C. albicans*, *Candida neoformans*, *C. glabrata*, *Candida apicola*, and *Trichosporon beigelii* [164]. The combination showed maximum synergy against *C. apicola*.

Santos et al. [165] related synergistic antifungal activity of an ethanol extract of *Hyptis martiusii* in combination with metronidazole against *C. albicans*, *C. krusei*, and *C. tropicalis*. Avijgan et al. [166] reported a potent synergistic effect between an *Echinophora platyloba* ethanolic extract and itraconazole or FLZ against isolates of *C. albicans* from vaginal secretions of patients with recurrent vulvovaginitis, significantly lowering the concentrations of both substances.

A combination between thymol and nystatin was found to have synergistic effects against *Candida* species [137], reducing the MICs of both products by 87.4%. Synergism was observed between a water insoluble fraction from *U. tomentosa* bark and terbinafine, as well as between it and FLZ against seven resistant isolates of *C. glabrata* and *C. krusei* [167]. Synergistic effects led to cell damage, and authors demonstrated, through differential scanning calorimetry and infrared analysis, that intermolecular interactions between the extract components and either terbinafine or FLZ occurring outside the cell wall are likely responsible for synergistic effects observed between substances.

Subfraction combinations of *Terminalia catappa*, *Terminalia mantaly*, and *Monodora tenuifolia* showed synergistic interactions against *C. albicans*, *C. glabrata*, *C. parapsilosis*, and *C. neoformans* isolates. Synergistic combination between *M. tenuifolia* and *T. mantaly* subfractions also showed fungicidal effects against most tested strains [168].

The combination therapy with curcumin and fluconazole was the most effective among the treatments tested against *Cryptococcus gattii*. The association was able to reduce the fungal burden and damage on lung tissues of infected mice and to eliminate the fungal burden in the brain, enhancing the survival of mice with *C. gattii*-induced cryptococcosis [120].

Methanolic extract of *Buchenavia tetraphylla* is a great source of antimicrobial compounds and enhanced the action of FLZ against different *C. albicans* isolates from vaginal secretions as well as azole-resistant isolates. The extract increased the action of FLZ in most strains through additive (20% of strains) or synergistic (60% of strains) effects [169].

Kumari et al. [170] investigated the effect of six essential oil compounds sourced from oregano oil (carvacrol), cinnamon oil (cinnamaldehyde), lemongrass oil (citral), clove oil (eugenol), peppermint oil (menthol), and thyme oil (thymol) against three infectious forms: planktonic cells, biofilm formation, and preformed biofilm of *C. neoformans* and *C. laurentii*. The anti-biofilm activity of the tested compounds was in the order thymol > carvacrol > citral > eugenol = cinnamaldehyde > menthol. The three most potent compounds thymol, carvacrol, and citral showed best anti-biofilm activity at a much lower concentration against *C. laurentii*. In the presence of these potent compounds, assays revealed the absence of extracellular polymeric matrix, reduction in cellular density, and alteration in the surface morphology of biofilm cells. In addition they were the most efficient in terms of human safety in keratinocyte-*Cryptococcus* spp. co-culture infection model suggesting that thymol, carvacrol, and citral can be further exploited as cost-effective and nontoxic anti-cryptococcal drugs.

The lectin pCramoll from *Cratylia mollis*, a native forage plant endemic to the semiarid region of Brazil (caatinga biome), showed an immunomodulatory effect and a synergism in combination with fluconazole, increasing the survival of animals with cryptococcosis caused by *C. gattii* and improving aspects of morbidity present in the progression of cryptococcosis [171].

Thymol exhibited synergistic effects when combined with fluconazole against clinical species of *Candida*, enhancing the antifungal potential of the drug and decreasing the concentration required for the effect [172]. Zaidi et al. [173] found that methanolic extract of leaves of *Ocimum sanctum* in combination with fluconazole showed higher antifungal potential and synergistic activity against resistant *Candida* spp. than methanolic extract or fluconazole when used alone.

Essential oils were also recently proposed to increase drug effectiveness. *Lavandula* and *Rosmarinus* essential oils were selected as antiproliferative agents to compound lipid nanoparticles for clotrimazole delivery in treatment of *Candida* skin infections. Authors confirmed the potential anti-*Candida* activity of the selected oils due to their interaction with membrane permeabilization. In addition, in vitro studies against *Candida albicans*, *Candida krusei*, and *Candida parapsilosis* showed an increase of the antifungal activity of clotrimazole-loaded nanoparticles prepared with *Lavandula* or *Rosmarinus*, thus confirming that nanostructured lipid carriers (NLC) containing these essential oils represent a promising strategy to improve drug effectiveness against topical candidiasis [174].

A novel therapeutic strategy that has been adopted is photodynamic therapy (PDT). It is based on the interaction between a nontoxic photosensitizer and a safe source of visible light at a low intensity; the combination of these two factors in the presence of oxygen leads to the development of reactive oxygen species (ROS) which are toxic and cause oxidative damage to microorganism cells [175]. Curcumin associated with LED light was an efficient strategy against biofilms of *C. dubliniensis* isolates [176]. The uptake of CUR by yeast cells and its penetration through the biofilm were accompanied by confocal laser scanning microscopy. Daliri et al. [177] have assessed the effect of curcumin- and methyl blue-mediated PDT in combination with different laser exposure parameters on *C. albicans* colonies. They verified that the 460-nm laser in combination with CUR has the maximum antifungal efficiency against *C. albicans*.

Although we have described herein many in vitro studies examining synergistic effects among potential antifungal biomolecules and traditional antifungal agents, the mechanisms underlying these synergistic effects are poorly understood. Randomized and controlled analyses have been performed with the objective of verifying the efficacy and risks of using traditional antifungal combinations; however, the results are poor and contradictory. High cost to conduct these strategies, reduced number of clinical cases, and the existence of confusing variables are factors that contribute to the obtaining of vague and non-reproducible results.

Therefore, it is extremely relevant to examine carefully possible synergism between new phytocompounds and conventional antimycotic drugs in order to obtain more insight. Understanding the cellular action of each substance in the combination process is also a key step in inferring ways to employ strategy in the clinic. A lack of consensus in the medical clinic emphasizes the need to conduct further clinical trials using combinations of antifungals. The experiments and results addressed herein support further investigation of new plant constituents with antifungal properties and the efficacy of combination therapies involving phytocomponents and traditional antifungal agents as an important start for the development of unusual and original antifungal therapies.

6. Conclusions

The increase in *Candida* and *Cryptococcus* infections is alarming leading to high rates of morbidity and mortality worldwide. Concomitantly with the increase in fungal infections, species emerged, and the resistance phenomenon increased so that the available antifungal arsenal becomes irrelevant in the face of the problem. In addition, there are limitations manifested by some antifungal agents such as fungistatic character, severe toxicity, and renal dysfunction. Therefore, it is crucial to develop new drugs as alternative therapies that are potentially active against *Cryptococcus* and *Candida*. Plants are considered abundant and safe sources of phytochemicals endowed with many biological activities. Several polyphenols have been isolated and studied in relation to their anti-yeast and anti-virulence activities and may be useful in obtaining promising, efficient, and cost-effective drugs for the inhibition of *Candida* and *Cryptococcus* infections. Many phytosubstances are extremely effective in combination therapy with traditional or other phytochemicals, which can be further exploited to lead to novel drug therapies against recalcitrant infections.

Acknowledgements

Authors thank Ceuma University for their contribution to the work.

Conflict of interest

The authors declare no competing interests.

Author details

Cristina de Andrade Monteiro* and Julliana Ribeiro Alves dos Santos
Universidade Ceuma, Brazil

*Address all correspondence to: cristina.monteiro@ceuma.br

References

[1] Vallabhaneni S, Chiller TM. Fungal infections and new biologic therapies. Current Rheumatology Reports. 2016; **18**:29. DOI: 10.1007/s11926-016-0572-1

[2] Brown GD, Denning DW, Gow NAR, Levitz SM, Netea MG, White TC. Hidden killers: Human fungal infections. Science Translational Medicine. 2012;**4**:165rv13. DOI: 10.1126/scitranslmed.3004404

[3] López-Martínez R. Candidosis, a new challenge. Clinics in Dermatology. 2010; **28**:178-184. DOI: 10.1016/j.clindermatol.2009.12.014

[4] Gow NAR, Hube B. Importance of the *Candida albicans* cell wall during commensalism and infection. Current Opinion in Microbiology. 2012;**15**: 406-412. DOI: 10.1016/j.mib.2012.04.005

[5] Barchiesi F, Orsetti E, Osimani P, Catassi C, Santelli F, Manso E. Factors related to outcome of bloodstream infections due to *Candida parapsilosis* complex. BMC Infectious Diseases. 2016;**16**:387. DOI: 10.1186/s12879-016-1704-y

[6] Mitchell TG, Perfect JR. Cryptococcosis in the era of AIDS—100 years after the discovery of *Cryptococcus neoformans*. Clinical Microbiology Reviews. 1995;**8**(4):515-548

[7] Bovers M, Hagen F, Boekhout T. Diversity of the *Cryptococcus neoformans-Cryptococcus gattii* species complex. Revista Iberoamericana de Micologia. 2008;**25**(1):4-12

[8] Hurtado JCC, Fernandes P, Navarro F, Lovane M, Casas L, Quintó I, et al. Mortality due to *Cryptococcus neoformans* and *Cryptococcus gattii* in low-income settings: An autopsy study. Scientific Reports. 2019;**9**:7493. DOI: 10.1038/s41598-019-43941-w

[9] Molina-Leyva A, Ruiz-Carrascosa JC, Leyva-Garcia A, Husein-Elahmed H. Cutaneous *Cryptococcus laurentii* infection in an immunocompetent child. International Journal of Infectious Diseases. 2013;**17**:e1232-e1233. DOI: 10.1016/j.ijid.2013.04.017

[10] Franco C, Tomei F, Assalone P, Traficante D, Di Pilla G, Pepe C, et al. *Cryptococcus laurentii* diarrhea in a neoplastic patient. Case Reports in Oncological Medicine. 2015:216458, 2 pages. DOI: 10.1155/2015/216458

[11] Smith N, Sehring M, Chambers J, Patel P. Perspectives on non-*neoformans* cryptococcal opportunistic infections. Journal of Community Hospital Internal Medicine Perspectives. 2017;**7**(4): 214-217. DOI: 10.1080/20009666.2017.1350087

[12] Springer DJ, Chaturvedi V. Projecting global occurrence of *Cryptococcus gattii*. Emerging Infectious Diseases. 2010;**16**(10):14-20. DOI: 10.3201/eid1601.090369

[13] Negroni R. Cryptococcosis. Clinics in Dermatology. 2012;**30**:599-609. DOI: 10.1016/j.clindermatol.2012.01.005

[14] Cogliati M. Global molecular epidemiology of *Cryptococcus neoformans* and *Cryptococcus gattii*: An atlas of the molecular types. Scientifica (Cairo). 2013;**2013**:675213. DOI: 10.1155/2013/675213

[15] Chayakulkeeree M, Perfect JR. Cryptococcosis. Infectious Disease Clinics of North America. 2006;**20**(3): 507-544, v-vi

[16] Chaturvedi V, Chaturvedi S. *Cryptococcus gattii*: A resurgent fungal pathogen. Trends in Microbiology. 2011; **19**(11):564-571. DOI: 10.1016/j.tim.2011.07.010

[17] Marques SA, Bastazini I, Martins AL, Barreto JA, Barbieri D'Elia MP, Lastória JC, et al. Primary cutaneous cryptococcosis in Brazil: Report of 11 cases in immunocompetent and immunosuppressed patients. International Journal of Dermatology. 2012;**51**(7):780-784

[18] Chen SC, Meyer W, Sorrell TC. *Cryptococcus gattii* infections. Clinical Microbiology Reviews. 2014;**27**(4): 980-1024. DOI: 10.1111/j.1365-4632.2011.05298.x

[19] Zaragoza O. Basic principles of the virulence of *Cryptococcus*. Virulence. 2019;**10**(1):490-501. DOI: 10.1080/21505594.2019.1614383

[20] Rajasingham R, Smith RM, Park BJ, Jarvis JN, Govender NP, Chiller TM, et al. Global burden of disease of HIV-associated cryptococcal meningitis: An updated analysis. The Lancet Infectious Diseases. 2017;**17**(8):873-881. DOI: 10.1016/S1473-3099(17)30243-8

[21] de Oliveira Santos GC, Vasconcelos CC, Lopes AJO, de Sousa Cartágenes MDS, Filho AKDB, do Nascimento FRF, et al. Infections and therapeutic strategies: Mechanisms of action for traditional and alternative agents. Frontiers in Microbiology. 2018;**9**:1351. DOI: 10.3389/fmicb.2018.01351

[22] Pfaller MA, Diekema DJ, Gibbs DL, Newell VA, Ellis D, Tullio V, et al. Results from the ARTEMIS DISK global antifungal surveillance study, 1997–2007: A 10.5-year analysis of susceptibilities of *Candida* species to fluconazole and voriconazole as determined by CLSI standardized disk diffusion. Journal of Clinical Microbiology. 2010;**48**(4):1366-1377. DOI: 10.1128/JCM.02117-09

[23] Perfect JR, Dismukes WE, Dromer F, Goldman DL, Graybill JR, Hamill RJ, et al. Clinical practice guidelines for the management of cryptococcal disease: 2010 update by the infectious diseases society of America. Clinical Infectious Diseases. 2010;**50**(3):291-322

[24] Perfect JR, Bicanic T. Cryptococcosis diagnosis and treatment: What do we know now. Fungal Genetics and Biology. 2015;**78**:49-54. DOI: 10.1086/649858

[25] Mpoza E, Rhein J, Abassi M. Emerging fluconazole resistance: Implications for the management of cryptococcal meningitis. Medical Mycology Case Reports. 2018;**19**:30-32. DOI: 10.1016/j.mmcr.2017.11.004

[26] Razzaghi-Abyaneh M, Sadeghi G, Zeinali E, Alirezaee M, Shams-Ghahfarokhi M, Amani A, et al. Species distribution and antifungal susceptibility of *Candida* spp. isolated from superficial candidiasis in outpatients in Iran. Journal de Mycologie Médicale. 2014;**24**(2):e43-e50. DOI: 10.1016/j.mycmed.2014.01.004

[27] Terças AG, Monteiro AS, Moffa EB, Dos Santos JRA, de Sousa EM, Pinto ARB, et al. Phytochemical characterization of *Candida* species isolated from HIV-positive patients recruited at a public hospital in São Luís, Maranhão, Brazil. Frontiers in Microbiology. 2017;**8**:595. DOI: 10.3389/fmicb.2017.00298

[28] Stop neglecting fungi. Nature Microbiology. 2017;**2**:17120. DOI: 10.1038/nmicrobiol.2017.120

[29] Vázquez-González D, Perusquía-Ortiz AM, Hundeiker M, Bonifaz A. Opportunistic yeast infections: Candidiasis, cryptococcosis, trichosporonosis and geotrichosis. Journal der Deutschen Dermatologischen Gesellschaft. 2013;**11**(5):381-393; quiz 94. DOI: 10.1111/ddg.12097

[30] Mayer FL, Wilson D, Hube B. *Candida albicans* pathogenicity mechanisms. Virulence. 2013;**4**(2): 119-128. DOI: 10.4161/viru.22913

[31] Tsai PW, Chen YT, Hsu PC, Lan CY. Study of *Candida albicans* and its interactions with the host: A mini review. *Biomedicine*. 2013;**3**:51-64. DOI: 10.1016/j.biomed.2012.12.004

[32] Ferreira AV, Prado CG, Carvalho RR, Dias KS, Dias AL. *Candida albicans* and non-*C. albicans Candida* species: Comparison of biofilm production and metabolic activity in biofilms, and putative virulence properties of isolates from hospital environments and infections. Mycopathologia. 2013;**175** (3–4):265-272. DOI: 10.1007/s11046-013-9638-z

[33] Lewis LE, Bain JM, Lowes C, Gow NA, Erwig LP. *Candida albicans* infection inhibits macrophage cell division and proliferation. Fungal Genetics and Biology. 2012;**49**(9): 679-680. DOI: 10.1016/j.fgb.2012.05.007

[34] Kwamin F, Nartey NO, Codjoe FS, Newman MJ. Distribution of *Candida* species among HIV-positive patients with oropharyngeal candidiasis in Accra, Ghana. Journal of Infection in Developing Countries. 2013;7(1):41-45. DOI: 10.3855/jidc.2442

[35] Wächtler B, Citiulo F, Jablonowski N, Förster S, Dalle F, Schaller M, et al. *Candida albicans*-epithelial interactions: Dissecting the roles of active penetration, induced endocytosis and host factors on the infection process. PLoS One. 2012;7(5):e36952. DOI: 10.1371/journal.pone.0036952

[36] Kim J, Sudbery P. *Candida albicans*, a major human fungal pathogen. Journal of Microbiology. 2011;**49**(2):171-177. DOI: 10.1007/s12275-011-1064-7

[37] Li SY, Yang YL, Chen KW, Cheng HH, Chiou CS, Wang TH, et al. Molecular epidemiology of long-term colonization of *Candida albicans* strains from HIV-infected patients. Epidemiology and Infection. 2006; **134**(2):265-269. DOI: 10.1017/S0950268805004905

[38] Raman SB, Nguyen MH, Cheng S, Badrane H, Iczkowski KA, Wegener M, et al. A competitive infection model of hematogenously disseminated candidiasis in mice redefines the role of *Candida albicans* IRS4 in pathogenesis. Infection and Immunity. 2013;**81**(5): 1430-1438. DOI: 10.1128/IAI.00743-12

[39] Pfaller MA, Diekema DJ. Epidemiology of invasive candidiasis: A persistent public health problem. Clinical Microbiology Reviews. 2007;**20**(1): 133-163. DOI: 10.1128/CMR.00029-06

[40] Sardi JC, Scorzoni L, Bernardi T, Fusco-Almeida AM, Mendes Giannini MJ. *Candida* species: Current epidemiology, pathogenicity, biofilm formation, natural antifungal products and new therapeutic options. Journal of Medical Microbiology. 2013;**62**(Pt 1): 10-24. DOI: 10.1099/jmm.0.045054-0

[41] Guery BP, Arendrup MC, Auzinger G, Azoulay E, Borges Sá M, Johnson EM, et al. Management of invasive candidiasis and candidemia in adult non-neutropenic intensive care unit patients: Part I. Epidemiology and diagnosis. Intensive Care Medicine. 2009;**35**(1):55-62. DOI: 10.1007/s00134-008-1338-7

[42] Hise AG, Tomalka J, Ganesan S, Patel K, Hall BA, Brown GD, et al. An essential role for the NLRP3 inflammasome in host defense against the human fungal pathogen *Candida albicans*. Cell Host & Microbe. 2009; **5**(5):487-497. DOI: 10.1016/j.chom.2009.05.002

[43] Junqueira JC, Vilela SF, Rossoni RD, Barbosa JO, Costa AC, Rasteiro VM, et al. Oral colonization by yeasts in HIV-positive patients in Brazil. Revista do Instituto de Medicina Tropical de São Paulo. 2012;**54**(1):17-24. DOI: 10.1590/S0036-46652012000100004

[44] Li YY, Chen WY, Li X, Li HB, Li HQ, Wang L, et al. Asymptomatic oral yeast carriage and antifungal susceptibility profile of HIV-infected patients in Kunming, Yunnan Province of China. BMC Infectious Diseases. 2013;**13**:46. DOI: 10.1186/1471-2334-13-46

[45] Sant'Ana PL, Milan EP, Martinez R, Queiroz-Telles F, Ferreira MS, Alcântara AP, et al. Multicenter Brazilian study of oral *Candida* species isolated from AIDS patients. Memórias do Instituto Oswaldo Cruz. 2002;**97**:253-257. DOI: 10.1590/S0074-02762002000200019

[46] Kaur R, Dhakad MS, Goyal R, Haque A, Mukhopadhyay G. Identification and antifungal susceptibility testing of *Candida* species: A comparison of Vitek-2 system with conventional and molecular methods. Journal of Global Infectious Diseases. 2016;**8**(4):139-146. DOI: 10.4103/0974-777X.192969

[47] Paramythiotou E, Frantzeskaki F, Flevari A, Armaganidis A, Dimopoulos G. Invasive fungal infections in the ICU: How to approach, how to treat. Molecules. 2014;**19**(1):1085-1119. DOI: 10.3390/molecules19011085

[48] Álvares CA, Svidzinski TIE, Consolaro MEL. Candidíase vulvovaginal: fatores predisponentes do hospedeiro e virulência das leveduras. Jornal Brasileiro de Patologia e Medicina Laboratorial. 2007;**43**(5):319-327. DOI: 10.1590/S1676-24442007000500004

[49] Shoham S, Marr KA. Invasive fungal infections in solid organ transplant recipients. Future Microbiology. 2012;**7**(5):639-655. DOI: 10.2217/fmb.12.28

[50] Soll DR. *Candida* biofilms: Is adhesion sexy? Current Biology. 2008;**18**(16):R717-R720. DOI: 10.1016/j.cub.2008.07.014

[51] McMullan BJ, Sorrell TC, Chen SC. *Cryptococcus gattii* infections: Contemporary aspects of epidemiology, clinical manifestations and management of infection. Future Microbiology. 2013;**8**(12):1613-1631. DOI: 10.2217/fmb.13.123

[52] Goeb JL, Leon V, Kechid G. Cryptococcal meningitis with acute psychotic confusion in a sarcoid patient. Primary Care Companion to The Journal of Clinical Psychiatry. 2007;**9**(5):393-394

[53] Prakash PY, Sugandhi RP. Neuropsychiatric manifestation of confusional psychosis due to *Cryptococcus neoformans* var. grubii in an apparently immunocompetent host: A case report. Cases Journal. 2009;**2**:9084. DOI: 10.1186/1757-1626-2-9084

[54] Rodrigues ML. Funding and innovation in diseases of neglected populations: The paradox of Cryptococcal meningitis. PLoS Neglected Tropical Diseases. 2016;**10**(3):e0004429

[55] Dewar GJ, Kelly JK. *Cryptococcus gattii*: An emerging cause of pulmonary nodules. Canadian Respiratory Journal. 2008;**15**(3):153-157. DOI: 10.1371/journal.pntd.0004429

[56] Franco-Paredes C, Womack T, Bohlmeyer T, Sellers B, Hays A, Patel K, et al. Management of *Cryptococcus gattii* meningoencephalitis. Lancet Infectious Diseases. 2015;**15**(3):348-355. DOI: 10.1016/S1473-3099(14)70945-4

[57] Lin X, Heitman J. The biology of the *Cryptococcus neoformans* species complex. Annual Review of Microbiology. 2006;**60**:69-105

[58] Byrnes EJ, Li W, Ren P, Lewit Y, Voelz K, Fraser JA, et al. A diverse population of *Cryptococcus gattii* molecular type VGIII in southern Californian HIV/AIDS patients. PLoS Pathogens. 2011;7(9):e1002205. DOI: 10.1371/journal.ppat.1002205

[59] Silva FB, Santos JRN, da Silva LC, Gomes WC, Villis PCM, Gomes EDS, et al. Climate drivers of hospitalizations for mycoses in Brazil. Scientific Reports. 2019;**9**(1):6902. DOI: 10.1038/s41598-019-43353-w

[60] Sanguinetti M, Posteraro B, Lass-Flörl C. Antifungal drug resistance among *Candida* species: Mechanisms and clinical impact. Mycoses. 2015;**58** (Suppl 2):2-13. DOI: 10.1111/myc.12330

[61] Georgopapadakou NH. Antifungals: Mechanism of action and resistance, established and novel drugs. Current Opinion in Microbiology. 1998;**1**(5): 547-557. DOI: 10.1016/S1369-5274(98)80087-8

[62] Groll AH, Gea-Banacloche JC, Glasmacher A, Just-Nuebling G, Maschmeyer G, Walsh TJ. Clinical pharmacology of antifungal compounds. Infectious Disease Clinics of North America. 2003;**17**(1):159-191, ix. DOI: 10.1016/S0891-5520(02)00068-5

[63] Maubon D, Garnaud C, Calandra T, Sanglard D, Cornet M. Resistance of *Candida* spp. to antifungal drugs in the ICU: Where are we now? Intensive Care Medicine. 2014;**40**(9):1241-1255. DOI: 10.1007/s00134-014-3404-7

[64] Patil S, Rao RS, Majumdar B, Anil S. Clinical appearance of oral *Candida* infection and therapeutic strategies. Frontiers in Microbiology. 2015;**6**:1391. DOI: 10.3389/fmicb.2015.01391

[65] Prasad R, Shah AH, Rawal MK. Antifungals: Mechanism of action and drug resistance. In: Ramos J, Sychrová H, Kschischo M, editors. Yeast Membrane Transporter Advances in Experimental Medicine and Biology. Cham: Springer International Publishing; 2016. pp. 327-349

[66] Nett JE, Andes DR. Antifungal agents: Spectrum of activity, pharmacology, and clinical indications. Infectious Disease Clinics of North America. 2016;**30**(1):51-83. DOI: 10.1016/j.idc.2015.10.012

[67] Mesa-Arango AC, Scorzoni L, Zaragoza O. It only takes one to do many jobs: Amphotericin B as antifungal and immunomodulatory drug. Frontiers in Microbiology. 2012;**3**: 286. DOI: 10.3389/fmicb.2012.00286

[68] Seyedmousavi S, Rafati H, Ilkit M, Tolooe A, Hedayati MT, Verweij P. Systemic antifungal agents: Current status and projected future developments. In: Lion T, editor. Methods in Molecular Biology. New York, NY: Springer New York; 2017. pp. 107-139

[69] Carrillo-Muñoz AJ, Giusiano G, Ezkurra PA, Quindós G. Antifungal agents: Mode of action in yeast cells. Revista Española de Quimioterapia. 2006;**19**(2):130-139

[70] Spampinato C, Leonardi D. *Candida* infections, causes, targets, and resistance mechanisms: Traditional and alternative antifungal agents. BioMed Research International. 2013;**2013**: 204237. DOI: 10.1155/2013/204237

[71] Amaral AC, Felipe MS. Nanobiotechnology: An efficient approach to drug delivery of unstable biomolecules. Current Protein & Peptide Science. 2013;**14**(7):588-594. DOI: 10.2174/1389203711209070632

[72] Stiufiuc R, Iacovita C, Stiufiuc G, Florea A, Achim M, Lucaciu CM. A new class of pegylated plasmonic liposomes: Synthesis and characterization. Journal of Colloid and Interface Science. 2015;**437**: 17-23. DOI: 10.1016/j.jcis.2014.09.023

[73] Souza AC, Amaral AC. Antifungal therapy for systemic mycosis and the nanobiotechnology era: Improving efficacy, biodistribution and toxicity. Frontiers in Microbiology. 2017;**8**:336. DOI: 10.3389/fmicb.2017.00336

[74] Bayhan GI, Garipardic M, Karaman K, Akbayram S. Voriconazole-associated visual disturbances and hallucinations. Cutaneous and Ocular Toxicology. 2016; **35**(1):80-82. DOI: 10.3109/15569527.2015.1020544:1-3

[75] Rautemaa R, Ramage G. Oral candidosis—Clinical challenges of a biofilm disease. Critical Reviews in Microbiology. 2011;**37**(4):328-336. DOI: 10.3109/1040841X.2011.585606

[76] Fothergill AW, Sutton DA, McCarthy DI, Wiederhold NP. Impact of new antifungal breakpoints on antifungal resistance in *Candida* species. Journal of Clinical Microbiology. 2014; **52**(3):994-997. DOI: 10.1128/JCM.03044-13

[77] Chowdhary A, Randhawa HS, Sundar G, Kathuria S, Prakash A, Khan Z, et al. In vitro antifungal susceptibility profiles and genotypes of 308 clinical and environmental isolates of *Cryptococcus neoformans* var. *grubii* and *Cryptococcus gattii* serotype B from North-Western India. Journal of Medical Microbiology. 2011;**60** (Pt 7):961-967. DOI: 10.1099/jmm.0.029025-0

[78] Ferreira GF, Santos DA. Heteroresistance and fungi. Mycoses. 2017;**60**(9):562-568. DOI: 10.1111/myc.12639

[79] Basso LR, Gast CE, Bruzual I, Wong B. Identification and properties of plasma membrane azole efflux pumps from the pathogenic fungi *Cryptococcus gattii* and *Cryptococcus neoformans*. The Journal of Antimicrobial Chemotherapy. 2015;**70**(5):1396-1407. DOI: 10.1093/jac/dku554

[80] Paul S, Doering TL, Moye-Rowley WS. *Cryptococcus neoformans* Yap1 is required for normal fluconazole and oxidative stress resistance. Fungal Genetics and Biology. 2015;**74**:1-9. DOI: 10.1016/j.fgb.2014.10.015

[81] Kanafani ZA, Perfect JR. Antimicrobial resistance: Resistance to antifungal agents: Mechanisms and clinical impact. Clinical Infectious Diseases. 2008;**46**(1):120-128. DOI: 10.1086/524071

[82] Gamaletsou MN, Walsh TJ, Sipsas NV. Invasive fungal infections in patients with hematological malignancies: Emergence of resistant pathogens and new antifungal therapies. Turkish Journal of Haematology. 2018; **35**(1):1-11. DOI: 10.4274/tjh.2018.0007

[83] Sheikh N, Jahagirdar V, Kothadia S, Nagoba B. Antifungal drug resistance in *Candida* species. European Journal of General Medicine. 2013;**10**:254-258

[84] Perlin DS. Mechanisms of echinocandin antifungal drug resistance. Annals of the New York Academy of Sciences. 2015;**1354**:1-11. DOI: 10.1111/nyas.12831

[85] Katiyar S, Pfaller M, Edlind T. *Candida albicans* and *Candida glabrata* clinical isolates exhibiting reduced echinocandin susceptibility. Antimicrobial Agents and Chemotherapy. 2006;**50**(8):2892-2894. DOI: 10.1128/AAC.00349-06

[86] Perlin DS. Current perspectives on echinocandin class drugs. Future Microbiology. 2011;**6**(4):441-457. DOI: 10.2217/fmb.11.19

[87] Sanglard D, Odds FC. Resistance of *Candida* species to antifungal agents: Molecular mechanisms and clinical consequences. Lancet Infectious Diseases. 2002;**2**(2):73. DOI: 10.1016/S1473-3099(02)00181-0

[88] Espinel-Ingroff A. Mechanisms of resistance to antifungal agents: Yeasts and filamentous fungi. Revista Iberoamericana de Micología. 2008;**25**(2):101-106

[89] Papon N, Noël T, Florent M, Gibot-Leclerc S, Jean D, Chastin C, et al.

Molecular mechanism of flucytosine resistance in *Candida lusitaniae*: Contribution of the FCY2, FCY1, and FUR1 genes to 5-fluorouracil and fluconazole cross-resistance. Antimicrobial Agents and Chemotherapy. 2007;**51**(1):369-371. DOI: 10.1128/AAC.00824-06

[90] Global action fund for fungal infection (GAFFI). Global Action Fund for Fungal Infections (GAFFI) [Internet]. 2014. Available from: http://www.gaffi.org/

[91] Martins N, Barros L, Santos-Buelga C, Henriques M, Silva S, Ferreira IC. Evaluation of bioactive properties and phenolic compounds in different extracts prepared from *Salvia officinalis* L. Food Chemistry. 2015;**170**:378-385. DOI: 10.1016/j.foodchem.2014.08.096

[92] Chen WC, Liou SS, Tzeng TF, Lee SL, Liu IM. Wound repair and anti-inflammatory potential of *Lonicera japonica* in excision wound-induced rats. BMC Complementary and Alternative Medicine. 2012;**12**:226. DOI: 10.1186/1472-6882-12-226

[93] Agyare C, Bempah SB, Boakye YD, Ayande PG, Adarkwa-Yiadom M, Mensah KB. Evaluation of antimicrobial and wound healing potential of *Justicia flava* and *Lannea welwitschii*. Evidence-based Complementary and Alternative Medicine. 2013;**2013**:632927. DOI: 10.1155/2013/632927

[94] Pereira AM, Hernandes C, Pereira SI, Bertoni BW, França SC, Pereira PS, et al. Evaluation of anticandidal and antioxidant activities of phenolic compounds from *Pyrostegia venusta* (Ker Gawl.) Miers. Chemico-Biological Interactions. 2014;**224**:136-141. DOI: 10.1016/j.cbi.2014.10.023

[95] Nordin MA, Wan Harun WH, Abdul Razak F, Musa MY. Growth inhibitory response and ultrastructural modification of oral-associated candidal reference strains (ATCC) by *Piper betle* L. extract. International Journal of Oral Science. 2014;**6**(1):15-21. DOI: 10.1038/ijos.2013.97

[96] Isa AI, Awouafack MD, Dzoyem JP, Aliyu M, Magaji RA, Ayo JO, et al. Some *Strychnos spinosa* (Loganiaceae) leaf extracts and fractions have good antimicrobial activities and low cytotoxicities. BMC Complementary and Alternative Medicine. 2014;**14**:456. DOI: 10.1186/1472-6882-14-456

[97] Gonçalves MJ, Piras A, Porcedda S, Marongiu B, Falconieri D, Cavaleiro C, et al. Antifungal activity of extracts from *Cynomorium coccineum* growing wild in Sardinia island (Italy). Natural Product Research. 2015;**29**(23):2247-2250. DOI: 10.1080/14786419.2014.1000892

[98] Akroum S. Antifungal activity of acetone extracts from *Punica granatum* L., *Quercus* suber L. and *Vicia faba* L. Journal de Mycologie Médicale. 2017;**27**(1):83-89. DOI: 10.1016/j.mycmed.2016.10.004

[99] da Silva AR, de Andrade Neto JB, da Silva CR, Campos RS, Costa Silva RA, Freitas DD, et al. Berberine antifungal activity in fluconazole-resistant pathogenic yeasts: Action mechanism evaluated by flow cytometry and biofilm growth inhibition in *Candida* spp. Antimicrobial Agents and Chemotherapy. 2016;**60**(6):3551-3557. DOI: 10.1128/AAC.01846-15

[100] Correia AF, Silveira D, Fonseca-Bazzo YM, Magalhães PO, Fagg CW, da Silva EC, et al. Activity of crude extracts from Brazilian cerrado plants against clinically relevant *Candida* species. BMC Complementary and Alternative Medicine. 2016;**16**:203. DOI: 10.1186/s12906-016-1164-3

[101] Terças AG, Monteiro AS, Moffa EB, Dos Santos JRA, de Sousa EM, Pinto ARB, et al. Phytochemical

characterization of *Terminalia catappa* Linn. extracts and their antifungal activities against *Candida* spp. Frontiers in Microbiology. 2017;**8**:595. DOI: 10.3389/fmicb.2017.00595

[102] Bottari NB, Lopes LQ, Pizzuti K, Filippi dos Santos Alves C, Corrêa MS, Bolzan LP, et al. Antimicrobial activity and phytochemical characterization of *Carya illinoensis*. Microbial Pathogenesis. 2017;**104**:190-195. DOI: 10.1016/j.micpath.2017.01.037

[103] Shahzad M, Sherry L, Rajendran R, Edwards CA, Combet E, Ramage G. Utilising polyphenols for the clinical management of *Candida albicans* biofilms. International Journal of Antimicrobial Agents. 2014;**44**(3): 269-273. DOI: 10.1016/j.ijantimicag.2014.05.017

[104] Daglia M. Polyphenols as antimicrobial agents. Current Opinion in Biotechnology. 2012;**23**:174-181. DOI: 10.1016/j.copbio.2011.08.007

[105] Lopes G, Pinto E, Salgueiro L. Natural products: An alternative to conventional therapy for dermatophytosis? Mycopathologia. 2017;**182**:143-167. DOI: 10.1007/s11046-016-0081-9

[106] Seleem D, Pardi V, Murata RM. Review of flavonoids: A diverse group of natural compounds with anti-*Candida albicans* activity *in vitro*. Archives of Oral Biology. 2017;**76**:73-83. DOI: 10.1016/j.archoralbio.2016.08.030

[107] Herrera CL, Alvear M, Barrientos L, Montenegro G, Salazar LA. The antifungal effect of six commercial extracts of Chilean propolis on *Candida* spp. Ciencia e Investigación Agraria. 2010;**37**(1):75-84. DOI: 10.4067/S0718-16202010000100007

[108] Mulaudzi RB, Ndhlala AR, Kulkarni MG, Van Staden J. Pharmacological properties and protein binding capacity of phenolic extracts of some Venda medicinal plants used against cough and fever. Journal of Ethnopharmacology. 2012;**143**(1): 185-193. DOI: 10.1016/j.jep.2012.06.022

[109] Serpa R, França EJ, Furlaneto-Maia L, Andrade CG, Diniz A, Furlaneto MC. *In vitro* antifungal activity of the flavonoid baicalein against *Candida* species. Journal of Medical Microbiology. 2012;**61**(12):1704-1708. DOI: 10.1099/jmm.0.047852-0

[110] Duval A, Avérous L. Characterization and physicochemical properties of condensed tannins from *Acacia catechu*. Journal of Agricultural and Food Chemistry. 2016;**64**: 1751-1760. DOI: 10.1021/acs.jafc.5b05671

[111] dos Santos C, Vargas A, Fronza N, Dos Santos JHZ. Structural, textural and morphological characteristics of tannins from *Acacia mearnsii* encapsulated using sol-gel methods: Applications as antimicrobial agents. Colloids and Surfaces B: Biointerfaces. 2016;**151**: 26-33. DOI: 10.1016/j.colsurfb.2016.11.041

[112] Yoshida T, Amakura Y, Yoshimura M. Structural features and biological properties of ellagitannins in some plant families of the order Myrtales. International Journal of Molecular Sciences. 2010;**11**:79-106. DOI: 10.3390/ijms11010079

[113] Reddy MK, Gupta SK, Jacob MR, Khan SL, Ferreira D. Antioxidant, antimalarial and antimicrobial activities of tannin-rich fractions, ellagitannins and phenolic acids from *Punica granatum* L. Planta Medica. 2007;**73**: 461-467. DOI: 10.1055/s-2007-967167

[114] Yamaguchi MU, Garcia FP, Cortez DA, Ueda-Nakamura T, Filho BP, Nakamura CV. Antifungal effects of ellagitannin isolated from leaves of *Ocotea odorifera* (Lauraceae). Antonie

Van Leeuwenhoek. 2011;**99**:507-514. DOI: 10.1007/s10482-010-9516-3

[115] Klewicka E, Sójka M, Klewicki R, Kołodziejczyk K, Lipinska L, Nowak A. Ellagitannins from raspberry (*Rubus idaeus* L.) fruit as natural inhibitors of *Geotrichum candidum*. Molecules. 2016; **21**:908. DOI: 10.3390/molecules21070908

[116] Montagner C, Souza SM, Groposoa C, Monacheb FD, Smaniaa EFA, Smania A Jr. Antifungal activity of coumarins. Zeitschrift für Naturforschung. 2008; **63**:21-28. DOI: 10.1515/znc-2008-1-205

[117] Lu M, Li T, Wan J, Li X, Yuan L, Sun S. Antifungal effects of phytocompounds on *Candida* species alone and in combination with fluconazole. International Journal of Antimicrobial Agents. 2017;**49**(2): 125-136. DOI: 10.1016/j.ijantimicag.2016.10.021

[118] Navarro-García VM, Rojas G, Avilés M, Fuentes M, Zepeda G. *In vitro* antifungal activity of coumarin extracted from *Loeselia mexicana* Brand. Mycoses. 2011;**54**(5):e569-e571. DOI: 10.1111/j.1439-0507.2010.01993.x

[119] Raut JS, Shinde RB, Chauhan NM, Karuppayil SM. Phenylpropanoids of plant origin as inhibitors of biofilm formation by *Candida albicans*. Journal of Microbiology and Biotechnology. 2014;**24**(9):1216-1225. DOI: 10.4014/jmb.1402.02056

[120] Da Silva DL, Magalhães TF, Dos Santos JRA, De Paula TP, Modolo LV, De Fatima A, et al. Curcumin enhances the activity of fluconazole against *Cryptococcus gattii*-induced cryptococcosis infection in mice. Journal of Applied Microbiology. 2015;**120**: 41-48. DOI: 10.1111/jam.12966. ISSN 1364–5072

[121] Ferreira FD, Mossini SAG, Ferreira FMD, Arroteia CC, Costa CL, Nakamura CV, et al. The inhibitory effects of *Curcuma longa* L. essential oil and curcumin on *Aspergillus flavus* link growth and morphology. Scientific World Journal. 2013;**1**:1-6. DOI: 10.1155/2013/343804

[122] Alalwan H, Rajendran R, Lappin DF, Combet E, Shahzad M, Robertson D, et al. The anti-adhesive effect of curcumin on *Candida albicans* biofilms on denture materials. Frontiers in Microbiology. 2017;**8**:659. DOI: 10.3389/fmicb.2017.00659

[123] Paul S, Mohanram K, Kannan I. Antifungal activity of curcumin-silver nanoparticles against fluconazole-resistant clinical isolates of *Candida* species. Ayu. 2019;**39**(3):182-186. DOI: 10.4103/ayu.AYU_24_18

[124] Teodoro GR, Gontijo AVL, Salvador MJ, Tanaka MH, Brighenti FL, Delbem ACB, et al. Effects of acetone fraction from *Buchenavia tomentosa* aqueous extract and gallic acid on *Candida albicans* biofilms and virulence factors. Frontiers in Microbiology. 2018; **9**:647. DOI: 10.3389/fmicb.2018.00647

[125] Zabka M, Pavela R. Antifungal efficacy of some natural phenolic compounds against significant pathogenic and toxinogenic filamentous fungi. Chemosphere. 2013;**93**:1051-1056

[126] De Toledo LG, Ramos MADS, Spósito L, Castilho EM, Pavan FR, Lopes EDO, et al. Essential oil of *Cymbopogon nardus* (L.) Rendle: A strategy to combat fungal infections caused by *Candida* species. International Journal of Molecular Sciences. 2016;**17**(8):E1252. DOI: 10.3390/ijms17081252

[127] Mondello F, De Bernardis F, Girolamo A, Salvatore G, Cassone A. *In vitro* and *in vivo* activity of tea tree oil against azole-susceptible and -resistant human pathogenic yeasts. Journal of Antimicrobial Chemotherapy. 2003;**51**: 1223-1229. DOI: 10.1093/jac/dkg202

[128] Sharifzadeh A, Khosravi AR, Shokri H, Sharafi G. Antifungal effect of *Trachyspermum ammi* against susceptible and fluconazole-resistant strains of *Candida albicans*. Journal de Mycologie Médicale. 2015;**25**(2): 143-150. DOI: 10.1016/j.mycmed.2015.03.008

[129] Gavanji S, Zaker SR, Nejad ZG, Bakhtari A, Bidabadi ES, Larki B. Comparative efficacy of herbal essences with amphotericin B and ketoconazole on *Candida albicans* in the in vitro condition. Integrative Medicine Research. 2015;**4**: 112-118. DOI: 10.1016/j.imr.2015.01.003

[130] He X, Ma Y, Yi G, Wu J, Zhou L, Guo H. Chemical composition and antifungal activity of *Carica papaya* Linn. seeds essential oil against *Candida* spp. Letters in Applied Microbiology. 2017; **64**(5):350-354. DOI: 10.1111/lam.12711

[131] Minooeianhaghighi MH, Sepehrian L, Shokri H. Antifungal effects of *Lavandula binaludensis* and *Cuminum cyminum* essential oils against *Candida albicans* strains isolated from patients with recurrent vulvovaginal candidiasis. Journal de Mycologie Médicale. 2017; **27**(1):65-71. DOI: 10.1016/j.mycmed.2016.09.002

[132] Abu-Darwish MS, Cabral C, Gonçalves MJ, Cavaleiro C, Cruz MT, Zulfiqar A, et al. Chemical composition and biological activities of *Artemisia judaica* essential oil from southern desert of Jordan. Journal of Ethnopharmacology. 2016;**191**:161-168. DOI: 10.1016/j.jep.2016.06.023

[133] Köse YB, İşcan G, Göger F, Akalın G, Demirci B, Başer KHC. Chemical composition and biological activity of *Centaurea baseri*: New species from Turkey. Chemistry & Biodiversity. 2016;**13**(10):1369-1379. DOI: 10.1002/cbdv.201600070

[134] Sánchez MA, Turina AV, García DA, Nolan MV, Perillo MA. Surface activity of thymol: Implications for an eventual pharmacological activity. Colloids and Surfaces B: Biointerfaces. 2004;**34**:77-86. DOI: 10.1016/j.colsurfb.2003.11.007

[135] De Lira Mota KS, de Oliveira Pereira F, de Oliveira WA, Lima IO, de Oliveira Lima E. Antifungal activity of *Thymus vulgaris* L. essential oil and its constituent phytochemicals against *Rhizopus oryzae*: Interaction with ergosterol. Molecules. 2012;**17**:14418-14433. DOI: 10.3390/molecules171214418

[136] Romero AL, Romero RB, Silva EL, Diniz SPSS, Oliveira RR, Vida JB. Composição química e atividade do óleo essencial de *Origanum vulgare* sobre fungos fitopatogênicos. UNOPAR Científica Ciências Biológicas e da Saúde. 2012;**14**:231-235. DOI: 10.17921/2447-8938.2012v14n4p25p

[137] de Castro RD, de Souza TMP, Bezerra LM, Ferreira GL, Costa EM, Cavalcanti AL. Antifungal activity and mode of action of thymol and its synergism with nystatin against *Candida* species involved with infections in the oral cavity: An in vitro study. BMC Complementary and Alternative Medicine. 2015;**15**:417. DOI: 10.1186/s12906-015-0947-2

[138] Zore GB, Thakre AD, Jadhav S, Karuppayil SM. Terpenoids inhibit *Candida albicans* growth by affecting membrane integrity and arrest of cell cycle. Phytomedicine. 2011;**18**: 1181-1190. DOI: 10.1016/j.phymed.2011.03.008

[139] Fan JT, Kuang B, Zeng GZ, Zhao SM, Ji CJ, Zhang YM, et al. Biologically active arborinane-type triterpenoids and anthraquinones from *Rubia yunnanensis*. Journal of Natural Products. 2011; **74**(10):2069-2080. DOI: 10.1021/np2002918

[140] Mileva M, Krumova E, Miteva-Staleva J, Kostadinova N, Dobreva A,

Galabov AS. Chemical compounds, in vitro antioxidant and antifungal activities of some plant essential oils belonging to rosaceae family. Proceeding of the Bulgarian Academy of Sciences. 2014;**67**:1363-1368

[141] Dalleau S, Cateau E, Bergès T, Berjeaud JM, Imbert C. *In vitro* activity of terpenes against *Candida* biofilms. International Journal of Antimicrobial Agents. 2008;**31**(6):572-576. DOI: 10.1016/j.ijantimicag.2008.01.028

[142] Khan A, Ahmad A, Akhtar F, Yousuf S, Xess I, Khan LA, et al. Induction of oxidative stress as a possible mechanism of the antifungal action of three phenylpropanoids. FEMS Yeast Research. 2011;**11**:114-122. DOI: 10.1111/j.15671364.2010.00697.x

[143] De Jesus Faria T, Ferreira RS, Yassumoto L, De Souza JRP, Ishikawa NK, De Melo Barbosa A. Antifungal activity of essential oil isolated from *Ocimum gratissimum* L. (eugenol chemotype) against phytopathogenic fungi. Brazilian Archives of Biology and Technology. 2006;**49**:867-871

[144] Zhou L, Zhang Z, Wei M, Xie Y, He S, Shi H, et al. Evaluation of the antifungal activity of individual and combined monoterpenes against *Rhizopus stolonifer* and *Absidia coerule*. Environmental Science and Pollution Research. 2019;**26**:7804-7809. DOI: 10.1007/s11356-019-04278-z

[145] Newman DJ, Cragg GM. Natural products as sources of new drugs over the 30 years from 1981 to 2010. Journal of Natural Products. 2012;**75**:311-335. DOI: 10.1021/np200906s

[146] Johnson MD, Perfect JR. Use of antifungal combination therapy: Agents, order, and timing. Current Fungal Infection Reports. 2010;**4**:87-95

[147] Lewis RE, Kontoyiannis DP. Rationale for combination antifungal therapy. Pharmacotherapy. 2001;**21**: 149S-164S. DOI: 10.1592/phco.21.12.149S.34505

[148] Bink A, Pellens K, Cammue B, Thevissen K. Antibiofilm strategies: How to eradicate *Candida* biofilms. The Open Mycology Journal. 2011;**5**:29-38. DOI: 10.2174/1874437001105010029

[149] Santos JRA, Ribeiro NQ, Bastos RW, Holanda RA, Silva LC, Queiroz ER, et al. High-dose fluconazole in combination with amphotericin B is more efficient than monotherapy in murine model of cryptococcosis. Scientific Reports. 2017;7:4661

[150] Serena C, Fernández-Torres B, Pastor FJ, Trilles L, Lazéra MS, Nolard N, et al. In vitro interactions of micafungin with other antifungal drugs against clinical isolates of four species of *Cryptococcus*. Antimicrobial Agents and Chemotherapy. 2005;**49**:2994-2996. DOI: 10.1128/AAC.49.7.2994-2996.2005

[151] Espinel-Ingroff A. Novel antifungal agents, targets or therapeutic strategies for the treatment of invasive fungal diseases: A review of the literature (2005–2009). Revista Iberoamericana de Micología. 2009;**26**:15-22. DOI: 10.1016/S1130-1406(09)70004-X

[152] Katragkou A, McCarthy M, Meletiadis J, Hussain K, Moradi PW, Strauss GE, et al. In vitro combination therapy with isavuconazole against *Candida* spp. Medical Mycology. 2017; **55**:859-868. DOI: 10.1093/mmy/myx006

[153] Chen YL, Lehman VN, Averette AF, Perfect JR, Heitman J. Posaconazole exhibits in vitro and in vivo synergistic antifungal activity with caspofungin or FK506 against *Candida albicans*. PLoS One. 2013;**8**(3):e57672. DOI: 10.1371/journal.pone.0057672

[154] Chaturvedi V, Ramani R, Andes D, Diekema DJ, Pfaller MA, Ghannoum MA, et al. Multilaboratory testing of

two-drug combinations of antifungals against *Candida albicans*, *Candida glabrata*, and *Candida parapsilosis*. Antimicrobial Agents and Chemotherapy. 2011;**55**(4):1543-1548. DOI: 10.1128/AAC.01510-09

[155] Barchiesi F, Falconi DFL, Scalise G. *In vitro* activities of terbinafine in combination with fluconazole and itraconazole against isolates of *Candida albicans* with reduced susceptibility to azoles. Antimicrobial Agents and Chemotherapy. 1997;**41**(8):1812-1814

[156] Perea S, Gonzalez G, Fothergill AW, Sutton DA, Rinaldi MG. *In vitro* activities of terbinafine in combination with fluconazole, itraconazole, voriconazole, and posaconazole against clinical isolates of *Candida glabrata* with decreased susceptibility to azoles. Journal of Clinical Microbiology. 2002; **40**(5):1831-1833. DOI: 10.1128/JCM.40.5.1831-1833.2002

[157] Cui J, Ren B, Tong Y, Dai H, Zhang L. Synergistic combinations of antifungals and anti-virulence agents to fight against *Candida albicans*. Virulence. 2015;**6**(4):362-371. DOI: 10.1080/21505594.2015.1039885

[158] Pappas PG, Kauffman CA, Andes DR, Clancy CJ, Marr KA, Ostrosky-Zeichner L, et al. Clinical practice guideline for the management of candidiasis: 2016 update by the Infectious Diseases Society of America. Clinical Infectious Diseases. 2016;**62**:e1-e50. DOI: 10.1093/cid/civ1194

[159] Rees CA, Baoa R, Zegansa ME, Cramera RA. Natamycin and voriconazole exhibit synergistic interactions with non-antifungal ophthalmic agents against *Fusarium* species ocular isolates. Antimicrobial Agents and Chemotherapy. 2019. DOI: 10.1128/AAC.02505-18

[160] Mukherjee PK, Sheehan DJ, Hitchcock CA, Ghannoum MA. Combination treatment of invasive fungal infections. Clinical Microbiology Reviews. 2005;**18**:163-194. DOI: 10.1128/CMR.18.1.163-194.2005

[161] Carrillo-Muñoz AJ, Finquelievich J, Tur-Tur C, Eraso E, Jauregizar N, Quindós G, et al. Combination antifungal therapy: A strategy for the management of invasive fungal infections. Revista Española de Quimioterapia. 2014;**27**(3):141-158

[162] Musiol R, Mrozek-Wilczkiewicz A, Polanski J. Synergy against fungal pathogens: Working together is better than working alone. Current Medicinal Chemistry. 2014;**21**(7):870-893

[163] Tangarife-Castaño V, Correa-Royero C, Zapata-Londoño B, Duran C, Stanshenko E, Mesa-Arango AC. Anti-*Candida albicans* activity, cytotoxicity and interaction with antifungal drugs of essential oils and extracts from aromatic and medicinal plants. Infectio. 2011;**15**:160-167. DOI: 10.1016/S0123-9392(11)70080-7

[164] Chanda S, Rakholiya K, Dholakia K, Baravalia Y. Antimicrobial, antioxidant, and synergistic properties of two nutraceutical plants: *Terminalia catappa* L. and *Colocasia esculenta* L. Turkish Journal of Biology. 2013;**37**:81-91. DOI: 10.3906/biy-1203-41

[165] Santos KKA, Matias EFF, Sobral-Souza CE, Tintino SR, Morais-Braga MFB, Guedes GMM, et al. Trypanocide, cytotoxic, and anti-*Candida* activities of natural products: *Hyptis martiusii* Benth. European Journal of Integrative Medicine. 2013;**5**:427-431. DOI: 10.1016/j.eujim.2013.06.001

[166] Avijgan M, Mahboubi M, Nasab MM, Nia EA, Yousefi H. Synergistic activity between *Echinophora platyloba* DC ethanolic extract and azole drugs against clinical isolates of *Candida albicans* from women suffering chronic recurrent vaginitis. Journal de

Mycologie Médicale. 2014;**24**:112-116. DOI: 10.1016/j.mycmed.2014.01.116

[167] Moraes RC, Lana AJD, Kaiser S, Carvalho AR, Oliveira LFS, Fuentefria AM, et al. Antifungal activity of *Uncaria tomentosa* (Willd.) D.C. against resistant non-albicans *Candida* isolates. Industrial Crops and Products. 2015;**69**:7-14. DOI: 10.1016/j.indcrop.2015.01.033

[168] Ngouana TK, Mbouna CDJ, Kuipou RMT, Tchuenmogne MAT, Zeuko'o EM, Ngouana V, et al. Potent and synergistic extract combinations from *Terminalia catappa, Terminalia mantaly* and *Monodora tenuifolia* against pathogenic yeasts. Medicine. 2015;**2**:220-235. DOI: 10.3390/medicines2030220

[169] Cavalcanti Filho JR, Silva TF, Nobre WQ, Oliveira De Souza LI, Silva E Silva FCS, Figueiredo RC, et al. Antimicrobial activity of *Buchenavia tetraphylla* against *Candida albicans* strains isolated from vaginal secretions. Pharmaceutical Biology. 2017;**55**(1): 1521-1527. DOI: 10.1080/13880209.2017.1304427

[170] Kumari P, Mishra R, Arora N, Chatrath A, Gangwar R, Roy P, et al. Antifungal and anti-biofilm activity of essential oil active components against *Cryptococcus neoformans* and *Cryptococcus laurentii*. Frontiers in Microbiology. 2017;**8**:2161. DOI: 10.3389/fmicb.2017.02161

[171] Jandu JJB, Costa MC, Santos JRA, Andrade FM, Magalhães TFF, Gomes AG, et al. Treatment with pCramoll alone and in combination with fluconazole provides therapeutic benefits in *C. gattii* infected mice. Frontiers in Cellular and Infection Microbiology. 2017;**7**:211. DOI: 10.3389/fcimb.2017.00211

[172] Sharifzadeh A, Khosravi AR, Shokri H, Shirzadi H. Potential effect of 2-isopropyl-5-methylphenol (thymol) alone and in combination with fluconazole against clinical isolates of *Candida albicans*, *C. glabrata* and *C. krusei*. Journal de Mycologie Médicale. 2018;**28**(2):294-299. DOI: 10.1016/j.mycmed.2018.04.002

[173] Zaidi KU, Shah F, Parmar R, Thawani V. Anticandidal synergistic activity of *Ocimum sanctum* and fluconazole of azole resistance strains of clinical isolates. Journal de Mycologie Médicale. 2018;**28**(2):289-293. DOI: 10.1016/j.mycmed.2018.04.004

[174] Carbone C, Teixeira MC, Sousa MC, Martins-Gomes C, Silva AM, Souto EMB, et al. Clotrimazole-loaded Mediterranean essential oils NLC: A synergic treatment of *Candida* skin infections. Pharmaceutics. 2019;**11**:231. DOI: 10.3390/pharmaceutics11050231

[175] Konopka K, Goslinski T. Photodynamic therapy in dentistry. Journal of Dental Research. 2007;**86**: 694-707

[176] Sanitá PV, Pavarina AC, Dovigo LN, Ribeiro APD, Andrade MC, Mima E. Curcumin-mediated anti-microbial photodynamic therapy against *Candida dubliniensis* biofilms. Lasers in Medical Science. 2018;**33**:709-717. DOI: 10.1007/s10103-017-2382-8

[177] Daliria F, Azizia A, Goudarzib M, Lawafc S, Rahimid A. *In vitro* comparison of the effect of photodynamic therapy with curcumin and methylene blue on *Candida albicans* colonies. Photodiagnosis and Photodynamic Therapy. 2019;**26**: 193-198. DOI: 10.1016/j.pdpdt.2019.03.017

Chapter 12

Modulation of Edible Plants on Hepatocellular Carcinoma Induced by Aflatoxin B_1

Peeradon Tuntiteerawit, Tichakorn Singto, Anupon Tadee and Supatra Porasuphatana

Abstract

Aflatoxin B_1 (AFB_1) is one of the major causes of liver cancer especially hepatocellular carcinoma that has high incidence and mortality rate in many countries. Owing to the climate that is suitable for fungal growth, the avoidance of AFB_1 exposure from agricultural product contamination is too difficult. This up-to-date review aims to collect insight on how edible plants attenuate AFB_1-toxicity. Cruciferous vegetables, green tea, purple rice, turmeric, green vegetables, ginger, *Dialium guineense*, *Parkia biglobosa*, carotenoid-rich fruits and vegetables, Allii Fistulosi Bulbus, and rosemary have reported their capabilities to alleviate AFB_1-toxicity though several mechanisms. All these plants showed anti-genotoxic activity while some of them are able to reduce hepatotoxicity, liver cancer, and oxidative stress and modulate metabolism enzymes induced by AFB_1. Furthermore, a few edible plants could handle AFB_1 in pre-exposure phase including anti-AFB_1 biosynthesis and AFB_1 absorption. Although the detoxification mechanisms of AFB_1 activated by various edible plants have been investigated in pre-clinical study for a decade, clinical trial is still rarely clarified. Further study associating with a protective effect on AFB_1 toxicity still needs to be carried out especially in the clinical study.

Keywords: mycotoxins, liver cancer, natural products, detoxification, chemoprevention

1. Introduction

Liver cancer is the third most common cause of cancer mortality worldwide especially in developing countries. Epidemiological studies among all continents found that Asia and Africa have higher incidence rate than western world [1]. Hepatocellular carcinoma (HCC), arise due to excessive growth of abnormal liver cells, is most commonly found among all liver cancer types [2]. Four main potential causes of HCC have been identified as viral infection (chronic hepatitis B and C), metabolic syndrome (diabetes and nonalcoholic fatty liver disease), immune-related disease (autoimmune hepatitis), and toxic substances (alcohol and aflatoxins) [3].

Aflatoxin B_1 (AFB_1) is a noxious carcinogen produced by certain fungi *Aspergillus flavus* and *A. parasiticus* which mostly contaminate in agricultural products such as rice, chili, and peanuts. AFB_1 is a Class 1 carcinogen classified by the International Agency for Research on Cancer (IARC), suggesting sufficient evidence of carcinogenicity caused by AFB_1 in both animals and human [4]. Consequently, it is considered as a serious contaminant in many foodstuffs.

Once the AFB_1 is absorbed through human body, it is metabolized at the liver site by phase I metabolizing enzymes including hydroxylation, hydration, demethylation, and epoxidation. Nontoxic metabolites are resulted from hydroxylation, hydration, and demethylation while the reactive metabolite, AFB_1-8,9-epoxide, is resulted from epoxidation [3, 4]. AFB_1-8,9-epoxide is the genotoxic form and can react efficiently with DNA at the N7 site of guanine to form AFB_1 adduct. This adduct can adversely affect DNA sequence and genetic materials. However, human defensive mechanisms are able to detoxify AFB_1 toxicity through phase II metabolism enzymes. AFB_1 can be converted into excretable forms after binding with glutathione and glucuronic acid generated by specific enzyme, glutathione S-transferase (GST) and UDP-glucuronosyltransferase (UGT), respectively [5–7]. Besides acute toxicity such as hepatic necrosis, bile drug proliferation, edema, and lethargy could also be observed after exposure to high dose of AFB_1 [8].

Regarding the current situation, there are many ways to avoid the risk of AFB_1-induced liver cancer as determined by two main periods, pre- and post-harvest period and exposure period [6]. During harvest time, several techniques are used for controlling and reducing the chance of harmful effects resulted from AFB_1: cultivation of AFB_1 tolerance plants, biocontrol using competitive fungi, irrigation, and insecticide. For exposure period, most researches aim to determine the effects of several foods or supplementary foods that are capable of decreasing AFB_1-induced toxicity. For example, oltipraz, a synthetic derivative of natural compound originated from cruciferous vegetables, is reported on its capacity to reduce AFB_1 toxicity. In addition, green tea polyphenol and chlorophyllin (a derivative of chlorophyll found in green leafy vegetables) are also stated. These natural compounds have a potential against AFB_1-induced hepatocarcinogenicity by decreasing the absorption of AFB_1, controlling metabolic pathway, and increasing AFB_1 excretion [6, 9]. To update the involvement of edible plants as chemoprevention for AFB_1, this review is aimed to emphasize the mechanistic alleviation of AFB_1-induced liver toxicity by polyphenol-containing plants.

2. Effects of edible plants on toxicity induced by AFB_1

2.1 Cruciferous vegetables

Cruciferous vegetables belong to *Brassica* genus, Brassicaceae family which are usually known as broccoli, Brussels sprouts, cabbage, cauliflower, kale, and radishes and commonly used for food consumption. They are not only rich sources of fibers, vitamins, and carotenoids as their important components, but also contain higher glucosinolate content than other vegetables [10]. Glucosinolates are secondary metabolites in cruciferous veggies and can be divided into three classes based on their structure: aliphatic glucosinolates, indole glucosinolates, and aromatic glucosinolates.

Nearly 200 types of glucosinolates have been reported in scientific literature, especially glucobrassicin and glucoraphanin. These two compounds can be transformed into hydrolysis products such as isothiocyanates, sulforaphane (SF), and indole-3-carbinol (I3C) by β-thioglucosidase (myrosinase) enzyme when plant cells are damaged. This mechanism could also be processed by bacteria in the gastrointestinal tract [11, 12].

The studies of anticancer effects of glucosinolates and their hydrolysis products revealed that numerous existing compounds also had anticancer mechanism against various types of cancers. For instance, the presence of sulforaphane could suppress carcinogen and prevent DNA adduct (a biomarker of AFB_1 exposure) directly through an inhibition of phase I metabolism enzymes. At the same time, it induces phase II metabolism enzymes which play an important role in converting carcinogens to the inactive metabolites and excreting from the body. Their hydrolysis products exhibit an ability to scavenge the free radicals, inhibit inflammation and angiogenesis, and also induce an apoptosis of cancer cells [11].

Previous studies investigated the effects of bioactive compounds such as I3C and 1-cyano-2 hydroxy-3 butene (Crambene), derivatives of glucosinolate group found in cruciferous veggies, on HCC occurrence. Glucosinolates did not only respond for abnormal liver cells, but they also enhance AFB_1 detoxification in the rat model. Pre-exposure to the high-dose combination of I3C and Crambene (0.15 and 0.165%, respectively) protected the liver cells effectively more than low-dose combinations and single exposure [13]. Risk reduction of liver cancer could also be observed in rainbow trout when pre-exposed to I3C at the dose 2000 ppm prior to AFB_1; however, the adverse effects and increase of liver cancer incidence were reported when the exposure sequence was reversed [14]. In addition, further studies revealed a dose-dependent relationship between I3C dose after exposure to AFB_1 and the incidence of liver cancer and other cancer types [15]. Thus, it could be summarized that the incidence of liver cancer is induced by AFB_1 relating to timing of I3C exposure. Pre-exposure to I3C prior to AFB_1 reduced the liver cancer incidence, but post-exposure reversely raised the liver cancer incidence [15]. Accordingly, subsequent mechanistic studies indicated an induction of I3C on phase I and II metabolism enzyme activities [16]. Continuous exposure to I3C might enhance phase II enzyme activity, so the absorbed AFB_1 would be excreted rapidly. In contrast, pre-exposure to AFB_1 triggered the adverse effects such as DNA abnormality and increase of liver cancer risk. The explanation was that pre-exposure to AFB_1 generates AFB_1-8,9-epoxide and this reactive metabolite would be more activated when treated later with I3C. In addition, I3C could be able to induce both phase I and II metabolism enzyme activities, thus AFB_1-8,9-epoxide was more generated as a result of activation of phase I metabolism. Although phase II enzyme was also stimulated, it was not enough to eliminate AFB_1.

Not only I3C is frequently reported, but other glucosinolate derivatives like SF and H-1,2-dithiole-3-thione (D3T) are also stated. For example, while rats were pre-exposed to these derivatives, AFB_1-DNA adduct in rat's liver was reduced due to an increase of GST activity, a phase II detoxification enzyme for AFB_1 [17]. Likewise, other previous studies reported that SF could competitively inhibit CYP1A2 in human liver cells [16], causing a decrease of AFB_1-DNA adduct. Remarkably, upregulation of gene expression-related tissue repairing system and number of hepatocytes were observed after induction of SF [18].

The current epidemiological and clinical studies revealed that only lung, colorectal, breast, prostate, and pancreatic cancers were given the positive response to glucosinolates while animal model showed the effective inhibition of liver cancer and other cancer types through various mechanisms. Nevertheless, randomized clinical trial of glucosinolates on liver cancer showed different results [11, 19]; comparison between broccoli sprout extract treatments and control group was studied simultaneously. After treatment, AFB_1-DNA adducts were clearly determined. The results indicated that no significant difference was observed among tested groups on AFB_1-DNA adduct level ($p = 0.68$). On the contrary, an inverse linear correlation of dithiocarbamates, a metabolite of sulforaphane, and AFB_1-DNA adduct excretion was noted ($p = 0.002$, $R = 0.31$). It can be implied that exposure to glucosinolates might decrease AFB_1-induced toxicity [20]. Besides, various compounds of glucosinolates have the potential to increase excretion of many carcinogens through glutathione *S*-transferase stimulation. Once the GST was stimulated, carcinogenicity and risk of diseases in human were also decreased [21].

2.2 Green tea

Green tea, *Camellia sinensis,* is a beverage that contains high contents of phenolic compounds at approximately 30% of dry weight. One of the major phenolic compounds in green tea is catechin, particularly epigallocatechin gallate (EGCG), epigallocatechin (EGC), epicatechin gallate (ECG), and epicatechin (EC) [22].

Recent studies have demonstrated the positive effects of green tea on many diseases and adverse human health conditions such as coronary artery disease, oral heath, bone integrity, thermoregulation balance, and kidney stones. Furthermore, an association between green tea consumption and the incidence of many types of cancer has also been reported such as oral and pharynx, esophageal, gastric, colorectal, bladder, prostate, breast, lung, skin, leukemia, pancreatic, and liver cancers [23, 24]. Various research methods including preclinical studies (*in vitro* and *in vivo*), epidemiology, and clinical trial were used to investigate the effect of crude green tea extract or single compound like EGCG on many types of cancer [25]. Overall, an anti-cancer mechanism of green tea extracts against cancer cells was evidently elucidated. Green tea extracts were able to induce apoptosis of cancer cells through inhibiting nuclear factor kappa light chain enhancer of activated B cells (NF-κB) activity and B-cell lymphoma extra-large (Bcl-xL) mRNA expression. Besides, the reduction of angiogenesis of cancer cells was also resulted by green tea extracts through inhibition of vascular endothelial growth factor (VEGF) expression [26].

Previous studies have been reported on several protective ways against AFB_1-induced liver cancer from the exposure to catechin compounds and green tea extracts. For example, the reduction of chromosome aberration in rat bone marrow cells was observed after pre-exposure with green tea or EGCG for 24 hours prior to AFB_1 [27]. Besides, hepatic nuclear AFB_1-DNA binding and glutathione S-transferase placental form (GST-P) positive single hepatocyte, specific markers of hepatocarcinogenic potential in the rat model, were also reduced after pre-exposure with green tea extracts for 2–4 weeks prior to AFB_1 [28]. Similarly, the levels of GST-P and γ-glutamyl transpeptidase positive hepatic foci induced by AFB_1 and carbon tetrachloride were reduced during pre- or co-treatment with green tea extracts. Furthermore, the inhibition of hepatocarcinogenesis was also observed [29].

The studies of green tea against AFB_1-induced human liver cancer are still currently limited, and most reports have been retrieved from China. As some Chinese commonly consume food contaminated with AFB_1, the risk of HCC is higher than other regions. A clinical study demonstrated a protective effect of 500 and 1000 mg/day green tea polyphenol (GTP) on hepatocarcinogenesis in 124 HCC patients who presented with HBsAg and aflatoxin-albumin adducts. Results showed that 8-hydroxydeoxyguanosine (8-OHdG) level, an oxidative DNA damage biomarker originating in urine specimens, significantly decreased ($p = 0.007$) during co-exposure with GTP for 3 months [30]. Besides, AFB_1-albumin adducts (AFB_1-AA) and AFB_1-mercapturic acid (AFB_1-NAC) level in blood and urine specimens of volunteers were compared among 500 and 1000 mg GTP treatment group and control group. This result revealed a reduction of AFB_1-AA level, an indicator of AFB_1 exposure, for both 500 and 1000 mg GTP treatment groups within 3 months. This reduction was strongly related to dose and duration of GTP exposure ($p = 0.049$). Furthermore, AFB_1-NAC, an indicator of AFB_1 elimination activated by phase II metabolism enzymes, significantly increased ($p < 0.001$) in both treatment groups related to dose and duration of GTP exposure as well

($p < 0.001$). Therefore, it could be summarized that GTP effectively modulated AFB_1 biotransformation by inhibition of phase I metabolism enzymes as can be seen from the reduction of AFB_1-AA. GTP also has an induction effect to phase II metabolism enzymes which transform AFB_1–8,9-epoxide to AFB_1-NAC [31].

Furthermore, results from a meta-analysis investigating the effect of green tea extracts on HCC and other liver diseases also showed that regular green tea drinkers had a lower incidence of HCC than nonregular drinkers approximately 26% ($R = 0.74$, 95% CI = 0.56–0.97, $p = 0.027$). Although there were some inconsistent results in this study ($I^2 = 80.1\%$, $p = 0.000$), no publication bias was detected and no data from one study significantly influenced the final conclusion [25].

2.3 Purple rice

Anthocyanins, members of flavonoid groups, are mostly found in blue, purple, orange, and red vegetables. Anthocyanins in plants play a vital role in attraction of bugs for pollination and insect resistance [32]. Pharmacologically, purple corn extracts have been known for its anti-diabetic and antiadipogenic effects, anti-prostate carcinogenesis, and others [33–35] while blue butterfly pea flower has a definite potential anti-inflammatory effect [36]. Furthermore, anthocyanin-rich plants were shown to protect neurodegenerative and also cardiovascular disease [37].

Purple rice bran (*Oryza sativa* L. var. *indica*) contained flavonoids and anthocyanins approximately 53 and 2 mg/g, respectively. Both compounds were reported to reduce AFB_1-induced toxicity, and they were capable of inhibiting mutagenicity in *Salmonella typhimurium* strains TA98 and TA100 [38]. In animal model, rats were pre-treated with purple rice bran extracts for a month before exposure to AFB_1. Then, the expression of CYP450 including CYP1A2 and CYP3A was investigated; both of them have an identical role in transforming AFB_1 to AFB_1-epoxide. The results showed that the extracts could not only inhibit the expression of CYP1A2 and CYP3A, but also increase the expression of GST and UGT which encouraged AFB_1 excretion. Further in *in vivo* studies, the genotoxicity was evaluated by micronucleus assay, and the result showed lower micronucleus formation in extract-pretreated group than AFB_1 treated alone, confirming the capability of purple rice bran extract on the prevention of AFB_1-induced genotoxicity [39].

Apart from purple rice bran extract, other anthocyanin-rich plants are also studied for their effects on AFB_1-induced cytotoxicity. For instance, *Lannea microcarpa,* a tropical African plant, has been studied for its activities against hepatotoxicity, DNA fragmentation, and oxidative stress induced by AFB_1. Before exposure to AFB_1, animals were pre-exposed with *Lannea microcarpa* extracts for 6 months. Results showed that hepatotoxicity, DNA fragmentation, and oxidative stress was lower in extract-pretreated group when compared to AFB1-treated group [40].

2.4 Turmeric

Turmeric is a flowering plant widely used as a food ingredient in South Asia for a long period of time. It has been also applied in pharmacognosy field as a powerful anti-inflammatory resulting from rheumatoid arthritis, bruise, epilepsy, abdominal pain or discomfort, and asthma [41]. An *in vivo* study of turmeric clearly showed the anticancer properties of turmeric on liver, skin, and colorectal cancers. It has a strong potential to inhibit cancer cell growth through stimulating apoptosis and inhibiting phase I metabolism enzymes. It can also stimulate phase II metabolism enzyme activities which play an important role in converting reactive metabolites to excretable forms. Also, turmeric exhibits the antioxidant capacity which can effectively detoxify oxidative stress [42].

Curcumin is a major active component of turmeric. It belongs to curcuminoid group and commonly found in 2–8%. Previous *in vivo* studies investigated the effects of turmeric and curcumin on AFB_1-induced toxicity, and results showed that turmeric and curcumin decreased AFB_1-adduct formation, biomolecule damage, and hepatotoxicity [43–46], and it also inhibited acute toxicity through disturbing the lysis of erythrocytes [47]. During AFB_1 metabolism, free radicals generated by AFB_1 could be readily inhibited by turmeric and curcumin via decreasing lipid peroxidation and enhancing glutathione content. Likewise, they could activate several antioxidant enzymes such as glutathione peroxidase (GPx), superoxide dismutase (SOD), catalase (CAT), GST, and UGT which play a fundamental role in converting AFB_1 to excretable forms [43–46].

Turmeric is found to be capable of reducing both AFB_1-induced toxicity and HCC. Besides, it could also stimulate apoptosis of liver cancer cells through a mitochondria-dependent pathway and accumulation of calcium ions within the cells [48]. Turmeric showed the protective effect against AFB_1-induced liver cancer in animal model by inhibition of metastasis and growth factor expression related to the progression of angiogenesis [49].

2.5 Green vegetables

Chlorophyll (chla), a main component of green vegetables, consists of a porphyrin ring structure where magnesium is the central atom of the ring. Chla is important for plants' photosynthesis pathway and used as food additives. One of the characteristics of chla is almost insoluble in water while chlorophyllin (CHL), a derivative of chla, is completely soluble. CHL can be transformed into water-soluble form by saponification, a reaction that magnesium central atom is replaced with copper. *In vivo* and clinical studies in pharmacological researches of both chla and CHL revealed that they provided the therapeutic uses such as wound healing, anti-inflammation, anti-oxidation, anti-mutagenesis, and anti-carcinogenesis [50, 51].

Previous studies on the protective effects of chla and CHL on AFB_1 toxicity indicated that both compounds could reduce absorption of AFB_1 from apical to basolateral sides in Caco-2 cell line [52]. Accordingly, a crossover clinical trial demonstrated that chla and CHL exposure could reduce maximum concentration (C_{max}) and area under the curves (AUC) of AFB_1 compared to untreated group [53]. These findings suggest that chla and CHL have a strong potential to decrease AFB_1 absorption. The effects of chla and CHL co-exposure with AFB_1 have also been studied in animal model by emphasizing on antioxidant activities. Both bioactive compounds are capable of reducing AFB_1 toxicity through enhancing the expression of glutathione level and several antioxidant enzyme activities such as GPx, SOD, and CAT [54].

A recent study investigated the effects of CHL on AFB_1-induced hepatotoxicity and incidence of carcinogenesis in animal model. Exposure with CHL reduced hepatotoxicity and incidence of liver cancer [54, 55]. In a clinical study, a randomized controlled trial reported that daily exposure with CHL for 4 months decreased AFB_1-N7-guanine level in urine compared to placebo group [56].

Several studies were in agreement that chla and CHL reduce AFB_1-induced liver cancer through decreasing AFB_1 absorption in digestive tract contributing to the decrease of AFB_1 bioavailability. Besides, chla and CHL are the powerful antioxidants which effectively lower AFB_1-induced oxidative stress. These two compounds not only reduce hepatotoxicity, but also incidence of liver cancer. Thus, the consumption of green vegetables is one of the alternatives to reduce toxicity caused by consuming AFB_1-contaminated foods.

2.6 Ginger (*Zingiber officinale* Roscoe)

Ginger (*Zingiber officinale* Roscoe) contained high content of phenolic compounds in which 6-gingeerol and 6-shogaol are main constitutions [57]. Ginger plays a critical role as hepatoprotective effects through antioxidant mechanism; for example, liver injury by administration of country-made liquor (CML) and iron-induced nonalcoholic fatty liver disease (NAFLD) [58] and liver cirrhosis induced by carbon tetrachloride [59]. It was also reported to show the protective effects against AFB_1-induced toxicity.

In in vitro model of AFB_1-treated HepG2 cells, ginger extract-pretreated cells exhibited higher percent cell viability and lower intracellular ROS production and DNA strand break when compared to AFB_1 treatment alone. In Wistar rats, pretreatment with ginger extract also increased the activities of antioxidant enzymes: GPx, GST, CAT, and SOD, decreased malondialdehyde (MDA) level, and increased reduced glutathione (GSH) content. Co-incubation with ginger extract along with AFB_1 also showed a hepatoprotective effect as seen by the lower level of serum enzymes: alanine aminotransferase (ALT), aspartate transaminase (AST), alkaline phosphatase (ALP), and lactate dehydrogenase (LDH). Moreover, fat droplets and hepatocyte infiltration with macro-vesicles in liver induced by AFB_1 were normalized when pre-treated with ginger extract, clearly showing the effectiveness of ginger on AFB1-induced hepatotoxicity [57].

Mechanism of ginger extract to reduce AFB_1-induced hepatotoxicity was demonstrated by *in vivo* study. The expression of nuclear factor-E2-related factor 2 (Nrf2), a redox-responsive transcription factor, was increased when pre-treated with ginger extract. Nrf2 was translocated into the nucleus to regulate the antioxidant response element (ARE) which is the promotor of detoxification and antioxidant genes. Moreover, administration of ginger extract induced the expression of heme oxygenase 1 (HO-1) which is associated with the normalization of redox status [57]. Therefore, ginger extract could reduce AFB_1-induced hepatotoxicity in both *in vitro* and *in vivo* through antioxidant activities controlled by the function of Nrf2 and HO-1.

2.7 Plants in family Fabaceae

2.7.1 *Dialium guineense*

Dialium guineense is a fruit-bearing tree known as the velvet tamarind. Their bark, leaves, seeds, and fruit showed biological properties such as antimicrobial activities, anti-infectious diseases, and wound-healing [60, 61]. Extract from *Dialium guineense* showed ROS scavenging activities and could normalize the levels of enzyme biomarkers of hepatotoxicity: ALP, AST, and AST induced by AFB_1. Furthermore, treatment of velvet tamarind extract before AFB_1 exposure increased the antioxidant activities of various enzymes including SOD, GPx, GR, CAT, GSH, and oxidized glutathione (GSSH), decreased lipid peroxidation, protein carbonyl and DNA fragmentation. *In vitro* and *in vivo* experiments have also confirmed the protective effects of velvet tamarind extract against hepatotoxicity induced by AFB_1 via antioxidant properties [62].

2.7.2 *Parkia biglobosa*

Parkia biglobosa, known as the African locust bean tree (ALBT), is a perennial tree legume growing in West Africa. Several parts of ALBT (bark, leaves, pods, stem, and fruit pulp) showed medicinal properties such as antimicrobial activities, antihypertensive effects, antidiabetic activity, antidiarrheal activity, and others [63]. Pulp extract of ALBT exhibited abilities against antioxidant imbalance induced by AFB_1. When pretreated in animal model, pulp extract of ALBT were capable of inducing SOD, CAT, GPx, GR, and glucose-6-phosphate dehydrogenase (G6PD) activities and increasing GSH and GSSG content. In addition, pretreatment with pulp extract of ALBT reduced lipid peroxidation products, protein carbonyl, and DNA fragmentation induced by AFB_1. AFB_1 treatment also resulted in decrease of hepatocellular enzyme activities: ALP, ALT, and AST compared to control while the pretreatment with pulp extract of ALBT increased these enzyme activities in a dose-dependent manner. Accordingly, antioxidant imbalance and hepatotoxicity induced by AFB_1 were able to be alleviated by pretreatment with pulp extract of ALBT [64].

2.8 Carotenoid-rich fruits and vegetables

Carotenoids, natural plant pigments giving the color of fruits and vegetables, are responsible for the red, orange, and yellow colors in mangoes, corns, carrots, pumpkins, tomatoes, etc. More than 700 different carotenoids have long been characterized and classified as two main groups regarding their basic functional group [65]. Xanthophylls, yellow or orange-yellow pigments, are found widely in nature and the majority of their structure consists of oxygen as the core element such as lutein and zeaxanthin. Carotenes, one of another division of carotenoids, are hydrocarbon compounds without other functional groups including α-carotene, β-carotene, and lycopene [66]. Both xanthophylls and carotenes are almost known as fat-soluble compounds dissolved well in petroleum, ether, chloroform, and hexane but carotenes seem to be more soluble in these nonpolar aliphatic solvents compared to xanthophylls; some are water-soluble [67]. Carotenoids have a potential role as a provitamin A compound which can be converted within the body to vitamin A, and they are broadly accepted as free radical antioxidants inhibiting several types of cancers [68, 69].

Several carotenoids like β-carotene, canthaxanthin, lycopene, and cryptoxanthin were studied on the mitigation of AFB_1-induced mutagenesis in bacterial mutation assay. Mutagenesis was inhibited by the addition of all carotenoids, except lycopene, and cryptoxanthin was shown to be the most potent inhibitor among all tested carotenoids [70]. The comparison of both ionone rings, α and β type of carotenoids, was observed through suspended disc culture. The α-ionone ring carotenoids, α-carotene, lutein, or α-ionone, showed more inhibition of AF biosynthesis than β-ionone ring, and the existence of hydroxyl groups on the rings seemed to lessen the inhibition capacity [71].

Previous study demonstrated the effects of antioxidants β-carotene and lycopene on AFB_1-induced hepatotoxicity. The result showed the presence of lycopene followed by the addition of AFB_1 increased cell viability at approximately 14%, while pretreatment with β-carotene had the highest increase in cell survival up to 54%. Both carotenoids recovered mitochondrial dehydrogenase (MD) activity up to 85%, upregulated *p53* gene expression in AFB_1-exposed cells, and decrease in AFB_1-N7-guanine adducts. These results clearly showed that both β-carotene and lycopene could prevent AFB_1-induced toxicity in HepG2 cells [69].

Lycopene, a strong free radical scavenger having the greatest ability to cope with the singlet oxygen compared to the other carotenoids, can alleviate AFB_1-induced oxidative stress through the conjugation of the p-electron system with several reactive oxygen species. It can protect DNA, proteins, and lipid damages against the carcinogenesis onset contributed to its numerous conjugated double bonds, high lipophilicity, and acyclic structure [72]. Regarding several scientific publications, lycopene has been confirmed as the carotenoid that exhibited robust positive effects on AFB_1 toxicities via several pathways.

2.9 Allii Fistulosi Bulbus

Allium plants like garlic and onion are well-known in Asian countries as food ingredients and remedial foods. They have been documented as medicinal foods worldwide due to their pharmacological properties. *Allium fistulosum* (*A. fistulosum*), a perennial herb in *Allium* genus, has been commonly utilized as appetite inducer and medication against cold symptoms [73]. Also, it has ability to activate the immune response and antihypertensive effect as well as antioxidant defense system. The consumption of *A. fistulosum* extract increased estrogen level, mediated the conversion of testosterone to estrogen, and conducted hormone balance in female rats resulting in the enhancement of ovarian function [74]. The extract is able to downregulate the accumulation of lipid in HepG2 cells without cytotoxic effect and fatty acid gene synthesis. Similarly, mice fed high-fat, high-sucrose diet displayed an increase in body weight, hepatic weight, and fat accumulation in hepatocytes, but these adverse effects were attenuated by extract supplementation [75].

The effects of Allii Fistulosi Bulbus (VEAF) extract on cytotoxicity and oxidative stress caused by AFB_1 exposure were observed in HepG2 cells. Preincubation with VEAF followed by the addition of AFB_1 obviously enhanced cell viability. It inhibited oxidative stress through declining ROS level and TBAR content induced by AFB_1 and promoting GSH level. The determination of 8-OHdG, an indicator of oxidative damage on DNA, was then investigated. The result showed the inhibitory effect in VEAF treatment group up to 59.1% suppression compared to AFB_1-treated group. This evidence proved the alleviating potential of VEAF on AFB_1-induced oxidative stress resulting in cytoprotection against AFB_1 toxicity [76].

Quercetin, flavonol, is one of the major bioactive compounds in *Allium* plants. It shows the potential to scavenge free radical and improve health effects, that is, aging, allergy, angioprotective properties, anti-inflammatory, anti-cancer, anti-obesity, arthritis, asthma, diabetes, etc. [77]. For AFB_1 biosynthesis in *Aspergillus flavus*, quercetin notably decreased AFB_1 production (51%) in corn flour supplemented with quercetin at 48-hour incubation. Quercetin has an ability to inhibit the expression of necessary enzymes for AFB_1 biosynthesis such as acetyl CoA synthetase, esterase, and O-methyl transferase A and involves in the MAPK pathway which is the major pathway to form AFB_1. Quercetin, therefore, has the ability to be an anti-aflatoxigenic agent [78]. Quercetin also inhibited proliferation of *Aspergillus flavus* and its AFB1-biosynthesis through regulating the expression of development-related genes and aflatoxin production-related genes [79].

In HepG2 cells, quercetin decreased AFB1-induced cytotoxicity and ROS production and increased GSH content while *in vivo* study showed enhanced antioxidant activities and reduced lipid peroxidation [80]. After AFB_1 consumption, quercetin depicted the prevention of genotoxicity caused by AFB_1 in rat liver microsomes. Co-incubation with quercetin significantly decreased micronuclei formation compared to treated with AFB_1 alone ($p < 0.05$) [81]. Corresponding to another study, serum cytokines, procollagen III, and nitric oxide were significantly reduced during co-administration with quercetin and AFB_1 ($p < 0.05$). Quercetin also upregulated the antioxidant enzymes that may affect the decrease of DNA fragmentation and apoptosis [82]. Likewise, the administration between

AFB1-contaminated diet in rat resulted in a decrease of total proteins and RNA content and fatty acid synthase (Fas) and tumor necrosis factor (TNF) gene expression in the liver tissue caused by AFB_1 while co-administration with quercetin normalized these parameters [83].

Even though numerous studies revealed the hepatoprotective effects of quercetin against xenobiotic-induced cellular toxicity, low bioavailability of quercetin absorbed into circulation is the remarkable barrier [84]. One of the supreme strategies widely used is nanoformulation. Quercetin nanoparticles not only demonstrated a noteworthy reduction of AFB_1-induced cell death, but it also suppressed the liver toxicity caused by AFB_1 including ROS formation, lipid peroxidation, mitochondrial membrane potential collapse, and GSH depletion. In addition, both quercetin and quercetin nanoparticles significantly enhanced the function of hepatic enzymes (AST, ALT, and ALP) and hepatic antioxidant enzymes (SOD, CAT, and GPx) ($p < 0.05$). Interestingly, quercetin nanoparticles showed higher effects than quercetin [84]. These result reflexes an inhibiting ability of AFB_1 toxicity by administration of quercetin AFB_1.

AFB_1 also caused increase of cytotoxicity in a bovine mammary epithelial cell line. The pre-incubation with quercetin affected to increase cell viability, AFM_1 biosynthesis (low toxic metabolite of AFB_1), GSH content, and mRNA level of glutathione S-transferase alpha 1 (GSTA1) which are important for AFB_1 detoxification [85].

2.10 Rosemary plant (*Rosmarinus officinalis L.*)

Rosemary plant (*Rosmarinus officinalis L.*), naturally found in the western Mediterranean region, has been widely used as a food additive. As it contains high polyphenolic contents, it shows many pharmacological properties such as antioxidant activity and antimicrobial and antimycotic properties, etc. [86]. Previous study proved that the growth of *Aspergillus flavus and A. parasiticus* were significantly inhibited by 4% commercial rosemary essential oil from 28.2 to 59.5% and 41.5 to 52.4%, respectively [87]. Apart from antimycotic properties, dose-dependent exposure of carnosic acid—major polyphenolic compound in rosemary plants—clearly decreased cell death caused by 10 μM AFB_1. Pre-treatment to carnosic acid also reduced the production of ROS and the concentration of 8-OH-deoxyguanine, clearly confirming an involvement of carnosic acid in the protection of cytotoxicity induced by AFB_1 [88]. Furthermore, both rosemary extract and its active components (carnosol and carnosic acid) exhibited a potent inhibition of DNA adduct formation. They not only inhibit phase I metabolizing enzymes but also induce phase II metabolizing enzymes such as GST that promote the cellular defensive mechanism against AFB_1 [89].

3. Conclusion

Consumption of AFB1-contaminated food is the current major cause of HCC in many countries. Many studies aim to lower AFB1-induced toxicity particularly the utilization of edible plants as protective foods. This review proposed the edible plants which could alleviate AFB1-induced toxicity and concluded the possible mitigation of AFB1 toxicities through several related pathways (**Table 1** and **Figure 1**). Although the detoxification mechanism of AFB1 activated by various plants has been investigated in a pre-clinical study for a decade, clinical trial is still rarely clarified. Further investigation on a risk reduction of AFB1 still needs to be carried out especially in the clinical study.

Plants	Reference	Protective effects							
		Inhibit AFB_1 biosynthesis	Inhibit AFB_1 absorption	Anti-oxidant	Anti-genotoxicity	Reduce cytotoxicity	Modulate metabolism enzymes	Inhibit hepatotoxicity	Decrease liver cancer
Cruciferous vegetables	[10]			/			/		/
	[14]				/				/
	[15]								/
	[16]			/			/		
	[17]				/		/		
	[20]				/				
	[21]						/		
Green tea	[25]								/
	[27]				/				
	[28]								/
	[29]								/
	[30]			/					
	[31]*						/		
Purple rice	[38]				/				
	[39]				/		/		
	[40]			/	/			/	
Turmeric	[43]			/				/	
	[44]			/	/		/	/	
	[45]			/				/	
	[46]			/				/	
	[47]					/			
	[48]					/			/
	[49]			/			/		/

Plants	Reference	Protective effects							
		Inhibit AFB_1 biosynthesis	Inhibit AFB_1 absorption	Anti-oxidant	Anti-genotoxicity	Reduce cytotoxicity	Modulate metabolism enzymes	Inhibit hepatotoxicity	Decrease liver cancer
Green vegetables	[52]		/						
	[53]		/						
	[54]			/				/	/
	[55]			/				/	/
	[56]				/				
Ginger	[57]			/	/	/		/	
Dialium guineense	[62]			/	/			/	
Parkia biglobosa	[64]			/	/				
Carotenoid-rich fruits and vegetables	[69]				/	/			/
	[70]				/				
	[71]	/							
	[72]			/			/		
Allii Fistulosi Bulbus	[76]			/		/	/		
	[78]	/							
	[79]	/							
	[80]			/		/			
	[81]				/		/		
	[82]*			/	/				
	[83]**			/	/			/	
	[84]			/		/		/	
	[85]			/		/	/		

Plants	Reference	Protective effects							
		Inhibit AFB_1 biosynthesis	Inhibit AFB_1 absorption	Anti-oxidant	Anti-genotoxicity	Reduce cytotoxicity	Modulate metabolism enzymes	Inhibit hepatotoxicity	Decrease liver cancer
Rosemary	[87]	/							
	[88]			/		/			
	[89]				/		/		

*Alleviate serum cytokine and procollagen III, NO.
**Alleviate content of nucleic acid of liver tissue.

Table 1.
The protective effects of edible plants against AFB_1-induced toxicity.

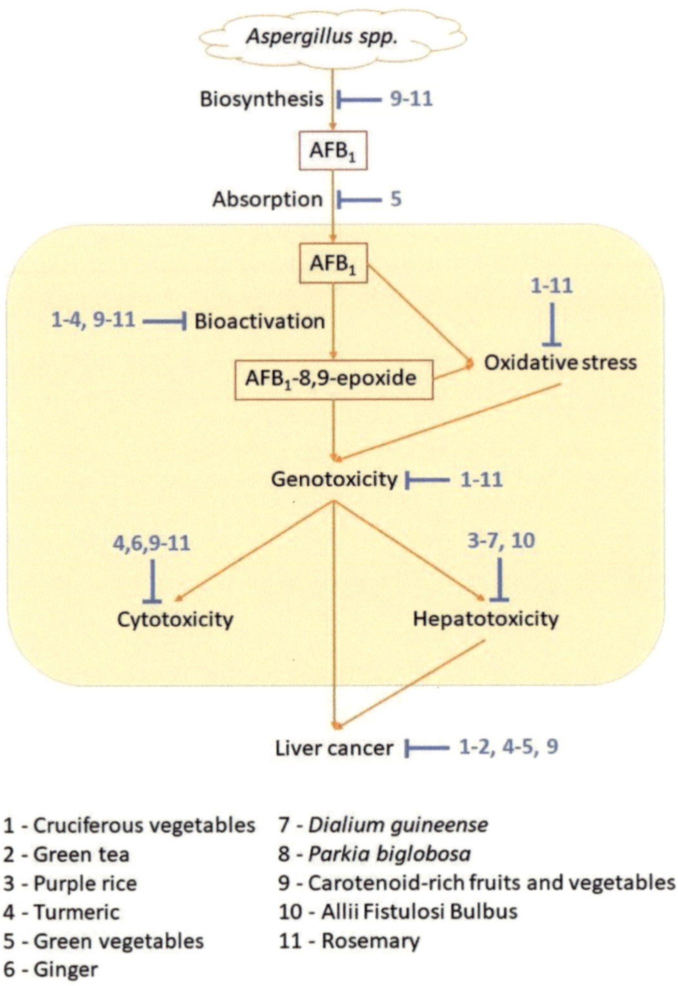

Figure 1.
Protective effects of edible plants against AFB_1-induced toxicity.

Acknowledgements

This work was funded and supported by Faculty of Pharmaceutical Sciences, Khon Kaen University, Khon Kaen, Thailand.

Conflict of interest

Authors declare no conflict of interest.

Abbreviations

AFB_1	Aflatoxin B_1
AFB_1-AA	AFB1-albumin adducts
AFB_1-NAC	AFB1-mercapturic acid
ALBT	African locust bean tree
ALP	Alkaline phosphatase

ALT	Alanine aminotransferase
ARE	Antioxidant response element
AST	Aspartate transaminase
AUC	Area under the curves
Bcl-xL	B-cell lymphoma-extra large
C_{max}	Maximum concentration
CAT	Catalase
CHL	Chlorophyllin
chla	Chlorophyll
CML	Country-made liquor
D3T	H-1,2-dithiole-3-thione
EC	Epicatechin
ECG	Epicatechin gallate
EGC	Epigallocatechin
EGCG	Epigallocatechin gallate
Fas	Fatty acid synthase
G6PD	Glucose-6-phosphate dehydrogenase
GPx	Glutathione peroxidase
GSH	Reduced glutathione
GSSH	Oxidized glutathione
GST	Glutathione S-transferase
GSTA1	Glutathione S-transferase alpha 1
GST-P	Glutathione S-transferase placental form
GTP	Green tea polyphenol
HCC	Hepatocellular carcinoma
HO-1	Heme oxygenase 1
I3C	Indole-3-carbinol
IARC	International Agency for Research on Cancer
LDH	Lactate dehydrogenase
MD	Mitochondrial dehydrogenase
MDA	Malondialdehyde
NAFLD	Nonalcoholic fatty liver disease
NF-κB	Nuclear factor kappa light chain enhancer of activated B cells
Nrf2	Nuclear factor-E2-related factor 2
8-OHdG	8-hydroxydeoxyguanosine
ROS	Reactive oxygen species
SF	Sulforaphane
SOD	Superoxide dismutase
TNF	Tumor necrosis factor
UGT	UDP-glucuronosyltransferase
VEAF	Allii Fistulosi Bulbus
VEGF	Vascular endothelial growth factor

Author details

Peeradon Tuntiteerawit, Tichakorn Singto, Anupon Tadee
and Supatra Porasuphatana*
Faculty of Pharmaceutical Sciences, Division of Pharmacognosy and Toxicology,
Khon Kaen University, Khon Kaen, Thailand

*Address all correspondence to: psupatra@kku.ac.th

References

[1] International Agency for Research on Cancer. Cancer Today: Estimated Number of Deaths in 2018, Worldwide, Both Sexes, all Ages. 2018. Available from: http://gco.iarc.fr/today/online-analysis [Accessed: March 30, 2019]

[2] Balogh J, Iii D, Asham E, Burroughs S, Boktour M, Saharia A, et al. Hepatocellular carcinoma: A review. J Hepatocell Carcinoma. 2016;**3**:41-53. DOI: 10.2147/JHC.S61146

[3] Chitapanarux T, Phornphutkul K. Risk factors for the development of hepatocellular carcinoma in Thailand. Journal of Clinical and Translational Hepatology. 2015;**3**(3):182-188. DOI: 10.14218/JCTH.2015.00025

[4] International Agency for Research on Cancer. Chemical Agents and Related Occupations. Lyon: WHO; 2009. 599 p

[5] Moudgil V, Redhu D, Dhanda S, Singh J. A review of molecular mechanisms in the development of hepatocellular carcinoma by aflatoxin and hepatitis B and C viruses. Journal of Environmental Pathology, Toxicology and Oncology. 2013;**32**(2):165-175. DOI: 10.1615/JEnvironPatholToxicolOncol.2013007166

[6] Kensler TW, Roebuck BD, Wogan GN, Groopman JD. Aflatoxin: A 50-year odyssey of mechanistic and translational toxicology. The Journal of Toxicological Sciences. 2011;**120**(S1):S28-S48. DOI: 10.1093/toxsci/kfq283

[7] Bbosa GS, Kitya D, Odda J, Ogwal-Okeng J. Aflatoxins metabolism. Health (Irvine Calif). 2013;**05**(10):14-34

[8] Kumar P, Mahato DK, Kamle M, Mohanta TK, Kang SG. Aflatoxins: A global concern for food safety, human health and their management. Frontiers in Microbiology. 2017;7:2170. DOI: 10.3389/fmicb.2016.02170

[9] Lavkor I, Var I. The control of aflatoxin contamination at harvest, drying, pre-storage ad storage periods in peanut: The new approach. In: Abdulra'Uf L, editor. Aflatoxin Control, Analysis, Detection and Health Risks. Rijeka: IntechOpen; 2017. pp. 45-65. DOI: 10.5772/intechopen.68675

[10] Sanlier N, Guler SM. The benefits of Brassica vegetables on human health. Journal of Human Health Research. 2018;**1**(1):104

[11] Kensler TW, Egner PA, Wang JB, Zhu YR, Zhang BC, Lu PX, et al. Chemoprevention of hepatocellular carcinoma in aflatoxin endemic areas. Gastroenterology. 2004;**127** (5 Suppl):310-318. DOI: 10.1053/j.gastro.2004.09.046

[12] Herr I, Büchler MW. Dietary constituents of broccoli and other cruciferous vegetables: Implications for prevention and therapy of cancer. Cancer Treatment Reviews. 2010;**36**(5):377-383. DOI: 10.1016/j.ctrv.2010.01.002

[13] Wallig MA, Heinz-Taheny KM, Epps DL, Gossman T. Synergy among phytochemicals within crucifers: Does it translate into chemoprotection? The Journal of Nutrition. 2018;**135**(12):2972S-2977S. DOI: 10.1093/jn/135.12.2972S

[14] Dashwood RH, Arbogast DN, Fong AT, Pereira C, Hendricks JD, Bailey GS. Quantitative inter-relationships between aflatoxin B1 carcinogen dose, indole-3-carbinol anti-carcinogen dose, target organ DNA adduction and final tumor respose. Carcinogenesis. 1989;**19**(1):175-181. DOI: 10.1093/carcin.10.1.175

[15] Dashwood RH. Indole-3-carbinol: Anticarcinogen or tumor promoter in brassica vegetables? Chemico-Biological

Interactions. 1998;**110**(1-2):1-5. DOI: 10.1016/s0009-2797(97)00115-4

[16] Latté KP, Appel KE, Lampen A. Health benefits and possible risks of broccoli—An overview. Food and Chemical Toxicology. 2011;**49**(12):3287-3309. DOI: 10.1016/j.fct.2011.08.019

[17] Wiriyachan N, Fiala JLA, Croy RG, Egner PA, Wogan GN, Ruchirawat M, et al. Sulforaphane-mediated reduction of aflatoxin B1-N7-guanine in rat liver DNA: Impacts of strain and sex. Toxicological Sciences. 2011;**121**(1):57-62. DOI: 10.1093/toxsci/kfr026

[18] Techapiesancharoenkij N, Fiala JLA, Navasumrit P, Robert G, Wogan GN, Groopman JD, et al. Sulforaphane, a cancer chemopreventive agent, induces pathways associated with membrane biosynthesis in response to tissue damage by aflatoxin B1. Toxicology and Applied Pharmacology. 2015;**282**(1):52-60. DOI: 10.1016/j.taap.2014.11.004

[19] Higdon JV, Delage B, Williams DE, Dashwood RH. Cruciferous vegetables and human cancer risk: Epidemiologic evidence and mechanistic basis. Pharmacological Research. 2007;**55**(3):224-236. DOI: 10.1016/j.phrs.2007.01.009

[20] Kensler TW, Chen JG, Egner PA, Fahey JW, Jacobson LP, Stephenson KK, et al. Effects of glucosinolate-rich broccoli sprouts on urinary levels of aflatoxin-DNA adducts and phenanthrene tetraols in a randomized clinical trial in He Zuo township, Qidong, people's Republic of China. Cancer Epidemiology, Biomarkers and Prevention. 2005;**14**(11 I):2605-2613. DOI: 10.1158/1055-9965.EPI-05-0368

[21] Kensler TW, Ng D, Carmella SG, Chen M, Jacobson LP, Muñoz A, et al. Modulation of the metabolism of airborne pollutants by glucoraphanin-rich and sulforaphane-rich broccoli sprout beverages in Qidong, China. Carcinogenesis. 2012;**33**(1):101-107. DOI: 10.1093/carcin/bgr229

[22] Venkateswara B, Sirisha K, Chava V. Green tea extract for periodontal health. Journal of Indian Society of Periodontology. 2011;**15**(1):18-22. DOI: 10.4103/0972-124X.82258

[23] Khlangwiset P, Wu F. Costs and efficacy of public health interventions to reduce aflatoxin–induced human disease. Food Additives and Contaminants. Part A, Chemistry, Analysis, Control, Exposure and Risk Assessment. 2010;**27**(7):998-1014. DOI: 10.1080/19440041003677475

[24] Yuan JM, Sun C, Butler LM. Tea and cancer prevention: Epidemiological studies. Pharmacological Research. 2011;**64**(2):123-135. DOI: 10.1016/j.phrs.2011.03.002

[25] Yin X, Yang J, Li T, Song L, Han T, Yang M, et al. The effect of green tea intake on risk of liver disease: A meta analysis. International Journal of Clinical and Experimental Medicine. 2015;**8**(6):8339-8346

[26] Sonoda J, Narumi K, Akio K, Toshiro M. Green tea catechins-pharmacokinetic properties and health beneficial effects. Pharmaceutica Analytica Acta. 2015;**6**(2):1-6. DOI: 10.4172/2153-2435.1000333

[27] Ito Y, Ohnishi S, Fujie K. Chromosome aberrations induced by aflatoxin B1 in rat bone marrow cells in vivo and their suppression by green tea. Mutation Research. 1989;**222**:253-261. DOI: 10.1016/0165-1218(89)90141-9

[28] Qin G, Gopalan-Kriczky P, Su J, Ning Y, Lotlikar P. Inhibition of aflatoxin B1-induced initiation of hepatocarcinogenesis in the rat by green tea. Cancer Letters. 1997;**112**(2):149-154. DOI: 10.1016/S0304-3835(96)04568-5

[29] Qin G, Ning Y, Lotlikar P. Chemoprevention of aflatoxin B1-initiated and carbon tetrachloride-promoted hepatocarcinogenesis in the rat by green tea. Nutrition and Cancer. 2000;**38**(2):215-222. DOI: 10.1207/S15327914NC382_11

[30] Zhang Z-Q, Liang Y, Wang J-S, Yu J, Zhang L, Wang K, et al. Phase IIa chemoprevention trial of green tea polyphenols in high-risk individuals of liver cancer: Modulation of urinary excretion of green tea polyphenols and 8-hydroxydeoxyguanosine. Carcinogenesis. 2005;**27**(2):262-268. DOI: 10.1093/carcin/bgi147

[31] Tang L, Tang M, Xu L, Luo H, Huang T, Yu J, et al. Modulation of aflatoxin biomarkers in human blood and urine by green tea polyphenols intervention. Carcinogenesis. 2008;**29**(2):411-417. DOI: 10.1093/carcin/bgn008

[32] Kong JM, Chia LS, Goh NK, Chia TF, Brouillard R. Analysis and biological activities of anthocyanins. Phytochemistry. 2003;**64**(5):923-933. DOI: 10.1016/S0031-9422(03)00438-2

[33] Park JH, Wang Z, Choi MK, Lim SS, Lee J-Y, Huang B, et al. Anti-diabetic effect of purple corn extract on C57BL/KsJ db/db mice. Nutrition Research and Practice. 2015;**9**(1):22-29. DOI: 10.4162/nrp.2015.9.1.22

[34] Reyna GS, Torres AG, Valenzuela R, Rincón-Cervera MÁ, Villanueva-Espinoza ME. Adding a purple corn extract in rats supplemented with chia oil decreases gene expression of SREBP-1c and retains Δ5 and Δ6 hepatic desaturase activity, unmodified the hepatic lipid profile. Prostaglandins, Leukotrienes and Essential Fatty Acids. 2018;**132**:1-7. DOI: 10.1016/j.plefa.2018.03.005

[35] Long N, Suzuki S, Sato S, Naiki-Ito A, Sakatani K, Shirai T, et al. Purple corn color inhibition of prostate carcinogenesis by targeting cell growth pathways. Cancer Science. 2013;**104**(3):298-303. DOI: 10.1111/cas.12078

[36] Nair V, Bang WY, Schreckinger E, Andarwulan N, Cisneros-Zevallos L. Protective role of ternatin anthocyanins and quercetin glycosides from butterfly pea (*Clitoria Ternatea* Leguminosae) blue flower petals against lipopolysaccharide (lps)-induced inflammation in macrophage cells. Journal of Agricultural and Food Chemistry. 2015;**63**(28):6355-6365. DOI: 10.1021/acs.jafc.5b00928

[37] Tsuda T. Dietary anthocyanin-rich plants: Biochemical basis and recent progress in health benefits studies. Molecular Nutrition and Food Research. 2012;**56**(1):159-170. DOI: 10.1002/mnfr.201100526

[38] Punvittayagul C, Sringarm K, Chaiyasut C, Wongpoomchai R. Mutagenicity and antimutagenicity of hydrophilic and lipophilic extracts of Thai northern purple rice. Asian Pacific Journal of Cancer Prevention. 2014;**15**(21):9517-9522. DOI: 10.7314/APJCP.2014.15.21.9517

[39] Suwannakul N, Punvittayagul C, Jarukamjorn K, Wongpoomchai R. Purple rice bran extract attenuates the aflatoxin B1-induced initiation stage of hepatocarcinogenesis by alteration of xenobiotic metabolizing enzymes. Asian Pacific Journal of Cancer Prevention. 2015;**16**(8):3371-3376. DOI: 10.7314/APJCP.2015.16.8.3371

[40] Ajiboye TO, Raji HO, Muritala HF, Ojewuyi OB, Yakubu MT. Anthocyanin extract of *Lannea microcarpa* fruits stall oxidative rout associated with aflatoxin B1 hepatocarcinogenesis. Food Bioscience. 2013;**4**:58-67. DOI: 10.1016/j.fbio.2013.09.002

[41] World Health Organization. WHO Monographs on Selected Medicinal Plants. Vol. 1. Malta: WHO; 1999. 295 p

[42] Sharma RA, Gescher AJ, Steward WP. Curcumin: The story so far. European Journal of Cancer. 2005;**41**(13):1955-1968. DOI: 10.1016/j.ejca.2005.05.009

[43] El-Agamy DS. Comparative effects of curcumin and resveratrol on aflatoxin B 1-induced liver injury in rats. Archives of Toxicology. 2010;**84**(5):389-396. DOI: 10.1007/s00204-010-0511-2

[44] Nayak S, Sashidhar RB. Metabolic intervention of aflatoxin B1 toxicity by curcumin. Journal of Ethnopharmacology. 2010;**127**(3): 641-644. DOI: 10.1016/j.jep.2009.12.010

[45] El-Bahr SM. Effect of curcumin on hepatic antioxidant enzymes activities and gene expressions in rats intoxicated with aflatoxin B1. Phytotherapy Research. 2015;**29**(1):134-140. DOI: 10.1002/ptr.5239

[46] Sharma V, Sharma C, Pracheta P, Sharma S. Protective potential of *Curcuma longa* and curcumin on aflatoxin B1 induced hepatoxicity in swiss albino mice. The Medical Journal of Cairo Univerisity. 2011;**1**(3):116-122

[47] Mathuria N, Verma RJ. Aflatoxin induced hemolysis and its amelioration by turmeric extracts and curcumin in vitro. Acta Poloniae Pharmaceutica. Drug Research. 2007;**64**(2):165-168

[48] Wang WH, Chiang IT, Ding K, Chung JG, Lin WJ, Lin SS, et al. Curcumin-induced apoptosis in human hepatocellular carcinoma j5 cells: Critical role of ca(+2) -dependent pathway. Evidence-based Complementary and Alternative Medicine. 2012;**2012**:512907. DOI: 10.1155/2012/512907

[49] Darvesh AS, Aggarwal BB, Bishayee A. Curcumin and liver cancer: A review. Current Pharmaceutical Biotechnology. 2011;**13**(1):218-228. DOI: 10.2174/138920112798868791

[50] Mishra VK, Bacheti RK, Husen A. Medicinal Uses of Chlorophyll: A critical overview. In: Hua L, Elisa S editor. Chlorophyll: Structure, function and medicinal uses. Nova Science Publishers; 2011. pp. 177-196

[51] Tumolo T, Lanfer-Marquez UM. Copper chlorophyllin: A food colorant with bioactive properties? Food Research International. 2012;**46**(2):451-459. DOI: 10.1016/j.foodres.2011.10.031

[52] Mata JE, Yu Z, Gray JE, Williams DE, Rodriguez-Proteau R. Effects of chlorophyllin on transport of dibenzo(a,l)pyrene, 2-amino-1-methyl-6-phenylimidazo-[4,5-b] pyridine, and aflatoxin B1 across Caco-2 cell monolayers. Toxicology. 2004;**196**(1-2):117-125. DOI: 10.1016/j.tox.2003.11.008

[53] Jubert C, Mata J, Bench G, Dashwood R, Pereira C, Tracewell W, et al. Effects of chlorophyll and chlorophyllin on low-dose aflatoxin B1 pharmacokinetics in human volunteers. Cancer Prevention Research. 2009;**2**(12):1015-1022. DOI: 10.1158/1940-6207.CAPR-09-0099

[54] Kumar M, Verma V, Nagpal R, Kumar A, Behare PV, Singh B, et al. Anticarcinogenic effect of probiotic fermented milk and chlorophyllin on aflatoxin-B1-induced liver carcinogenesis in rats. The British Journal of Nutrition. 2012;**107**(7):1006-1016. DOI: 10.1017/S0007114511003953

[55] Abdel-Latif MS, Elmeleigy KM, Aly TAA, Khattab MS, Mohamed SM. Pathological and biochemical evaluation of coumarin and chlorophyllin against aflatoxicosis in rat. Experimental and Toxicologic Pathology. 2017;**69**(5):285-291. DOI: 10.1016/j.etp.2017.01.014

[56] Egner P, Wang J, Zhu Y, Zhang B, Wu Y, Zhang Q, et al. Chlorophyllin intervention reduces aflatoxin—DNA

adducts in individuals at high risk for liver cancer. 2001;**98**(25):14601-14606. DOI: 10.1073/pnas.251536898

[57] Vipin A, Raksha R, Kurrey NK, Anu A, Venkateswaran G. Protective effects of phenolics rich extract of ginger against aflatoxin B1-induced oxidative stress and hepatotoxicity. Biomedicine and Pharmacotherapy. 2017;**91**:415-424. DOI: 10.1016/j.biopha.2017.04.107

[58] Bhandari U, Shamsher AA, Pillai KK, Khan MSY. Antihepatotoxic activity of ginger ethanol extract in rats. Pharmaceutical Biology. 2003;**41**(1):68-71. DOI: doi.org/10.1076/phbi.41.1.68.14697

[59] Abd-Allah GA, El-Bakry KA, Bahnasawy MH, El-Khodary ESR. Protective effects of curcumin and ginger on liver cirrhosis induced by carbon tetrachloride in rats. International Journal of Pharmacology. 2016;**12**(4):361-369. DOI: 10.3923/ijp.2016.361.369

[60] Ogu GI, Ezcadila J, Ehiobu JM. Antioxidant and antimicrobial activities of *Dialium guineense* (wild) leaf extract. Pharmacy and Pharmacology Research. 2013;**1**(1):1-7

[61] Okeke NC, Udeani TK, Onyebuchi UL. Wound-healing and antimicrobial properties of dichloromethane fraction of *Dialium guineense* (Wild) fruit coat. Research in Pharmaceutical Sciences. 2016;**11**(3):219-226

[62] Adeleye AO, Ajiboye TO, Iliasu GA, Abdussalam FA, Balogun A, Ojewuyi OB. Phenolic extract of *Dialium guineense* pulp enhances reactive oxygen species detoxification in aflatoxin B1 hepatocarcinogenesis. Journal of Medicinal Food. 2014;**17**(8):875-885. DOI: 10.1089/jmf.2013.0157

[63] Dedehou V, Olounladé P, Adenilé D, Alowanou G, Azando E, Hounzangbé-Adoté M. A review on medicinal plants of *Parkia biglobosa* (Mimosaceae -Fabaceae) and *Pterocarpus erinaceus* (Leguminosae - Papilionoidea). Journal of Medicinal Plants Studies. 2016;**4**(6):132-137

[64] Ajiboye TO, Adeleye AO, Salau AK, Ojewuyi OB, Adigun NS, Sabiu S, et al. Phenolic extract of *Parkia biglobosa* fruit pulp stalls aflatoxin B1-mediated oxidative rout in the liver of male rats. Revista Brasileira de Farmacognosia. 2014;**24**:668-676. DOI: 10.1016/j.bjp.2014.10.010

[65] Kotake-Nara E, Nagao A. Absorption and metabolism of xanthophylls. Marine Drugs. 2011;**9**:1024-1037. DOI: 10.3390/md9061024

[66] Saini RK, Nile SH, Park SW. Carotenoids from fruits and vegetables: Chemistry, analysis, occurrence, bioavailability and biological activities. Food Research International. 2015;**76**:735-750. DOI: 10.1016/j.foodres.2015.07.047

[67] Widomska J, Zareba M, Subczynski WK. Can xanthophyll-membrane interactions explain their selective presence in the retina and brain? Food. 2016;**5**(1):E7. DOI: 10.3390/foods5010007

[68] Haskell MJ. Provitamin A carotenoids as a dietary source of vitamin A. In: Tanumihardjo S, editor. Carotenoids and Human Health. New Jersey: Humana Press; 2013. pp. 249-260. DOI: 10.1007/978-1-62703-203-2_15

[69] Reddy L, Odhav B, Bhoola K. Aflatoxin B1-induced toxicity in HepG2 cells inhibited by carotenoids: Morphology, apoptosis and DNA damage. Biological Chemistry. 2006;**387**(1):87-93. DOI: 10.1515/BC.2006.012

[70] He Y, Campbell TC. Effects of carotenoids on aflatoxin B1-induced

mutagenesis in *S. typhimurium* TA 100 and TA 98. Nutrition and Cancer. 1990;**13**(4):243-253. DOI: 10.1080/01635589009514066

[71] Norton RA. Effect of carotenoids on aflatoxin B1 synthesis by *Aspergillus flavus*. Phytopathology. 1997;**87**(8):814-821. DOI: 10.1094/PHYTO.1997.87.8.814

[72] Yilmaz S, Kaya E, Kisacam M. The effect on oxidative stress of aflatoxin and protective effect of lycopene on aflatoxin damage. In: Abdulra'uf LB, editor. Aflatoxin-Control, Analysis, Detection and Health Risks. Rijeka: InTech; 2017. pp. 67-88. DOI: 10.5772/intechopen.69321

[73] Fukaya M, Nakamura S, Nakagawa R, Kinka M, Nakashima S, Matsuda H. Cyclic sulfur-containing compounds from *Allium fistulosum* 'Kujou'. Journal of Natural Medicines. 2019;**73**(2):397-403. DOI: 10.1007/s11418-018-1272-0

[74] Lee YH, Yang H, Lee SR, Kwon SW, Hong EJ, Lee HW. Welsh onion root (*Allium fistulosum*) restores ovarian functions from ltrozole induced-polycystic ovary syndrome. Nutrients. 2018;**10**(10):E1430. DOI: 10.3390/nu10101430

[75] Hwang JT, Shun EJ, Chung MY, Park JH, Chung S, Choi HK. Ethanol extract of *Allium fistulosum* inhibits development of non-alcoholic fatty liver disease. Nutrition Research and Practice. 2018;**12**(2):110-117. DOI: 10.4162/nrp.2018.12.2.110

[76] Lee JK, Choi EH, Lee KG, Chun HS. Alleviation of aflatoxin B1-induced oxidative stress in HepG2 cells by volatile extract from Allii fistulosi bulbus. Life Sciences. 2005;**77**(23):2896-2910. DOI: 10.1016/j.lfs.2005.03.028

[77] D'Andrea G. Quercetin: A flavonol with multifaceted therapeutic applications? Fitoterapia. 2015;**106**:256-271. DOI: 10.1016/j.fitote.2015.09.018

[78] Tiwari S, Shankar J. Integrated proteome and HPLC analysis revealed quercetin-mediated inhibition of aflatoxin B1 biosynthesis in *Aspergillus flavus*. 3 Biotech. 2018;**8**(1):47. DOI: 10.1007/s13205-017-1067-0

[79] Li XM, Li ZY, Wang YD, Wang JQ, Yang PL. Quercetin inhibits the proliferation and aflatoxins biosynthesis of *Aspergillus flavus*. Toxins (Basel). 2019;**11**(3):E154. DOI: 10.3390/toxins11030154

[80] Choi KC, Chung WT, Kwon JK, Yu JY, Jang YS, Park SM, et al. Inhibitory effects of quercetin on aflatoxin B1-induced hepatic damage in mice. Food and Chemical Toxicology. 2010;**48**(10):2747-2753. DOI: 10.1016/j.fct.2010.07.001

[81] Kohli E, Raj HG, Kumari R, Rohil V, Kaushik NK, Prasad AK, et al. Comparison of the prevention of aflatoxin B1-induced genotoxicity by quercetin and quercetin pentaacetate. Bioorganic and Medicinal Chemistry Letters. 2002;**12**(18):2579-2582. DOI: 10.1016/S0960-894X(02)00478-X

[82] Abdel-Wahhab MA, Aljawish A, El-Nekeety AA, Abdel-Aiezm SH, Abdel-Kader HAM, Rihn BH, et al. Chitosan nanoparticles and quercetin modulate gene expression and prevent the genotoxicity of aflatoxin B1 in rat liver. Toxicology Reports. 2015;**2**:737-747. DOI: 10.1016/j.toxrep.2015.05.007

[83] El-Nekeety AA, Abdel-Azeim SH, Hassan AM, Hassan NS, Aly SE, Abdel-Wahhab MA. Quercetin inhibits the cytotoxicity and oxidative stress in liver of rats fed aflatoxin-contaminated diet. Toxicology Reports. 2014;**1**:319-329. DOI: 10.1016/j.toxrep.2014.05.014

[84] Eftekhari A, Ahmadian E, Panahi-Azar V, Hosseini H, Tabibiazar M, Maleki Dizaj S. Hepatoprotective and free radical scavenging actions of quercetin nanoparticles on aflatoxin

B1-induced liver damage: In vitro/in vivo studies. Artificial Cells Nanomedicine and Biotechnology. 2018;**46**(2):411-420. DOI: 10.1080/21691401.2017.1315427

[85] Ghadiri S, Spalenza V, Dellafiora L, Badino P, Barbarossa A, Dall'Asta C, et al. Modulation of aflatoxin B1 cytotoxicity and aflatoxin M1 synthesis by natural antioxidants in a bovine mammary epithelial cell line. Toxicology In Vitro. 2019;**57**:174-183. DOI: 10.1016/j.tiv.2019.03.002

[86] Hernández MD, Sotomayor JA, Hernández Á, Jordán MJ. Rosemary (*Rosmarinus officinalis* L.) oils. In: Preedy VR, editor. Essential Oils in Food Preservation, Flavor and Safety. London: Elsevier; 2015. pp. 677-688. DOI: 10.1016/B978-0-12-416641-7.00077-8

[87] Mahmoud MA, El-Zaidy M, Al-Othman MR, ARM AE-A, Al-Gahtani MF. Efficacy of *Rosmarinus officinalis* essential oil on *Aspergillus flavus* and parasiticus. Journal of Pure and Applied Microbiology. 2014;**8**(8):185-190

[88] Costa S, Utan A, Speroni E, Cervellati R, Piva G, Prandini A, et al. Carnosic acid from rosemary extracts: A potential chemoprotective agent against aflatoxin B1. An in vitro study. Journal of Applied Toxicology. 2007;**27**(2):152-159. DOI: 10.1002/jat.1186

[89] Offord EA, Macé K, Avanti O, Pfeifer AM. Mechanisms involved in the chemoprotective effects of rosemary extract studied in human liver and bronchial cells. Cancer Letters. 1997;**114**(1-2):275-281. DOI: 10.1016/S0304-3835(97)04680-6